ELECTRICAL SAFETY-RELATED WORK PRACTICES

THIRD EDITION

NATIONAL JOINT APPRENTICESHIP AND TRAINING COMMITTEE

IBEW — NECA
ATTITUDE * SKILL * KNOWLEDGE
NJATC ®
FOR THE ELECTRICAL INDUSTRY
APPRENTICESHIP & TRAINING

JONES & BARTLETT
LEARNING

World Headquarters
Jones & Bartlett Learning
5 Wall Street
Burlington, MA 01803
978-443-5000
info@jblearning.com
www.jblearning.com

Jones & Bartlett Learning books and products are available through most bookstores and online booksellers. To contact Jones & Bartlett Learning directly, call 800-832-0034, fax 978-443-8000, or visit our website, www.jblearning.com.

Substantial discounts on bulk quantities of Jones & Bartlett Learning publications are available to corporations, professional associations, and other qualified organizations. For details and specific discount information, contact the special sales department at Jones & Bartlett Learning via the above contact information or send an email to specialsales@jblearning.com.

Production Credits
Chief Executive Officer: Ty Field
President: James Homer
Chief Product Officer: Eduardo Moura
Executive Publisher: Kimberly Brophy
Executive Acquisitions Editor—Fire and Electrical: William Larkin
Managing Editor: Carol B. Guerrero
Production Manager: Jenny L. Corriveau
Associate Marketing Manager: Jessica Carmichael

VP, Manufacturing and Inventory Control: Therese Connell
Composition: diacriTech
Cover Design: Kristin E. Parker
Text Design: Anne Spencer
Director of Photo Research and Permissions: Amy Wrynn
Cover Image: Courtesy of Salisbury by Honeywell
Printing and Binding: RR Donnelley
Cover Printing: RR Donnelley

Library of Congress Cataloging-in-Publication Data
National Joint Apprenticeship and Training Committee for the Electrical Industry.
Electrical safety-related work practices / National Joint Apprenticeship and Training Committee. – Third Edition.
    pages cm
Includes index.
ISBN 978-1-4496-4278-5 (pbk.)
1. Electrical engineering–United States–Safety measures. I. Title.
TK152.E543 2013
621.3028'9–dc23
                                                    2013023995

6048

Printed in the United States of America
17 16 15 14   10 9 8 7 6 5 4 3 2

Acknowledgments

Principal Writer

Palmer Hickman
Director of Safety, Codes and Standards
NJATC

Contributing Writers

Tim Crnko
Manager, Training & Tech Services
Cooper Bussmann®

Mary Capelli-Schellpfeffer, MD, MPA
CapSchell, Inc.

James R. White
Training Director
Shermco Industries

Text Contributors

The NJATC wishes to thank the following individuals for permission to reprint materials in this edition.

Steve Abbott
Stark Safety Consultants

Vince A. Baclawski
Technical Director, Power Distribution Products
National Electrical Manufacturers Association
(NEMA)

Dennis J. Berry
Secretary of the Corporation
Director of Licensing
National Fire Protection Association

**Electrical Construction & Maintenance (EC&M)
Magazine**
Penton Media

Jacqueline Hansson
IEEE

Stephen M. Lipster
Training Director
Columbus Joint Apprenticeship and Training
Committee for the Electrical Industry

Dennis K. Neitzel, CPE
Director Emeritus
AVO Training Institute, Inc.

Kristen Schmidt
Technical Services Director
InterNational Electrical Testing Association (NETA)

Contents

Foreword

Electrical Safety-Related Work Practices: NJATC's Guide based on *NFPA 70E*®

For years, many in the industry only considered electrical shock when contemplating worker protection and safe work practices. Today, however, the industry recognizes numerous additional hazards associated with work involving electrical hazards. These include, for example, the hazards associated with an arcing fault, including arc flash and arc blast. Consideration must be given to the devastating forces generated when molten copper expands to 67,000 times its original volume as it vaporizes, arc temperatures that can reach 35,000°F, pressures that can reach thousands of pounds per square foot, and shrapnel expelled from ruptured equipment at speeds that may exceed 700 miles per hour. Workers are exposed to these and other hazards even during a seemingly routine task such as voltage testing.

Working on circuits and equipment deenergized and in accordance with established lockout and tagout procedures has always been the primary safety-related work practice and a cornerstone of electrical safety. Only after it has been demonstrated that deenergizing is infeasible or would create a greater hazard, may equipment and circuit parts be worked on energized, and then only after other safety-related work practices, such as insulated tools and appropriate personal protective equipment, have been implemented. Examples of additional concerns that should be considered include worker, contractor, and customer attitude regarding energized work, comprehensiveness of an electrical safety plan, appropriate training, the role of overcurrent protective devices in electrical safety, equipment maintenance, and design and work practice considerations. These are a few of the issues that play an important role in worker safety, and are among the topics examined in this publication.

Electrical Safety-Related Work Practices has been developed in an effort to give those in the electrical industry a better understanding of a number of the hazards associated with work involving electrical hazards and the manner and conditions under which such work may be performed. These work practices and protective techniques have been developed over many years and are drawn from industry practice, national consensus standards, and federal electrical safety requirements. In many cases, these requirements are written in performance language. This publication also explores *NFPA 70E, Standard for Electrical Safety in the Workplace* as a means to comply with the electrical safety-related work practice requirements of the Occupational Safety and Health Administration.

Photo Contributors

The NJATC wishes to thank the following companies and individuals for submitting photos for inclusion in this edition.

Boltswitch
Jim Erickson
President

Cooper Bussmann®
Vincent J. Saporita
Vice President Technical Marketing and Services

DuPont Engineering
Daniel Doan

H. Landis Floyd, II
Corporate Electrical Safety Competency Leader
Principal Consultant—Electrical Safety & Technology

Eaton Corporation
Thomas A. Domitrovich, P.E., LEED AP BD+C
National Application Engineer

JoAnn Frank

electricaltrades.org
Stephen M. Lipster

Fluke Corporation
Leah Friberg

Ideal Industries, Inc.
Rob Conrad
Sales Training & Development Manager

Institute of Electrical and Electronics Engineers (IEEE)
Jacqueline Hansson

International Brotherhood of Electrical Workers (IBEW)
Jim Spellane

Klein Tools
Stephen Ratkovich

Lower Colorado River Authority
Corby Weiss

Milwaukee Electric Tool Corp.
Scott Teson
Director of Skilled Trades

National Electrical Contractors Association (NECA)
Michael J. Johnston
Executive Director Standards and Safety

National Fire Protection Association
Dennis J. Berry
Secretary of the Corporation
Director of Licensing

PACE Engineers Group Pty Ltd.
Robert Fuller

Salisbury by Honeywell
Brian McCauley, DOM

Schneider Electric Inc.
George B. Hendricks
Manager, Learning and Development Solutions

Service Electric Company
Mark Christian

Shermco Industries
James R. White
Training Director

Stark Safety Consultants
Steve Abbott

Westex Inc.
John Gaston
Sales and Marketing Manager

The NJATC also wishes to thank the following who participated in the photos that are courtesy of NECA:

Wilson Electric Co.
Ryan Hand
Michael Maffiolli

Morse Electric, Inc.
Brad Munda
Kyle Borneman

Larry McCrae, Inc.
Jeff Costello

J.P. Rainey Company, Inc.
Dave Ganther
Bill Inforzato

Carr and Duff, Inc.
George Novelli
Tom McCusker

Northern Illinois Electrical JATC
Todd Kindred

1 Electrical Safety Culture

OBJECTIVES

1. Recognize the important role that a safety culture plays for every person and in every organization, and understand how it affects worker exposure to electrical hazards.

2. Understand that a number of decisions are made before and during the time a worker is exposed to electrical hazards, and appreciate how decisions can reduce or eliminate electrical hazards.

3. Recognize the important role that understanding and complying with requirements plays in reducing and eliminating hazards.

REFERENCES

1. National Institute for Occupational Safety and Health (NIOSH) Fatality Assessment and Control Evaluation (FACE) program

2. Occupational Safety and Health Administration (OSHA) 29 CFR Part 1910

3. OSHA 29 CFR Part 1926

A 36-year-old electrician's helper was electrocuted in a fitting room of a department store located in a suburban shopping mall. The victim and a coworker were replacing the overhead fluorescent light tubes and ballast transformers in the fitting room. The employer had a written safety program that included lockout/tagout procedures and employed a safety coordinator who conducted weekly safety meetings.

Under the direction of the foreman, the victim and a second helper started work on replacing the ballasts. The foreman reportedly shut off the power to the lights by turning off and locking out the wall switch and then checked the lights with a circuit tester. This step deenergized all but one center light fixture, which was on a separate "night light" circuit that remained on. The foreman went to check the breaker but was unable to find the switch to shut off the remaining light. He confirmed that the victim had worked on live wires, and then told the victim to go ahead with the job.

The victim used a six-foot fiberglass ladder to reach the lights and began removing the tubes and ballast transformers. He was working on the ladder when he cut the energized black wire. The power entered though the victim's hands and exited to the grounded metal doorframe that he was leaning against. The victim's coworker, who was working in the stall beside him, heard the victim say, "Help me," and saw sparks flying from the wire. The coworker cut the black wire, breaking the contact and releasing the victim, who collapsed against the metal frame.

At this time, the foreman entered the room and helped move the victim to the floor. The store manager called 911; police arrived and began cardiopulmonary resuscitation (CPR). The victim was transported to the local hospital, where he was pronounced dead.

Source: For details of this case, see New Jersey FACE Investigation #95NJ080. Accessed May 29, 2012.

For additional information, visit qr.njatcdb.org Item #1184.

Introduction

Too often a culture exists in the workplace where workers are routinely allowed and expected to work on or near energized electrical circuits. This tendency to accept the risk of an electrical injury is unacceptable and must change.

This practice might be due to ignorance of laws that have been in place for decades, lack of knowledge of the severity of the hazards, or perhaps failure to realize how quickly a task situation might change and cause an energy release. It is less likely that workers, contractors, and facilities owners would allow energized work if everyone involved in the decision-making process fully understood the laws, requirements, hazards, true costs, and consequences associated with energized work.

Safety Culture

A false sense of security devalues safe practice. As a consequence, Electrical Workers may work on energized circuits owing to misperceptions of the risks involved. These paradigms are part of the electrical work culture, and may lead workers to take risks that are not in their best interest. Many do not understand the existing and potential hazards; others, who do understand these risks, do not realize how quickly a situation can change when things go wrong. The following list of statements reflects mindsets and attitudes that can lead to taking unnecessary risks:

- I don't care what the law says—I'm going to work it energized.
- I'm an Electrical Worker; working stuff hot is part of my job.
- That's what the customer expects. If my people won't do it, then they'll get another contractor.
- It's the office of the president of the company—you can't deenergize the circuit to change that ballast.
- You can't shut that assembly line down, because it will cost too much.
- There are people out of work looking for a job, so if I won't work it hot, someone else will.
- I've been doing it this way for 30 years, and nothing has ever happened to me.
- I know I should be wearing personal protective equipment (PPE), but it slows me down.
- There's no time to shut it down.
- That protective equipment is too expensive.
- What's the worst that can happen?
- It won't happen to me.

Far too many Electrical Workers believe that working on energized circuits is part of their job or is expected of them; in fact, such tasks are not part of routine electrical work. A tendency to work on or near electrical circuits while energized and accept the risk of an electrical injury creates an unacceptable culture. The need to change this mind-set must be recognized by all involved in the decision-making process. **See Figure 1-1**.

Figure 1-1. *Energized work is permitted only under limited circumstances as set forth by OSHA and NFPA 70E®.*

*NFPA-70E® is a registered trademark of the National Fire Protection Association, Quincy, MA.

Contractors have reported feeling pressured by their customers to work on energized equipment when a shutdown is warranted. Likewise, workers have reported feeling pressured by management to perform energized work when it is not justified. Workers who accept this risk expose themselves to injury or death. They also expose the contractors and their clients to undue risks of increased insurance premiums and loss of production. In many cases, customers may not understand the total costs and risks associated with energized work.

A well-informed client understands the hazards of energized work and the financial implications associated with an electrical incident. Equipment or circuits that are not permitted to be shut down for a few minutes ultimately might be shut down for days or weeks, or even longer, due to an unplanned event such as a dropped tool or a loose part falling into energized equipment and creating an unscheduled shutdown. A well-informed client is less likely to permit energized work, much less expect it.

Hazard Awareness and Recognition

A full understanding and recognition of existing and potential hazards is crucial to ensuring that an environment is electrically safe. The following must be done at a minimum:

- Eliminate the hazard.
- Develop and implement appropriate procedures.
- Develop, conduct, and implement training for qualified and unqualified persons.

- Deenergize and follow all of the necessary steps of the lockout/tagout program and establish an electrically safe work condition unless the employer demonstrates a true need for energized work.
- Develop and implement a hazard identification and risk assessment procedure.
- Engineer out the hazards or reduce them as far as is practicable.
- Provide adequate protection against hazards when the need for energized work is demonstrated.

A comparison can be made between the hazards of driving an automobile and the hazards associated with working on or near energized electrical equipment. Protective systems such as seat belts and air bags were developed to reduce the likelihood of injury or death; likewise, personal protective equipment (PPE) was developed to increase Electrical Worker safety. Such protections have a key limitation, however; they are effective only when they are actually used.

It is much the same with the hazards associated with working while exposed to electricity. Electrical Workers will continue to be exposed to electrical hazards if they do not take appropriate steps. Potential hazards in this environment include fire, falls and falling objects, electrical shock, and the hazards associated with arcing faults, including arc flash and arc blast. An *arcing fault* is a fault characterized by an electrical arc through the air. *Arc flash* is a dangerous condition caused by the release of energy in an electric arc, usually associated with electrical distribution equipment. **See Figure 1-2**. An arcing fault, for example, could be initiated by a dropped tool or by operation of equipment that has not been maintained properly. Electrical Workers may believe that the chance of such a lapse is unlikely; however, it may need to happen only once to result in injury or death. If used, protective systems, work practices, and protective equipment can reduce or eliminate exposure to the hazards. Workers may still suffer injury or death if circuits and equipment are not worked on in an electrically safe work condition. An *electrically safe work condition* is defined in *NFPA 70E* as follows:

Electrically safe work condition. A state in which an electric conductor or circuit part has been disconnected from energized parts, locked/tagged in accordance with established standards, tested to ensure the absence of voltage, and grounded if determined necessary

Reprinted with permission from NFPA-70E 2012, Electrical Safety in the Workplace Copyright 2011, National Fire Protection Association, Quincy, MA 02169. This reprinted material is not the complete and official position of the NFPA on the referenced subject, which is represented only by the standard in its entirety.

Figure 1-2. *Shock, arc flash, and arc blast subject workers to a number of hazards.*

From the Cooper Bussmann Safety BASICs Handbook *with permission from Cooper Bussmann.*

The hazard of electrical shock has been recognized since the dawn of electricity. The industry has evolved and made great strides to protect against electrical shock through the use of ground-fault circuit interrupters (GFCI) and rubber protective goods such as insulating gloves and blankets. A *GFCI* is defined in *NFPA 70E* as follows:

Ground-Fault Circuit Interrupter (GFCI). A device intended for the protection of personnel that functions to deenergize a circuit or portion thereof within an established period of time when a current to ground exceeds the values established for a Class A device. (Informational Note: Class A ground-fault circuit interrupters trip when the current to ground is 6 mA or higher and do not trip when the current to ground is less than 4 mA. For further information, see ANSI/UL 943, *Standard for Ground-Fault Circuit Interrupters.*)

Reprinted with permission from NFPA-70E 2012, Electrical Safety in the Workplace Copyright 2011, National Fire Protection Association, Quincy, MA 02169. This reprinted material is not the complete and official position of the NFPA on the referenced subject, which is represented only by the standard in its entirety.

These products are effective when used and maintained properly. Even with these advances, however, injury and death still occur from electrical shock. **See Figure 1-3**. The Bureau of Labor Statistics (BLS) data for electric shocks in nonfatal cases involving days away from work for the period 1992–2001 indicate that there were an average of 2,726 cases annually in private industry.

Arc flash and arc blast constitute lesser-known hazards; electrical burns happen frequently. The BLS data for nonfatal cases involving days away from work for the period 1992–2001 indicate an average of 1,710 electrical burns per year (peaking at 2,200 in 1995)

BACKGROUND

The following passage demonstrates one way that acceptance of risk and the culture in place today could have evolved. The current culture's outdated work practices might be the result of the way Electrical Workers were taught in the past. Readers should contemplate the guidance that was once given to Electrical Workers. The recommended practice at that time actually was to check for voltage using the fingers.

Historical Methods for Testing Voltage

As late as the mid-1900s, Electrical Workers performed testing for voltage procedures, employing a variety of less desirable techniques when viewed from today's perspective. On lower voltages typically found on bell, signal, and low-voltage control work, the presence of voltage (or pressure, as it commonly was called) could be tested using the "tasting method":

- A method of stripping the ends of the conductors from both sides of the circuit and placing the ends of these conductors a short distance apart on the tongue could be used to determine the presence of voltage.
- The "testee", or more appropriately, "tester" would experience a burning sensation followed by a slight salt taste. Depending on the amount of voltage present, holding one of the conductors in the bare hand and touching the other to the tongue could also be used. In this case, the body was acting as a voltage divider, lessening the burning sensation on the tongue.

Other variations of voltage testing included standing on wet ground when one end of the voltage system was grounded, while touching the tongue with the other terminal of the voltage source. This also was an "approved method." Individuals using this method often stated that once these test methods were performed, the end result was not often forgotten.

On higher voltages typically found in building power applications, the "finger method" was employed as an acceptable method of determining the presence of voltage in circuits of 250 volts or less:

- Electrical Workers would test the wires for voltage by touching the conductors to the ends of the fingers on one hand. Often, due to skin thickness, skin dryness, and calluses, the Electrical Worker would have to first lick the fingers to wet them to be able to sense the voltage being measured.

This method was billed as easy and convenient for determining whether live wires were present. The individual Electrical Worker's threshold for pain determined whether or not this was an acceptable method for everyday use. Some Electrical Workers supposedly had the ability, depending on the intensity of the sensation, to determine the actual voltage being tested.

Source: American Electrician's Handbook: A Reference Book for Practical Electrical Workers, *5th edition, by Terrell Croft (revised by Clifford C. Carr). Copyright ©* 1942 by The McGraw-Hill Companies, Inc. Reprinted by permission of The McGraw-Hill Companies, Inc.

Considering these historical aspects of electrical work, it is no wonder that today's culture often does not appreciate and implement safe work practices. Safe work practices today do not allow procedures such as the "tasting method" and the "finger method" to test for the presence or absence of voltage.

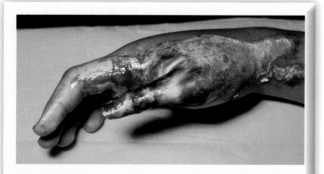

Figure 1-3. *The consequences of exposure to electrical hazards are often traumatic.*

© Charles Stewart & Associates

in private industry. That averages out to nearly one worker suffering the consequences of electrical burns every hour, based on a 40-hour work week. These data were instrumental in advancing electrical safety in general and *NFPA 70E* in particular during that time period.

BACKGROUND

The following passage contains excerpts from a procedure that once was considered an appropriate method of CPR. Like other practices and procedures, CPR has evolved over time. Clearly, electrical safety-related work practices, like all practices and procedures, require updating over time.

Cardiopulmonary Resuscitation: Accepted Practice in the Early 1900s

A review of several of the techniques used as methods of resuscitation show that medical technology has come a long way since the early 1900s.

The primary method of treating an individual who had experienced heart failure and/or respiratory arrest was simple. This primary method required two rescuers to perform the resuscitation procedure:

- After placing the victim on his or her back, one rescuer would grab and wiggle the victim's tongue, while the other rescuer would work the victim's arms back and forth to help induce breathing.

While possibly resuscitating some stricken individuals, a secondary approach was to be used should the first method fail:

- In cases where manual inflation of the lungs was attempted with no success, an attempt to cause the victim to gasp for air was performed. To initiate this gasping, the rescuers would insert two fingers into the victim's rectum, pressing them suddenly and forcibly towards the back of the individual.

Needless to say, it is not hard to understand why today's CPR methods provide more favorable results for both the victim and the rescuer.

Source: The Fire Underwriters of the United States, Standard Wiring: Electric Light and Power. H. G. Cushing Jr., New York, NY, 1911.

OSHA Tip

Occupational Safety and Health Administration (OSHA) 29 CFR 1910.333(a)(1)

Deenergized parts. Live parts to which an employee may be exposed shall be deenergized before the employee works on or near them, unless the employer can demonstrate that deenergizing introduces additional or increased hazards or is infeasible due to equipment design or operational limitations. Live parts that operate at less than 50 volts to ground need not be deenergized if there will be no increased exposure to electrical burns or to explosion due to electric arcs.

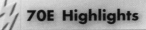

70E Highlights

An electrically safe work condition is defined in *NFPA 70E®* as "a state in which an electrical conductor or circuit part has been disconnected from energized parts, locked/tagged in accordance with established standards, tested to ensure the absence of voltage, and grounded if determined necessary."

Reprinted with permission from NFPA-70E 2012, Electrical Safety in the Workplace Copyright 2011, National Fire Protection Association, Quincy, MA 02169. This reprinted material is not the complete and official position of the NFPA on the referenced subject, which is represented only by the standard in its entirety.

Trauma Following Electrical Events

An examination of multihazard electrical incidents and the incident effects on survivors is provided in a paper delivered by Mary Capelli-Schellpfeffer, MD, MPA, of CapSchell, Inc., at the 2004 IEEE IAS Electrical Safety Workshop. Excerpts from Dr. Capelli-Schellpfeffer's paper are provided here.

The clinical spectrum of electrical incident effects on survivors ranges from the absence of any external physical signs to severe multiple trauma. Reported neuropsychiatric difficulties can vary from vague complaints seemingly unrelated to the injury event by their distance in time or apparent severity to effects consistent with anoxic brain injury accompanying an electrical trauma. In addition to physical limitations, complaints commonly described in electrical incident survivors include hearing loss, headache, memory changes, disorientation, slowing of mental processes, agitation, confusion, irritability, affective disorders, and post-traumatic stress disorder (PTSD; severe anxiety resulting from a traumatic experience).

The evaluation and treatment of electrical incident survivors can be variable, as there is little information available to provide rigorous decision making around the mental health care of these patients. Opinions differ about the nature and cause of patient symptoms, and the relationship between symptoms and factors like trauma severity, litigation, or premorbid personality. Not all survivors develop cognitive and emotional difficulties, and no consistent relationship has been established between characteristics, such as age, injury-related characteristics (e.g., voltage, current source, work error), and neuropsychological test performance.

Questions remain as to how electrical exposure affects central nervous system function. The pattern of neuropsychological effects suggests diffuse cerebral injury. Moreover, the effects of electrical incidents may produce emotional disturbance through damage to the limbic system or hypothalamic-pituitary axis. It is noteworthy that from the therapeutic perspective, it has been appreciated that the medical application of electric current in proximity to the brain during electroconvulsive therapy (ECT) affects mental status, psychiatric condition, and neuromuscular function. While the biologic mechanisms for the individual responses seen following ECT remain to be articulated, persistent alterations in patients' neuropsychiatric condition following ECT are well documented and in effect, often represent desired clinical outcomes.

Regarding electrical incident clinical effects in survivors, a study [Pliskin, Capelli-Schellpfeffer, et. al., 1998. Neuropsychological sequelae of electrical shock. *Journal of Trauma* 44 (4):709–715] analyzed the experience of the largest reported series of electrical injury survivors with neuropsychological complaints. All patients had peripheral electrical contacts (i.e., shock) with no evidence on history or examination of direct mechanical electrical contact with the head. A total of 45 males and 8 females were included in the final analysis. These individuals had a mean age of 38.5 years (range, 22 to 70 years) and a mean educational level of 13.1 years (range, 8 to 18 years).

The mean time between injury and completion of the measures used in this study was 11.2 months (range, 0.2 to 66.7 months). Twenty of the 53 patients were employed as electricians or line operators at the time of injury. There were also 7 mechanics or railroad workers, 5 office workers, 3 factory workers, 3 service technicians, 2 food service workers, 2 police officers, as well as 11 individuals with other occupations. Forty-four patients were injured on the job, and 9 were injured during nonvocational activities. At the time of follow-up contacts, 30 (56.6%) patients were working again, 18 (44.0%) patients were unemployed or retired, 1 patient was deceased, and 4 patients could not be contacted. Twenty-one of the 53 patients were injured by voltage sources less than 1000 volts (39.6%), and 27 patients sustained voltage exposures greater than 1000 volts (50.9%). Forty-four patients were hospitalized for observation or to receive initial treatment for their injuries (83.0%), while 9 patients were released after initial evaluation. Twenty patients underwent surgery for their injuries (37.7%), 32 patients received either nonsurgical treatment or no treatment, and the treatment history for 1 patient was unknown. Sixteen of the 53 patients (30.2%) sustained a loss of consciousness as a result of their electrical accident, and 4 patients (7.5%) experienced cardiac arrest. Twenty-nine patients complained of ringing in their ears (tinnitus) (18%), and 5 patients reported a loss of hearing (3%). In the 49 patients for whom complete data were available, there was no significant relationship between the reported neuropsychological symptoms, Beck Depression Inventory, self-rated memory complaints, and injury experience parameters, including voltage exposure, loss of consciousness, trauma severity, or litigation. Blast effects from the electrical incidents may help to explain why these survivors without external signs of electrical contact presented with nervous system or hearing impairment.

The potential for injury and death is directly related to the energy output from an electrical fault. However, it is the actual energy exposure (i.e., the energy transferred to the individual) that provokes a biologic effect. **See Figure 1-4.** Critical in predicting the extent of injury after an electrical incident are the quantity and form of energy transfer and the affected individual's biologic characteristics.

The dose or amount of energy transferred to an individual involved in an electrical incident can be conceptualized by considering the efficiency of the coupling between the energized source and the individual during the energy release. This "efficiency" is a function in part of the current, exposure duration, distance from the source, barriers used, surface area of the body exposed, and the material properties of biologic tissues, including the following:

- Tissue conductance
- Tissue impedance
- Tissue resistance
- Absorbance of human "biomaterials" (i.e., the water, lipids, fats, proteins, and minerals that constitute the body)

It is not simple to predict the possible energy transfer that may have occurred after an electrical incident. This information is routinely lacking during the immediate period of an incident investigation and may only be approximated later. Employees who may be at risk of an injury (while doing a job near energized equipment) also lack this information in concept and detail.

Arcing Faults Present Multiple Hazards

It is important to note that in an electrical incident, injury, or fatality can have different causes and are not limited to a "heat" injury. In general, use of the word "burn" is often misunderstood to mean "heat injury" alone. Electrical current flow absent heating can result in a "burn." Radiation injury can also create a "burn," as for example with ultraviolet radiation damage to the corneas. There are also traumatic effects from falls, crush, and shrapnel.

In the acute or rehabilitative care setting, the relationship between survivor symptoms and the multiple effects from electrical, thermal, acoustic, and radiation forces may not be apparent. However, the magnitude of the physical forces generated during an electrical incident explains in part the resemblance of some electrical incident survivors to those with mild traumatic brain injury. In particular, the acoustic component of the electrical incidents can be a mechanism for brain injury. As electrical flash/blast reportedly occurs in almost 75 percent of survived electrical incidents, the contribution of the explosion forces is presumed similar in its injury mechanism to the acceleration–deceleration event experienced by head trauma patients in motor vehicle accidents. In the triage setting, these effects may be obscured by the lack of physical signs readily recognizable by Emergency Medical Technicians (EMTs) or triage clinicians.

Knowledge of the nature of electrical incident effects is evolving, because the electrical system itself is evolving. As power density increases across more compact spaces, more fault energy is available in electrical installations. With installations designed to be compact, smaller spatial cushions exist, with less barrier protection emerging in innovative designs. The result may be less time and less physical distance through which an unintentional release of energy can be dissipated during an electrical incident. With higher power density and smaller, more compact spaces, risk of collateral damage from electrical incident effects increases, including injuries and fatalities. **See Figure 1-5.**

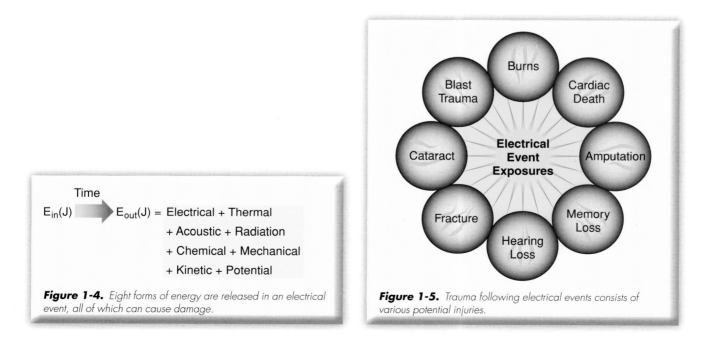

$$\text{Time}$$
$$E_{in}(J) \rightarrow E_{out}(J) = \text{Electrical} + \text{Thermal}$$
$$+ \text{Acoustic} + \text{Radiation}$$
$$+ \text{Chemical} + \text{Mechanical}$$
$$+ \text{Kinetic} + \text{Potential}$$

Figure 1-4. *Eight forms of energy are released in an electrical event, all of which can cause damage.*

Figure 1-5. *Trauma following electrical events consists of various potential injuries.*

High power density in a compact space has bomb-like potential, with characteristics of closed blast effects. For example, 1 megawatt (MW) is roughly equivalent to one stick of dynamite. One stick of dynamite is roughly 1/3 lb of TNT equivalent. So, a 100 MW scenario can be represented as loaded with 100 sticks of dynamite. Or, a 300 MW scenario can be represented as loaded with 100 lbs of TNT. A scenario in a closed space creates the possibility of closed blast effects, where the walls of an enclosure confine the blast, and energy can be dispersed only in one direction—toward the worker. This is unlike a scenario in an open space, such as a lineman working at the top of a utility pole in a vacant field, which allows energy to be dispersed in all directions (a free blast scenario).

With knowledge of the electrical power in a workplace scenario, hazards and possible electrical incident effects can be discussed in terms that are intuitively more obvious.

Adapted from: Facts on Trauma Following Electrical Events, *by Mary Capelli-Schellpfeffer, MD, MPA, CapSchell, Inc. Reported and presented at the 2004 IEEE IAS Electrical Safety Workshop held in Oakland, California. Reprinted with permission. Courtesy of Mary Capelli-Schellpfeffer, MD, MPA.*

Recognizing Limitations of PPE

In addition to understanding the effects of exposure to electrical hazards, it is important to recognize that electrical protective equipment provides limited protection against electrical hazards, much like seat belts and air bags provide limited protection from the hazards that could be encountered in an automobile accident. In much the same way that a hard hat could not be expected to protect a worker from a falling steel beam, arc-rated clothing should not be expected to always allow a worker to escape an incident unscathed. An *arc rating* is defined in *NFPA 70E* as follows:

Arc rating. The value attributed to materials that describes their performance to exposure to an electrical arc discharge. The arc rating is expressed in calories per square centimeter (cal/cm^2) and is derived from the determined value of the arc thermal performance value (ATPV) or energy of breakopen threshold (E_{BT}) (should a material system exhibit a breakopen response below the ATPV value). Arc rating is reported as either ATPV or E_{BT}, whichever is the lower value.

Informational Note No. 1: Arc-rated clothing or equipment indicates that it has been tested for exposure to an electric arc. Flame-resistant (FR) clothing without an arc rating has not been tested for exposure to an electric arc.

Informational Note No. 2: *Breakopen* is a material response evidenced by the formation of one or more holes in the innermost layer of arc-rated material that would allow flame to pass through the material.

Informational Note No. 3: ATPV is defined in ASTM F 1959–06 as the incident energy on a material or a multilayer system of materials that results in a 50 percent probability that sufficient heat transfer through the tested specimen is predicted to cause the onset of a second degree skin burn injury based on the Stoll curve, cal/cm^2.

70E Highlights

NFPA 70E includes informational notes. Per *NFPA 70E* Section 90.5(C), these materials are considered informational only and are not enforceable as requirements.

Informational Note No. 4: E_{BT} is defined in ASTM F 1959–06 as the incident energy on a material or a material system that results in a 50 percent probability of breakopen. Breakopen is defined as a hole with an area of $1.6 cm^2$ ($0.5 in^2$) or an opening of 2.5 cm (1.0 in.) in any dimension.

Reprinted with permission from NFPA-70E 2012, Electrical Safety in the Workplace Copyright 2011, National Fire Protection Association, Quincy, MA 02169. This reprinted material is not the complete and official position of the NFPA on the referenced subject, which is represented only by the standard in its entirety.

Although arc-rated apparel might provide a level of protection against a thermal event that it is rated for, many other hazards might be associated with an incident. Explosive effects, including shrapnel, could rip through protective clothing, a pressure wave could rupture eardrums, or the differential pressure that results from the wave might collapse lungs and damage other internal organs.

Understanding Requirements

It is safe to assume that not all energized work performed today falls within what the Occupational Safety and Health Administration (OSHA) recognizes as justification to work on an energized circuit. OSHA requires employers to furnish each employee with a place of employment free from recognized hazards that are causing or are likely to cause death or serious physical harm. Live parts to which an employee might be exposed must be deenergized before the employee works on or near them, unless the employer can demonstrate that deenergizing introduces additional or increased hazards or is infeasible due to equipment design or operational limitations.

It is worthwhile to consider why some laws are followed routinely while others, such as refraining from working on exposed energized electrical equipment, are often ignored. One factor that can lead to the performance of energized work is ignorance of the laws in effect. It is critical to recognize

70E Highlights

NFPA 70E Section 130.7(A) Informational Note No. 1:

The PPE requirements of 130.7 are intended to protect a person from arc flash and shock hazards. While some situations could result in burns to the skin, even with the protection selected, burn injury should be reduced and survivable. Due to the explosive effect of some arc events, physical trauma injuries could occur. The PPE requirements of 130.7 do not address protection against physical trauma other than exposure to the thermal effects of an arc flash.

Reprinted with permission from NFPA-70E 2012, Electrical Safety in the Workplace Copyright 2011, National Fire Protection Association, Quincy, MA 02169. This reprinted material is not the complete and official position of the NFPA on the referenced subject, which is represented only by the standard in its entirety.

the limitations on energized work. Both *NFPA 70E* and OSHA generally stipulate that energized work is permitted only where the employer can demonstrate that deenergizing introduces additional or increased hazards or where the task to be performed is infeasible in a deenergized state due to equipment design or operational limitations. **See Figure 1-6**. Equipment must be locked out and tagged out in accordance with

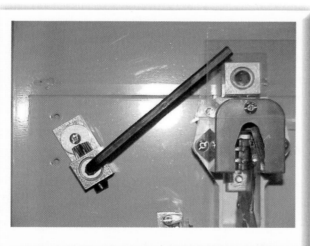

Figure 1-6. *Personnel and equipment could be severely damaged in the event of a mishap during energized work.*

established policy, unless the need to work energized is demonstrated.

Electrical Workers must be intimately familiar with their company's policy on working while exposed to electrical hazards. While it may be laudable to work on energized equipment only when "we absolutely have to," that may not be entirely possible. Energized work includes voltage testing; voltage testing is among the tasks that are infeasible to perform deenergized. As the OSHA and *NFPA 70E* requirements are fully explored in subsequent chapters, it will become apparent that the vast majority of work performed on energized equipment does not qualify as work allowed by OSHA and *NFPA 70E*.

Decisions

Many choices about how to perform a task are made long before a worker is placed before a hazard. Other decisions are made along the way. Many decisions are made just before, and even during, the performance of a task.

Many factors should be considered in creating a safe work environment. The following questions should be among those included in the development and implementation of a safety program, training, and a hazard identification and a risk assessment procedure:

- Has an electrical safety plan been developed and implemented?
- Has appropriate training been provided?
- Are safe work practices in place and understood, including lockout/tagout and placing equipment in an electrically safe work condition?
- Has the required protective equipment been provided?
- Has an attempt been made to reduce the potential worker exposure through work practice or design considerations, such as arc-resistant switchgear or remote switching?
- Was the overcurrent protection selected solely to protect the equipment or was worker protection considered as well?
- Was a current-limiting overcurrent protective device (OCPD) selected?
- Was the OCPD applied within its rating?
- If the OCPD was replaced, was the appropriate degree of current limitation applied?
- Has the OCPD been maintained properly?
- Has the impedance on the transformer on the supply side of the service changed?

These are among a few of the concerns that must be addressed when the hazard/risk analysis is conducted.

Appropriate Priorities

A corporation's management might say that it cannot afford to develop a safety plan, provide necessary training, or provide the appropriate PPE and insulated tools. In such a case, corporate priorities should be analyzed. While there may be other important financial priorities, crucial items such as appropriate training and PPE should always be a primary consideration.

Training and Personal Responsibility

Some Electrical Workers who fail to maintain a safety culture might argue that they were never trained or were never provided with the required protective equipment. It is possible that a worker might receive training in safe work practices, but then fail to implement that training. In such a case, there is no lack of training, but the worker makes the conscious decision not to wear protective equipment. Other barriers to working in an electrically safe work condition include a lack of a safety plan and, as discussed, a lack of understanding of the hazards.

Costs of Energized Work Versus Shutting Down

The true cost of an electrical injury or fatality must be considered before deciding that work must be performed on energized equipment. Consider the following:

- What is the true cost if something goes wrong?
- As mentioned earlier, a shutdown that "cannot" be scheduled could become an unscheduled shutdown. An unscheduled shutdown may ultimately cost more than a planned shutdown.
- Have the costs associated with human life, equipment, loss of production, insurance premium increases, potential exclusion from bid lists, corporate image, and worker morale been considered in the decision-making process?
- Can the equipment be shut down at night when very few, if any, people will be inconvenienced?
- Can equipment be shut down over the weekend?
- Can a shutdown be scheduled at some point?

Consider whether supervisors are rewarded for safety shortcuts. Are the costs associated with injuries and citations charged against the job, or are they considered overhead (a cost of doing business)? If such costs are not charged against the job, a job site where injuries occur may appear to be more profitable than it really is. A manager who is rewarded for safety shortcuts may not make decisions that are in the best interest of Electrical Workers or the company.

Time Pressure

Imagine an Electrical Worker operating out of a service truck who responds to a report of a transfer switch that malfunctioned at 3 a.m. at a nursing home. In such a case, the worker will be expected to get the power restored as soon as possible. An Electrical Worker must adjust the customer's understanding of "as soon as possible" to include working safely, and the added time that working safely could require. The Electrical Worker must know how to evaluate the magnitude of the hazards present, and how to make decisions about the job while simultaneously prioritizing electrical safety.

Employee Qualification

Electrical Workers must be qualified to perform the tasks to which they are assigned. An Electrical Worker must receive qualification in advance of responding to an emergency call. In such a case, items to consider include the following: Is the personal and other protective equipment in the truck? Is the protection adequate? Is the Electrical Worker qualified to make those decisions at the job site, or will someone from the engineering department need to be consulted as well? Has the worker been trained to know how to use the equipment properly and understand its limitations?

Implementation Choices

At some point, a decision is made on how to implement an electrical safety program, including implementation of safety-related work practices and personal and other protective equipment.

An examination of the estimated cost of implementing an arc-flash PPE program is provided in the article "Escaping Arc Danger" by H. Landis Floyd, published in the May/June 2008 issue of *IEEE Industry Applications Magazine*. The article explores the cost of arc-flash clothing implementation through three examples: (1) doing nothing, (2) selecting PPE using the tables in *NFPA 70E*, and (3) conducting hazard analysis and PPE selection based on the results of a detailed arc-flash hazard analysis. **See Figure 1-7.**

Figure 1-7. *Cost comparison (in U.S. $1000) of average 5-year costs for sample company for three options for an arc flash mitigation program.*

Method	PPE Costs	Analysis Costs	Injury Costs	Total Costs
No arc flash PPE	$0	$0	$20,800	$20,800
Two hazard level PPE choice	$1570	$100	$6150	$7820
Detailed analysis	$835	$2000	$150	$2985

Note that a comparison of the costs of implementing the three scenarios is summarized, which is Table 1 from the article.

Doan and Floyd [12] estimated the total cost of implementing an arc flash protective clothing by considering three options: do nothing, minimum compliance, and application of protective measures based on the state-of-the-art hazard analysis methods. This comparison is summarized in Table 1. This is based on the assumption that all recommendations are followed. In practice, human error and other factors can increase injury frequency and overall costs with any of these options.

The quality of implementing an arc hazard mitigation program can vary from doing nothing, to minimum compliance, to a state-of-the-art program with arc hazard analysis as the basis for a full range of control measures. A "do nothing" approach is very much out of step with regulatory requirements, and evolution in electrical safety knowledge, as previously noted, is likely the most costly long-term choice. Advancements in the arc-flash hazards mitigation in the mining industry have been hampered somewhat by the delay in recognizing the relevance of *NFPA 70E* to mining operations.

One of the options provided in *NFPA 70E* 2004 for selecting arc flash PPE is based on tables, which provide lists of common tasks, with appropriate arc flash protective equipment noted for each task. These tables can be useful, but they can also be misapplied. The explanatory footnotes accompanying the tables may be overlooked. These notes explain that the electrical system must have certain specifications for the tables to be applicable. Failure to assure that the electrical system meets these requirements can result in either underrated or overrated PPE. Both conditions can have serious consequences. Underrated PPE can result in serious injuries from arc hazard exposure. In this case, the thermal protection rating of the multilayer system of garments was less than the exposure, resulting in severe burns, even though the inner garment layer is relatively intact. On the other extreme, overrating of PPE can lead to unnecessary heat stress.

An approach that is based on a detailed arc hazard assessment enables the identification of exposures where engineering design or administrative controls can reduce the severity or frequency of exposure, reduce the frequency of potential arc flash events, and better assure that the PPE is appropriately rated for exposures.

Summary

The list of workers' statements addressing safety culture, as presented earlier in this chapter, might seem justified and realistic. It might be true that the worker is pressured to do a job quickly or that PPE seems inconvenient or uncomfortable to wear. Nevertheless, not wearing PPE increases the potential for serious injury and perhaps—and even worse—death. When a work situation is so inherently dangerous, issues such as time pressure are irrelevant. The need to ensure the worker's safety overrides any other concerns that, while real and difficult, are not as important.

A customer's needs, while important to a business, should not be prioritized over the need of the worker to remain alive and uninjured. Remember the family and friends who will suffer the consequences of these decisions if things go wrong. Whether an incident results in an injury or a fatality, family and friends suffer emotionally and financially. If the worker does survive, he or she often requires months or years of rehabilitation.

A person might get only one chance to make a decision that those left behind will regret for years, or a lifetime. What would a loved one recommend when asked if it is worth the risk of ignoring the rules "just this one time"? There are many reasons why risks are taken. Sometimes, it could be calculated risk; at other times, it might be uninformed risk. Whatever the reason, it is not likely worth the risk.

There is little question that, all too often, a culture is in place where workers are allowed and expected to work on energized equipment. Too often the existing culture supports a tendency to work routinely on or near energized electrical circuits. What is most important—that the worker performs the work without becoming injured or killed—happens only when workers, contractors, and customers become educated about the hazards and ways to properly handle them. Workers, contractors, and their customers must be made aware of the requirements that are in place, the hazards that exist, the decisions that can and should be made, and the true costs associated with an incident when things go wrong. It is a multifaceted challenge that requires a multifaceted education process and a multifaceted change in culture.

BACKGROUND

Introduction to the NIOSH FACE Program

The National Institute for Occupational Safety and Health's (NIOSH) Fatality Assessment and Control Evaluation (FACE) program is a research program designed to identify and study all fatal occupational injuries, including those of an electrical nature. The goal of the FACE program is to prevent occupational fatalities across the United States by identifying and investigating work situations at high risk for injury and then formulating and disseminating prevention strategies to those who can intervene in the workplace.

FACE Program's Two Components

NIOSH's in-house FACE began in 1982. Participating states voluntarily notify NIOSH of traumatic occupational fatalities resulting from targeted causes of death that have included confined spaces, electrocutions, machine-related fatalities, falls from elevation, and logging incidents. The program is currently focusing on deaths associated with machinery, deaths of youths younger than 18 years of age, and street/highway construction work-zone fatalities.

The FACE program began operating as a state program in 1989. Today, nine state health or labor departments have cooperative agreements with NIOSH for conducting surveillance, targeted investigations, and prevention activities at the state level using the FACE model.

FACE is a research program; investigators do not enforce compliance with state or federal occupational safety and health standards and do not determine fault or blame.

Primary Activities of the FACE Program

The primary activities of the FACE program include the following:

- Conducting surveillance to identify occupational fatalities
- Performing investigations of specific types of events to identify injury risks
- Developing recommendations designed to control or eliminate identified risks
- Making injury prevention information available to workers, employers, and safety and health professionals

On-Site Investigations

On-site investigations are essential for observing sites where fatalities have occurred and for gathering facts and data from company officials, witnesses, and coworkers. Investigators collect facts and data on what was happening just before, at the time of, and right after the fatal injury. These facts become the basis for writing investigative reports.

During the on-site investigations, facts and data are collected on items such as the following:

- Type of industry involved
- Number of employees in the company
- Company safety program
- The victim's age, sex, and occupation
- The working environment
- The tasks the victim was performing
- The tools or equipment the victim was using
- The energy release that results in fatal injury
- The role of management in controlling how these factors interact

Each day, on average, 16 workers die as a result of a traumatic injury on the job. Investigations conducted through the FACE program allow the identification of factors that contribute to fatal occupational injuries. This information is used to develop comprehensive recommendations for preventing similar deaths.

FACE Information and Reports

Surveillance and investigative reports are maintained by NIOSH in a database. NIOSH researchers use this information to identify new hazards and case clusters. FACE information may suggest the need for new research or prevention efforts or for new or revised regulations to protect workers. NIOSH publications are developed to highlight high-risk work situations and to provide safety recommendations. These reports are disseminated to targeted audiences and are available on the Internet through the NIOSH homepage or through the NIOSH publications office.

The names of employers, victims, and/or witnesses are not used in written investigative reports or included in the FACE database.

Adapted from: Fatality Assessment and Control Evaluation (FACE) program website. Accessed May 29, 2012.

For additional information, visit qr.njatcdb.org Item #1185.

REVIEW QUESTIONS

1. Often, a culture exists in the workplace where workers are routinely expected or allowed to work on or near energized electrical circuits. This practice might be due to ignorance of laws that have been in place for decades, lack of knowledge of the severity of the hazards, or perhaps failure to realize how __?__ a task situation might change and cause an energy release.
   a. inexplicably
   b. quickly
   c. rarely
   d. slowly

2. It is less likely that workers, contractors, and facility owners would allow energized work if __?__ involved in the decision-making process fully understood the laws and requirements, hazards and true costs, as well as consequences associated with energized work.
   a. builders
   b. employees
   c. employers
   d. everyone

3. To ensure that an environment is electrically safe, the following must be done at a minimum: develop, conduct, and implement __?__ for qualified and unqualified persons, develop and implement hazard identification and risk assessment procedures, and engineer out the hazards or reduce them as far as is practicable.
   a. guidelines
   b. quizzes
   c. study skills
   d. training

4. The industry has evolved and made great strides to protect personnel against electrical shock through the use of __?__ and rubber protective goods such as insulating gloves and blankets.
   a. arc-fault circuit interrupters (AFCI)
   b. circuit breakers
   c. fuses
   d. ground-fault circuit interrupters (GFCI)

5. In addition to physical limitations, complaints commonly described by survivors of electrical incidents include hearing loss and headache, memory changes and disorientation, confusion, and post-traumatic stress disorder.
   a. True
   b. False

6. With installations designed to be compact, smaller spatial cushions exist, with less barrier protection emerging in innovative designs. The result may be less time and less physical distance through which a(n) __?__ release of energy can be dissipated during an electrical incident.
   a. intentional
   b. large
   c. small
   d. unintentional

7. With higher power density and smaller, more compact spaces, the risk of collateral damage (including injuries and fatalities) from the effects of an electrical incident __?__ .
   a. decreases
   b. increases
   c. remains the same
   d. none of the above

8. Electrical protective equipment generally does not provide protection from __?__ .
   a. arc blast
   b. arc flash
   c. shock
   d. all of the above

9. An Electrical Worker must adjust the customer's understanding of "as soon as possible" while on a troubleshooting call to include working safely, and to the added time that working safely could require, how to evaluate the magnitude of the hazards present and how to make decisions about the job.
   a. True
   b. False

OBJECTIVES

1. Identify electrical hazards.
2. Explain the effects of current on the human body.
3. Describe an arcing fault event and the effects it can have on the human body.
4. Understand the role of overcurrent protective devices in arcing fault energy release.

REFERENCE

1. *NFPA 70E®*, 2012 Edition

CASE STUDY

A 19-year-old electrician's apprentice and a Journeyman Electrical Worker were installing two new switch boxes during an office building renovation project. The circuits in the room where the new switch boxes were being installed were deenergized, with the exception of the circuit to an existing metal switch box suspended by conduit from the ceiling. The circuit feeding the suspended switch box was energized by a 277-volt circuit from the adjacent room. A metal-sheathed cable fed this box and entered it through the box's side.

The Journeyman momentarily left the room and told the victim that "they would figure out how to wire the boxes" when he returned. Apparently, the victim thought the circuit feeding the suspended box was deenergized because it was in the same room as the new boxes being installed. While the victim was alone, he elected to disassemble the switch box suspended from the ceiling. He reached into the suspended box and cut the conductors from each of the four terminal connections in the box. Then, with his left hand, the victim pulled the metal sheathed conductor out of the switch box that he was holding in his right hand. The bare conductors must have contacted the box and/or his left hand. In turn, he provided a path to ground and was electrocuted. Burn marks found on the victim's right hand were consistent with the shape of the box.

The victim was found 14 feet from the switch box. Emergency medical services personnel responded and administered advanced cardiac life-support procedures. Attempts to resuscitate the victim were unsuccessful. The victim was pronounced dead on arrival at a nearby hospital.

For details of this case, see FACE Program Case 87-34. Accessed May 2, 2012.

For additional information, visit qr.njatcdb.org Item #1186.

Introduction

Electricity is pervasive in our modern infrastructure. Electrical Workers continuously install, maintain, and troubleshoot circuits, and both they and their customers often take electricity for granted. Electricity, however, remains a very dangerous hazard for people working on or near it. Even when electrical circuits do not directly pose serious shock or burn hazards by themselves, these circuits are found adjacent to circuits with potentially lethal levels of energy. A minor shock from a low energy circuit can cause a worker to drop a tool onto another circuit, resulting in a lethal arcing fault. Involuntary reaction to a shock can result in bruises, bone fractures, and even death from collisions or falls.

The following are recognized as examples of some common electrical hazards that can cause injury and even death while a person works on or near electrical equipment and systems:

- Electrical shock
- Arc flash
- Arc blast

Workplace Hazards

Electrical Workers face health and safety challenges on a daily basis. Potential dangers abound. As discussed in the chapter, "Electrical Safety Culture," potential reasons for the decision to expose oneself to dangers may include an incomplete understanding of the applicable requirements. Examples may include lack of understanding of the lockout/tagout program or poor retention of hazardous communication training. Workers may lack knowledge regarding how to read a Material Safety Data Sheet (MSDS) or be unable to recall the location and availability of the written hazardous communication program and the required list of hazardous chemicals. Potential hazards include laser equipment, paints and solvents, improperly built scaffolding, and an improperly designed excavation. Unfortunately, not all Electrical Workers exposed to these and other hazards fully appreciate the exposures they face, nor do they always know how best to avoid the many potential dangers that can cause injury and death.

The workplace must be evaluated to identify and eliminate all hazards. However, this course will focus primarily on the electrical hazards of shock, arc flash, and arc blast.

Electrical Shock

According to data compiled by the U.S. Department of Labor's Bureau of Labor Statistics (BLS), between the years of 2003 and 2009, electricity was the cause of death in 1,573 on-the-job fatalities. In addition, during the same period, the BLS estimates that 18,460 workers sustained nonfatal injuries in electrical incidents. Many of these nonfatal injuries took place in the manufacturing and construction industries.

Most Electrical Workers are aware of the danger of electrical shock, including electrocution. Historically, shock is the electrical hazard most prominently mentioned in the majority of electrical safety standards. In reality, few Electrical Workers truly understand that only a minimal amount of current is required to cause injury or death. The current drawn by a seven-and-a-half-watt, 120-volt lamp, passing across the chest, from hand to hand or from hand to foot, is enough to cause electrocution.

The effects of electric current on the human body depend on the following factors:

- Circuit characteristics (current, resistance, frequency, and voltage)
- Contact resistance and internal resistance of the body
- The current's pathway through the body, which is determined by the contact location and internal body chemistry
- Duration of the contact
- Environmental conditions that affect the body's contact resistance

OSHA Tip

Lockout is one of the main protective measures that can be taken to prevent workers from working on an energized circuit. This specific procedure involves placing a lockout device on an energy-isolating device, making it physically impossible to operate the equipment until the lockout device is removed.

OSHA defines lockout as follows:

lockout The placement of a lockout device on an energy-isolating device, in accordance with an established procedure, ensuring that the energy-isolating device and the equipment being controlled cannot be operated until the lockout device is removed.

OSHA Tip

Tagout is a protective measure in which a prominent warning device, such as a tag, is placed on the energy-isolating device, indicating that the equipment must not be operated. The tagout device warns that a worker might be injured if he or she operates the energy-isolating device.

OSHA defines tagout as follows:

tagout The placement of a tagout device on an energy-isolating device, in accordance with an established procedure, to indicate that the energy-isolating device and the equipment being controlled may not be operated until the tagout device is removed.

Skin Resistance

An integral concept for understanding the magnitude of currents possible in the human body is skin contact resistance. **See Figure 2-1**. The skin's resistance can change as a function of the moisture present in its external and internal layers, which can be altered by such factors as ambient temperatures, humidity, fear, and anxiety.

Body tissue, vital organs, blood vessels, and nerve (nonfat) tissue in the human body contain water and electrolytes and are highly conductive, offering only limited resistance to alternating electric current. As the skin is broken down by electrical current, resistance drops and current levels increase.

Consider an example of a person with hand-to-hand resistance of 1,000 ohms. The voltage determines the amount of current passing through the body.

While 1,000 ohms might appear to be low, even lower levels can occur. For example, an Electrical Worker wearing sweat-soaked noninsulating gloves on both hands while maintaining a full-hand grasp of an energized conductor and a grounded pipe or conduit would approach lower levels. Moreover, cuts, abrasions, or blisters on hands can negate skin resistance, leaving only internal body resistance to oppose current flow. A circuit in the range of 50 volts could be dangerous in this instance.

Caution

Even current as low as that drawn by a seven-and-a-half-watt, 120-volt lamp can be enough to cause electrocution.

Using Ohm's Law, the current (I, in amperes) in a circuit can be calculated based on the circuit voltage (V) and the resistance (R). *Ohm's Law* is the mathematical relationship between voltage, current, and resistance in an electric circuit; it states that current flowing in a circuit is proportional to electromotive force (voltage) and inversely proportional to resistance: I = E/R. Current (amperes) equals voltage (volts) divided by resistance (ohms), a relationship that may be expressed in equation form:

$$I \text{ (amperes)} = \frac{V \text{(volts)}}{R \text{ (ohms)}}$$

$$\text{Exmple 1: } I = \frac{480}{1,000} = 0.480 \text{ amp (480 mA)}$$

$$\text{Exmple 2: } I = \frac{120}{1,000} = 0.120 \text{ amp (120 mA)}$$

Electrical currents can cause muscles to lock up, resulting in the inability of a person to release his or her grip from the current source. The lowest current at which muscle lockup occurs is known as the let-go threshold current. The *let-go threshold* is the electrical current level at which the brain's electrical signals to muscles can no longer overcome the signals introduced by an external electrical system. Because these external signals lock muscles in the contracted position, the body may not be able to let go when the brain tells it to do so. At 60 hertz (Hz), most people have a "let-go" limit of 10 to 40 milliamperes (mA). **See Figure 2-2**.

Figure 2-1. *Human resistance values range for a variety of skin-contact conditions.*

Condition	Resistance (ohms)	
	Dry	Wet
Finger touch	40,000 to 1,000,000	4,000 to 15,000
Hand holding wire	15,000 to 50,000	3,000 to 6,000
Finger–thumb grasp	10,000 to 30,000	2,000 to 5,000
Hand holding pliers	5,000 to 10,000	1,000 to 3,000
Palm touch	3,000 to 8,000	1,000 to 2,000
Hand around 1½-inch pipe	1,000 to 3,000	500 to 1,500
Two hands around 1½-inch pipe	500 to 1,500	250 to 750
Hand immersed	N/A	200 to 500
Foot immersed	N/A	100 to 300
Human body, internal, excluding skin	200 to 1,000	

N/A: Not applicable
Data source: Kouwenhoven, W. B., and Milnor, W. R., Field Treatment of Electric Shock Cases—1, AIEE Trans. Power Apparatus and Systems, Volume 76, pp. 82–84, April 1957; discussion pp. 84–87.

Figure 2-2. *The effects of electrical shock vary according to current level (60 Hz AC).*

Response	60 Hz, AC Current (mA)
Tingling sensation	0.5 to 3
Muscle contraction and pain	3 to 10
Let-go threshold	10 to 40
Respiratory paralysis	30 to 75
Heart fibrillation; might clamp tight	100 to 200
Tissue and organs burn	More than 1,500

Data source: Kouwenhoven, W. B., and Milnor, W. R., Field Treatment of Electric Shock Cases—1, AIEE Trans. Power Apparatus and Systems, Volume 76, pp. 82–84, April 1957; discussion pp. 84–87.

Potential injury (current flow) also increases with time. A victim who cannot "let go" of a current source is much more likely to be electrocuted than someone whose reaction removes him or her from the circuit more quickly. A victim who is exposed for only a fraction of a second is less likely to sustain an injury.

Extent of Injury

The most damaging path for electrical current is through the chest cavity and head. **See Figure 2-3**. Any prolonged exposure to 60 hertz current of 10 milliamperes or more might be fatal. Fatal ventricular fibrillation of the heart (a state in which the heart does not contract, but instead twitches, resulting in stopping of its rhythmic pumping action) can be initiated by a current flow of 100 to 200 milliamperes. These injuries can cause fatalities resulting from direct paralysis of the respiratory system, failure of the rhythmic heart pumping action, or immediate heart stoppage.

During fibrillation, the victim might become unconscious. Alternatively, the individual may remain conscious, deny needing help, walk a few feet, and then collapse within a short time frame. Either situation can result in death within a few minutes or hours. Prompt medical attention is needed for anyone receiving electrical shock. Many people involved in electrical incidents can be saved, provided that they receive proper medical treatment, including cardiopulmonary resuscitation (CPR), quickly.

The extent of injury resulting from electrical shock is typically not immediately visible because the current flows through muscle tissue and organs and not through the skin, except at the entrance and exit points. Entrance and exit wounds are usually coagulated areas and may exhibit charring. Or, these areas might be missing, having "exploded" away from the body due to the level of energy present. The smaller the area of contact, the greater the heat produced. For a given current, damage in the limbs might be the greatest, due to the higher current flux per unit of cross-sectional area.

Within the body, the current can burn internal body parts in its path, yet leave the skin unaffected. This type of injury might be difficult to diagnose, as the only initial signs of injury are the entry and exit wounds. Damage to the internal tissues along the current path, while not apparent immediately, might cause delayed internal tissue swelling and irritation. Prompt medical attention can minimize possible loss of blood circulation. However, for some surviving shock victims, so much internal tissue is permanently damaged that amputation of an extremity is necessary to prevent death.

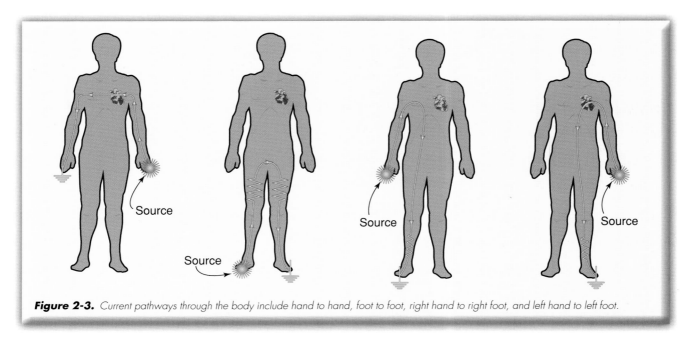

Figure 2-3. *Current pathways through the body include hand to hand, foot to foot, right hand to right foot, and left hand to left foot.*

Prevention

All electrocutions are preventable. A significant portion of Occupational Safety and Health Administration (OSHA) requirements is dedicated to electrical safety. Current OSHA regulations were promulgated many years ago; OSHA compliance is considered a minimum requirement for improving the safety of the workplace.

Several standards offer guidance regarding safe approach distances to minimize the possibility of shock from exposed electrical conductors of different voltage levels. One of the most recent, and perhaps the most authoritative, guidelines is presented in *NFPA 70E: Standard for Electrical Safety in the Workplace®*, in Section 130.4, Approach Boundaries to Energized Electrical Conductors or Circuit Parts. (*Electrical Safety in the Workplace* and *NFPA 70E* are registered trademarks of the National Fire Protection Association, Quincy, Massachusetts.) The requirements related to approach boundaries will be covered in a subsequent chapter.

Arcing Faults: Arc Flash and Arc Blast

Arcing Fault Basics

The unique aspect of an arcing fault is that the fault current flows through the air between conductors or a conductor and a grounded part. The arc has an associated arc voltage because of arc impedance. The product of the fault current and arc voltage in a concentrated area may result in a tremendous amount of energy that is released in several forms. **See Figure 2-4.**

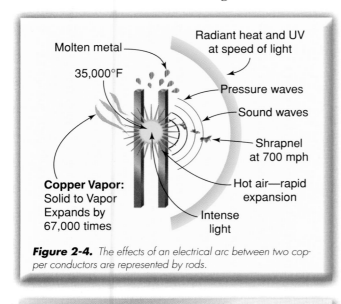

Figure 2-4. *The effects of an electrical arc between two copper conductors are represented by rods.*

Labels in figure:
- Molten metal
- 35,000°F
- Radiant heat and UV at speed of light
- Pressure waves
- Sound waves
- Shrapnel at 700 mph
- Hot air—rapid expansion
- Intense light
- **Copper Vapor:** Solid to Vapor Expands by 67,000 times

⊙SHA Tip

OSHA compliance is considered a minimum requirement.

The resulting energies can take the form of intense heat, brilliant light, and tremendous pressures. Intense heat from the arcing source travels at the speed of light. The temperature of the arc terminals can reach approximately 35,000°F, which is about four times as hot as the surface of the sun. The high arc temperature changes the state of conductors from solid to both hot molten metal and to metal vapor. The immediate vaporization of the conductors is an explosive change in state. Copper vapor expands to 67,000 times the volume of solid copper; thus a copper conductive component the size of a penny could expand as it vaporizes and disperse to the size of a refrigerator. Because of the expansive vaporization of conductive metal, a line-to-line or line-to-ground arcing fault can escalate into a three-phase arcing fault in less than a thousandth of a second.

The release of intense thermal energy superheats the immediate surrounding air. The air also expands in an explosive manner. The rapid vaporization of conductors and superheating of air result in a pressure wave of air and gases and a conductive plasma cloud that can engulf a person. In addition, the thermal shock and pressures can violently destroy circuit components. The pressure waves hurl the destroyed, fragmented components like shrapnel at high velocity; shrapnel fragments can be expelled in excess of 700 miles per hour, or about the speed at which shotgun pellets leave a gun barrel. Molten metal droplets at high temperatures typically are blown out from the event due to the pressure waves.

Testing has proved that the arcing fault current magnitude and time duration are among the most critical variables in determining the amount of energy released. Notably, the predictability of arcing faults and the energy released by an arcing fault are subject to significant variance. Among the variables that affect the outcome are the following:

- Available bolted short-circuit current
- The time the fault is permitted to flow (speed of the overcurrent protective device [OCPD]; *overcurrent protective device* is a general term that includes fuses, circuit breakers, and relays.)
- Arc-gap spacing
- Size of the enclosure (or no enclosure)
- Power factor of the fault
- System voltage
- Whether the arcing fault can sustain itself
- Type of system grounding scheme
- Distance of the worker's body parts from the arc

Typically, engineering data that the industry provides about arcing faults are based on specific values of these variables. For instance, for systems of 600 volts and less, much of the data have been gathered from testing on systems with an arc-gap spacing of 1.25 or 1.0 inches and incident energy determined at 18 inches from the point of the arcing fault. *Incident energy* is defined in *NFPA 70E* as follows:

> **Incident energy.** The amount of energy impressed on a surface, at a certain distance from the source, generated during an electrical arc event. One of the units used to measure incident energy is calories per centimeter squared (cal/cm²).
>
> *Reprinted with permission from NFPA-70E 2012, Electrical Safety in the Workplace Copyright 2011, National Fire Protection Association, Quincy, MA 02169. This reprinted material is not the complete and official position of the NFPA on the referenced subject, which is represented only by the standard in its entirety.*

⚡ 70E Highlights

NFPA 70E Section 130.7(A) Informational Note No. 1
The PPE requirements of 130.7 are intended to protect a person from arc flash and shock hazards. While some situations could result in burns to the skin, even with the protection selected, burn injury should be reduced and survivable. Due to the explosive effects of some arc events, physical trauma injuries could occur. The PPE requirements of 130.7 do not address protection against physical trauma other than exposure to the thermal effects of an arc flash.

Reprinted with permission from NFPA-70E 2012, Electrical Safety in the Workplace Copyright 2011, National Fire Protection Association, Quincy, MA 02169. This reprinted material is not the complete and official position of the NFPA on the referenced subject, which is represented only by the standard in its entirety.

Arc Flash

Arc flash is the hazard associated with the release of thermal energy during an arcing fault. In recent years, awareness of arc-flash hazards has been increasing. Each year, more than 2,000 people are admitted to burn centers in the United States with severe electrical burns. Electrical burns are considered extremely hazardous for a number of reasons. Direct contact with the circuit is not necessary to incur a serious, even deadly, burn. In fact, serious or fatal burns can occur at distances of more than 10 feet from the source of an arc flash. Ignition of flammable clothing worn by a worker is a cause of some of the most severe burns. Molten metal splatter or thermal energy emitted from an arcing fault ignites flammable clothing and before the clothing can be removed severe burns result.

The most common unit of measure used to quantify an arc-flash hazard is cal/cm². This unit of measurement is the amount of heat energy imposed on a surface area at a given distance from an arcing fault. Arc-flash suits and arc-rated clothing also have an arc rating expressed in cal/cm² and are designed to help protect workers from an arc-flash hazard.

NFPA 70E Annex D and the current edition of IEEE 1584, *IEEE Guide for Performing Arc Flash Hazard Calculations*, by the Institute of Electrical and Electronics Engineers, illustrate methods to estimate the amount of thermal energy (incident energy) available. An arc-flash hazard analysis allows a person to select the proper personal protective equipment (PPE). Incident energy analysis is discussed in detail in subsequent chapters.

Arc Blast

Arc blast is associated with the release of tremendous pressure that can occur during an arcing fault event. The worst arc blast hazards typically result from arcing faults that release high energy in a short-time duration. Various individuals and organizations in the industry have researched and continue to research ways to quantify the risks associated with arc blast. A number of the potential hazards associated with arc blast were covered in the discussion of arcing fault basics. However, there is little or no information at this time on arc-blast hazard risk assessment or on ways to protect workers from an arc-blast hazard. Arc-flash suits and other arc-rated clothing used to protect Electrical Workers from arc-flash hazards might not protect them from arc blast.

How Arcing Faults Can Affect Humans

Too few people recognize the extreme nature of electrical arcing faults, the potential for severe burns from arc flash, and the potential for injuries due to high pressures from arc blast. For Electrical Workers, the effects of an arcing fault can be devastating.

Burns are the most prevalent consequence of electrical incidents. These injuries can be due to either contact (shock hazard) or arc flash. Three basic types of such burns are distinguished:

- **Electrical burns due to current flow** Tissue damage (whether skin deep or deeper) occurs because the body is unable to dissipate the heat from the current flow through the body. The damage to tissue can be internal and initially not obvious from external examination. Typically, electrical burns are slow to heal and frequently result in amputation.

- **Arc burns by radiant or convective heat** Temperatures generated by electric arcs can burn flesh and ignite clothing at a distance of 10 feet or more.
- **Thermal contact burns (conductive heat)** These injuries are normally experienced from skin contact with the hot surfaces of overheated electric conductors or a person's clothing that ignites due to an arc flash.

Caution

It is not necessary to be in contact with the circuit to incur a serious burn. Serious or fatal burns can occur at distances of more than 10 feet from energized conductors.

Studies show that when skin temperature is as low as 110°F, the body's temperature equilibrium begins to break down in about six hours. At 158°F, a one-second duration is sufficient to cause total cell destruction. Skin at temperatures of 205°F for more than 0.1 second can cause incurable, third-degree burns. **See Figure 2-5**.

In addition to burn injuries, victims of arcing faults can experience damage to their sight, hearing, and lungs, as well as skeletal, respiratory, muscular, and nervous systems. The speed of an arcing fault event can be so rapid that the human system cannot react quickly enough for a worker to take corrective measures. The radiant thermal waves, high pressure waves, spewing of hot molten metal, intense light,

hurling shrapnel, and the hot conductive plasma cloud can be devastating in a fraction of a second. The intense thermal energy released can cause severe burns or ignite flammable clothing. Molten metal, when blown out from the circuit, can burn skin or ignite flammable clothing and cause serious burns over much of the body. The worker might gasp and inhale hot air and vaporized metal, sustaining severe injury to the respiratory system. The tremendous pressure blast from the vaporization of conducting materials and superheating of air can fracture ribs, collapse lungs, and knock a worker down or cause him or her to be thrown some distance.

It is important to realize that the time in which the arcing fault event runs its course might be only a fraction of a second. In a matter of approximately 0.001 second, a single-phase arcing fault can escalate to a three-phase arcing fault. Tremendous energy can be released in a few hundredths of a second. Humans cannot detect, comprehend, or react to events in these time frames.

Sometimes a greater respect for arcing fault and shock hazards is afforded to medium- and high-voltage systems. However, injury reports reveal that serious accidents are occurring at an alarming rate on systems of 600 volts or less, in part because of the high fault currents that are possible. Also, designers, managers, and workers tend not to take the same necessary precautions when designing or working on medium- and high-voltage systems.

Staged Arc-Flash Tests

An ad-hoc electrical safety working group within the IEEE Petroleum and Chemical Industry Committee conducted staged arc-flash tests to investigate arcing fault hazards. These tests and others are detailed in "Staged Tests Increase Awareness of Arc-Fault Hazards in Electrical Equipment" (IEEE Petroleum and Chemical Industry Conference Record, September 1997, pp. 313–322). One finding of this IEEE paper is that current-limiting OCPDs reduce damage and arc-fault energy (provided that the fault current is within the current-limiting range of the OCPD). To better assess the benefit of limiting the current of an arcing fault, it is important to note some key thresholds of injury for humans. **See Figure 2-6**. Results of the staged arc-flash tests were recorded by sensors on mannequins and can be compared to these thresholds.

Test 4, Test 3, and Test 1

The results of three of the electrical safety working group's tests are reviewed here, identified as Test 4, Test 3, and Test 1. All three of these tests were conducted on

Figure 2-5. *There is a relationship between skin temperature and tolerance.*

Skin Temperature	Duration	Damage Caused
110°F	6.0 hours	Cell breakdown begins
158°F	1.0 second	Total cell destruction
176°F	0.1 second	Curable (second-degree) burn
205°F	0.1 second	Incurable (third-degree) burn

Source: Bussmann Safety BASICs Handbook, Courtesy of Cooper Bussmann, Inc. 2004.

Figure 2-6. *There are several key thresholds for injury from an arcing fault.*

Threshold for Injury	Measurement
Just curable burn threshold	80°C/176°F (0.1 second)
Incurable burn threshold	96°C/205°F (just under the temperature where water will boil) for 0.1 second
Eardrum rupture threshold	720 lb/ft²
Lung damage threshold	1,728–2,160 lb/ft² (approximately the equivalent to having a compact car resting its weight on one's chest)
OSHA required ear protection threshold	85 decibel (db) for a sustained time period (Note: An increase of 3 db is equivalent to doubling the sound level.)

the same electrical circuit setup with an available bolted 3-phase, short-circuit current of 22,600 symmetrical root mean square (rms) amperes at 480 volts. **See Figure 2-7**.

In each case, an arcing fault was initiated in a size 1 combination motor controller enclosure with the door open, as if an Electrical Worker were performing work on the unit while energized or before it was placed in an electrically safe work condition. Test 4 and Test 3 were identical except for the OCPD protecting the circuit. In Test 4, a 640-ampere circuit breaker with a short-time delay is protecting the circuit; the circuit was cleared in six cycles. **See Figure 2-8**. In Test 3, 601-ampere (KRP-C-601SP), current-limiting fuses (Class L) are protecting the circuit; these fuses opened the fault current in less than one-half cycle and limited the current. In addition, the arcing fault was initiated on the line side of the branch circuit device in both Test 4 and Test 3 (the fault is on the feeder circuit, but

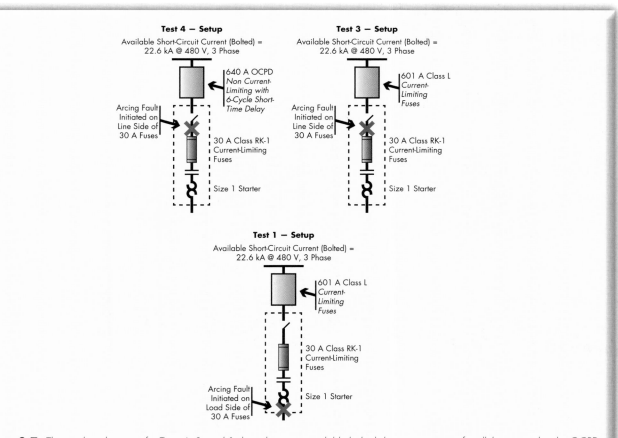

Figure 2-7. *The one-line diagrams for Tests 4, 3, and 1 show the same available bolted short-circuit current for all three tests, but the OCPDs differ and the point of initiation of the arcing fault differs.*

within the controller enclosure). **See Figure 2-9.** In Test 1, the arcing fault is initiated on the load side of the 30-ampere branch-circuit OCPD (LPS-RK 30SP) current-limiting fuses (Class RK1). **See Figure 2-10.** These fuses limited this fault current to a much lower amount and cleared the circuit in approximately one-fourth cycle or less.

The tests were filmed via high speed camera and still photos extracted. The results of the tests were recorded from the various sensors on the mannequin closest to the arcing fault. T1 and T2 recorded the temperature on the bare hand and neck, respectively. The hand with the T1 sensor was very close to the arcing fault. T3 recorded the temperature on the chest under the shirt. P1 recorded the pressure on the chest. Also, the sound level was measured at the ear. Some results "pegged the meter"—that is, the specific measurements were unable to be recorded because the actual level exceeded the range of the sensor/recorder setting. These values are shown as >, which indicates that the actual value exceeded the value given, but it is unknown how high a level the actual value attained.

The Role of Overcurrent Protective Devices in Electrical Safety

If an arcing fault occurs while a worker is in close proximity to the fault, the survivability of the worker is mostly dependent upon the following factors:

- The magnitude of the arcing fault current
- The characteristics of the OCPDs: time to clear the arcing current and ability to limit the current
- Precautions the worker has taken prior to the event, such as wearing PPE appropriate for the hazard

For additional information, visit qr.njatcdb.org Item #1223.

Figure 2-8. In Test 4, this staged test is protected by a circuit breaker having an intentional six-cycle (0.1 second) short-time delay (not a current-limiting OCPD).

From Cooper Bussmann Safety BASICs Handbook with permission from Cooper Bussmann.

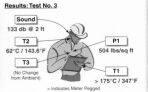

For additional information, visit qr.njatcdb.org Item #1224.

Figure 2-9. In Test 3, this staged test is protected by KRP-C-601SP Low-Peak® current-limiting fuses (Class L), which were in their current-limiting range and cleared in less than one-half cycle (0.0083 second).

From Cooper Bussmann Safety BASICs Handbook with permission from Cooper Bussmann.

For additional information, visit qr.njatcdb.org Item #1225.

Figure 2-10. In Test 1, this staged test is protected by LPS-RK-30SP, Low-Peak current-limiting fuses (Class RK1), which were in their current-limiting range and cleared in approximately one-fourth cycle or less (0.004 second).

From Cooper Bussmann Safety BASICs Handbook with permission from Cooper Bussmann.

The selection and performance of OCPDs play a significant role in electrical safety. Extensive tests and analysis by industry members have shown that the energy released during an arcing fault is related primarily to two characteristics of the OCPD protecting the affected circuit:

- The time it takes the OCPD to open. The faster the fault is cleared by the OCPD, the smaller the amount of energy released.
- The amount of fault current the OCPD lets through. Current-limiting OCPDs may reduce the current let-through (when the fault current is within the current-limiting range of the OCPD) and, therefore, can reduce the energy released.

The lower the amount of energy released, the better for both worker safety and equipment protection. The photos and recording sensor readings from the staged tests illustrate this point.

The following conclusions can be drawn from the staged tests:

1. Arcing faults can release tremendous amounts of energy in many forms in a very short period of time, as indicated by the measured values compared to key thresholds of injury for humans. Although the circuit in Test 4 was protected by a 640-ampere OCPD, it was a non-current-limiting device and took six cycles (0.1 second) to open.

2. The OCPD's characteristic can have a significant impact on the outcome. A 601-ampere, current-limiting OCPD protected the circuit in Test 3. The current that flowed was reduced (limited), and the clearing time was one-half cycle or less. This was a significant reduction compared to Test 4. **See Figure 2-11.** Compare the Test 3 and Test 4 results to the threshold for injury values, and note the difference in exposure. In addition, note that the results of Test 1 are significantly less than those in Test 4 and even those in Test 3. The reason is that Test 1 utilized a much smaller (30-ampere) current-limiting device.

Test 3 and Test 1 both show that there are benefits of using current-limiting OCPDs. Test 1 proves the point that the greater the current limitation, the more the arcing fault energy may be reduced. Both Test 3 and Test 1 utilized very current-limiting fuses, but the lower ampere-rated fuses limited the current more than the larger ampere-rated fuses. Note that the fault current must be in the current-limiting range of the OCPD to receive the benefit of the lower current let-through. **See Figure 2-12.**

Figure 2-11. *Three staged arc-flash tests are compared.*

Test	Protective Device Used	OCPD Clearing Time
Test 4	640-ampere, non-current-limiting device	Six cycles
Test 3	KRP-C 601SP, 601-ampere, current-limiting fuses (Class L)	Less than one-half cycle
Test 1	LPS-RK 30SP, 30-ampere, current-limiting fuses (Class RK1)	One-fourth cycle

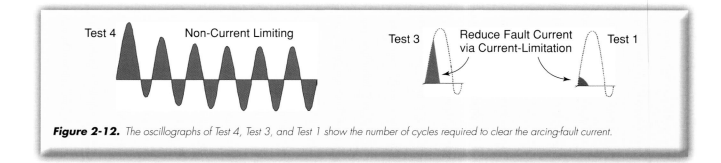

Figure 2-12. *The oscillographs of Test 4, Test 3, and Test 1 show the number of cycles required to clear the arcing-fault current.*

3. The shirt reduced the thermal energy exposure on the chest (the T3 sensor measured temperature under the shirt). This illustrates the benefit of workers wearing protective garments.

Summary

Recognizing the many hazards to which a worker might be exposed and understanding the severe consequences of that exposure are important steps in convincing everyone involved that much must be done to provide a workplace free from recognized hazards. In summary, the electrical hazards associated with working on or near exposed energized electrical conductors include the following:

- Electrical shock
- Arc flash
- Arc blast

Electrical shock hazard is related to the magnitude of current and the path through the human body. Arc-flash hazard is related to the arcing-current magnitude and time duration. Overcurrent protection can have a significant role in the level of arc-flash hazard. At this time, there is insufficient information to quantify arc-blast hazards and recommend appropriate PPE.

REVIEW QUESTIONS

1. Which of the following electrical hazards can potentially cause injury and even death to a person working on or near electrical equipment and systems?
   a. Arc blast
   b. Arc flash
   c. Electric shock
   d. All of the above

2. Few Electrical Workers understand that a minimal amount of current is required to cause injury or death due to shock hazard. What is the approximate range of current needed to cause the onset of heart fibrillation?
   a. 5 to 10 mA
   b. 10 to 20 mA
   c. 30 to 50 mA
   d. 100 to 200 mA

3. What does the severity of injury due to electric current flowing through the human body depend on?
   a. The amount of current through the body
   b. The current's pathway through the body
   c. Time duration of the contact
   d. All of the above

4. The most damaging path for electrical current is through the chest cavity and head. Ventricular fibrillation of the heart can cause fatalities resulting from _?_.
   a. contractions of the digestive system
   b. failure of the rhythmic heart pumping action
   c. indirect contact
   d. tingling sensation

5. All electrocutions are preventable. Which standard offers guidance regarding safe approach distances for different voltage levels to minimize the possibility of shock from exposed electrical conductors?
   a. NFPA 70
   b. NFPA 70E
   c. NFPA 72
   d. NFPA 99

6. The unique aspect of an arcing fault is that the resulting energies can be in the form of intense heat, brilliant light, and tremendous pressures. Intense heat from the arcing source travels at the speed of light. The temperature of the arc terminals can reach approximately _?_, or about four times as hot as the surface of the sun.
   a. 650°F
   b. 800°F
   c. 1,900°F
   d. 35,000°F

7. Incident energy is the amount of energy impressed on a surface, a certain distance from the source, generated during an electrical arc event. One of the units used to measure incident energy is _?_.
   a. amperes per square yard
   b. calories per centimeter squared (cal/cm²)
   c. pounds per square foot
   d. watts per cubic foot

8. Bare skin exposed to an arc-flash event can result in injury due to thermal conditions. Skin exposed to temperatures of _?_ for more than one-tenth of one second can cause incurable, third-degree burns.
   a. 110°F
   b. 158°F
   c. 176°F
   d. 205°F

9. If an arcing fault occurs while a worker is in close proximity to the fault, the survivability of the worker is mostly dependent upon which of the following factors?
   a. Characteristics of the overcurrent protective device
   b. Magnitude of the arcing current
   c. Precautions the worker has taken prior to the event
   d. All of the above

3 OSHA Considerations

CHAPTER OUTLINE

- OSHA: History, Application, Enforcement, and Responsibility
- The OSH Act and OSHA Standards and Regulations
- Safety and Health Regulations for Construction
- Occupational Safety and Health Standards

OBJECTIVES

1. Demonstrate an understanding of OSHA's history, application, enforcement, and responsibility.
2. Demonstrate an understanding of the OSH Act, the General Duty Clause, and OSHA Standards and Regulations.
3. Demonstrate an understanding of the regulations presented from Part 1926, Safety and Health Regulations for Construction.
4. Demonstrate an understanding of the regulations presented from Part 1910, Occupational Safety and Health Standards.

REFERENCES

1. *NFPA 70E®*, 2012 Edition
2. OSHA 29 CFR Part 1926
3. OSHA 29 CFR Part 1910
4. Occupational Safety and Health Act of 1970

CASE STUDY

A crew of electricians was working at a facility that was shut down for the July 4 holiday. The crew members had approximately one hour of work left to finish before they could go home and enjoy the holiday. The workers were pulling three sets of wiring from a source in the main plant to new electrical equipment in an addition to the facility. Two sets were for air conditioning, and one set was for a new lighting panel. Each set of wiring had its own breaker, which the foreman—but not the lead electrician—had locked out, modifying the normally followed lockout/tagout procedure. Normally, the employee performing the work would place his or her lockout/tagout equipment on the breakers and then remove the lockout/tagout equipment after the work was completed.

After completing connections for the new lighting panel, the lead electrician was getting ready to connect the wires for the air conditioning. As he pulled the wires into a junction box, he tapped the ends of the wires into his right hand to make them even. He was not wearing insulated gloves as he handled the wires and made the connections. At the same time, the lead electrician was ready for the breaker to the lighting panel to be turned on and instructed the foreman to throw the breaker to the "on" position. The foreman, thinking he should throw all three breakers to the "on" position, walked over to the breaker panel and removed his lockout/tagout on all three breakers. He then proceeded to throw all three to the "on" position. This action sent electricity through the wires into the lead electrician's hand, killing him.

Reportedly, the victim looked at his coworker, said "Help me," and then collapsed. Nearby workers called out to the foreman to contact emergency services, which he did immediately. While emergency services were en route, cardiopulmonary resuscitation was performed until paramedics arrived. Paramedics took the victim to a nearby hospital, where a physician notified the coroner, who declared the victim dead. The cause of death was electrocution.

Source: For details of this case, see FACE Investigation 03KY115. Accessed August 30, 2012.

For additional information, visit qr.njatcdb.org Item #1187.

Introduction

The Occupational Safety and Health Administration (OSHA) was established more than four decades ago when the U.S. Congress passed the Occupational Safety and Health Act (OSH Act) of 1970.

Its purpose, in part, is "to assure so far as possible every working man and woman in the nation safe and healthful working conditions and to preserve our human resources." Requirements were put into place that, when complied with, go a long way toward avoiding the hazards encountered in the workplace.

For additional information, visit qr.njatcdb.org Item #1188.

Electrical Workers are exposed to a number of hazards, and the work that they perform may be covered by any number of OSHA regulations. The primary focus of this chapter is on the regulations in Part 1926, Safety and Health Regulations for Construction, and Part 1910, Occupational Safety and Health Standards. The discussion here provides an overview of a number of performance-based OSHA regulations that may apply.

The information discussed in this chapter is not intended to be all-inclusive or the basis for an electrical safety program. Instead, the intent is to point out a number of the regulations that may apply. Part 1926 and Part 1910 regulations must be studied in their entirety for a full and complete look at these provisions and their application.

OSHA: History, Application, Enforcement, and Responsibility

OSHA publication 3302-02R 2012, entitled *All About OSHA* (published in 2012), includes the following general overview of basic topics related to OSHA's mission, what it covers, and how it operates.

For additional information, visit qr.njatcdb.org Item #1189.

On December 29, 1970, President Nixon signed the Occupational Safety and Health Act of 1970 (OSH Act) into law, establishing OSHA. Congress created OSHA to assure safe and healthful conditions for working men and women by setting and enforcing standards and providing training, outreach, education, and compliance assistance. Coupled with the efforts of employers, workers, safety and health professionals, unions, and advocates, OSHA and its state partners have dramatically improved workplace safety, reducing work-related deaths and injuries by more than 65%.

History

According to OSHA, in 1970, an estimated 14,000 workers were killed on the job—about 38 every day. For 2010, the Bureau of Labor Statistics reports this number fell to about 4,500, or about 12 workers per day. At the same time, U.S. employment has almost doubled to over 130 million workers at more than 7.2 million worksites. The rate of reported serious workplace injuries and illnesses has also dropped markedly, from 11 per 100 workers in 1972 to 3.5 per 100 workers in 2010.

OSHA's safety and health standards have prevented countless work-related injuries, illnesses, and deaths. Nevertheless, far too many preventable injuries and fatalities continue to occur. Significant hazards and unsafe conditions still exist in U.S. workplaces; each year more than 3.3 million working men and women suffer a serious job-related injury or illness. Millions more are exposed to toxic chemicals that may cause illnesses years from now.

In addition to the direct impact on individual workers, the negative consequences for America's economy are substantial. Occupational injuries and illnesses cost American employers more than $53 billion a year—over $1 billion a week—in workers' compensation costs alone. Indirect costs to employers, including lost productivity, employee training, and replacement costs, as well as time for investigations following injuries can more than double these costs. Workers and their families suffer great emotional and psychological costs, in addition to the loss of wages and the costs of caring for the injured, which further weakens the economy.

Application

Under the OSH law, employers are responsible for providing a safe and healthful workplace for their workers. The OSH Act covers most private sector employers and their workers, in addition to some public sector employers and workers in the 50 states and certain territories and jurisdictions under federal authority. OSHA covers most private sector employers and workers in all 50 states, the District of Columbia, and other U.S. jurisdictions either directly through Federal OSHA or through an OSHA-approved state plan. State plans are OSHA-approved job safety and health programs operated by individual states instead of Federal OSHA.

For additional information, visit qr.njatcdb.org Item #1190.

Enforcement

Enforcement plays an important part in OSHA's efforts to reduce workplace injuries, illnesses, and fatalities. When OSHA finds employers who fail to uphold their safety and health responsibilities, the agency takes strong, decisive actions. Inspections are initiated without advance notice, conducted using on-site or telephone and facsimile investigations, performed by highly trained compliance officers, and scheduled based on the following priorities:

- Imminent danger
- Catastrophes—fatalities or hospitalizations
- Worker complaints and referrals
- Targeted inspections—particular hazards, high injury rates
- Follow-up inspections

Responsibility

Employers have the responsibility to provide a safe workplace. Employers must provide their workers with a workplace that does not have serious hazards and follows all OSHA safety and health standards. Employers must find and correct safety and health problems. OSHA further requires that employers must first try to eliminate or reduce hazards by making feasible changes in working conditions rather than relying on personal protective equipment such as masks, gloves, or earplugs. Switching to safer chemicals, enclosing processes to trap harmful fumes, and using ventilation systems to clean the air are examples of effective ways to eliminate or reduce risks.

The OSH Act and OSHA Standards and Regulations

OSHA requirements are not recommendations; rather, the requirements set forth in the OSHA standards are law. By passing the Occupational Safety and Health Act of 1970, Congress authorized

Figure 3-1. *The Occupational Safety and Health Act of 1970 (OSH Act) was signed into law on December 29, 1970, establishing OSHA.*

enforcement of the standards developed under the Act. **See Figure 3-1.**

The OSH Act comprises 35 sections:
Section 1: Introduction
Section 2: Congressional Findings and Purpose
Section 3: Definitions
Section 4: Applicability of This Act
Section 5: Duties
Section 6: Occupational Safety and Health Standards
Section 7: Advisory Committees; Administration
Section 8: Inspections, Investigations, and Recordkeeping
Section 9: Citations
Section 10: Procedure for Enforcement
Section 11: Judicial Review
Section 12: The Occupational Safety and Health Review Commission
Section 13: Procedures to Counteract Imminent Dangers
Section 14: Representation in Civil Litigation
Section 15: Confidentiality of Trade Secrets
Section 16: Variations, Tolerances, and Exemptions
Section 17: Penalties
Section 18: State Jurisdiction and State Plans
Section 19: Federal Agency Safety Programs and Responsibilities
Section 20: Research and Related Activities

For additional information, visit qr.njatcdb.org Item #1191.

The publication *All About OSHA* also discusses standards that may be applicable in the workplace. OSHA's Construction, General Industry, Maritime, and Agriculture standards protect workers from a wide range of serious hazards. Examples of OSHA standards include requirements for employers to take the following steps:

- Provide fall protection
- Prevent trenching cave-ins
- Prevent exposure to some infectious diseases
- Ensure the safety of workers who enter confined spaces
- Prevent exposure to harmful chemicals
- Put guards on dangerous machines
- Provide respirators or other safety equipment
- Provide training for certain dangerous jobs in a language and vocabulary that workers can understand

Employer and Employee Duties

Employers must comply with the General Duty Clause of the OSH Act, which is found in Section 5, Duties. This clause requires employers to keep their workplaces free of serious recognized hazards and is generally cited when no specific OSHA standard applies to the hazard. Note also Section 5(b), which outlines employee responsibility.

Standards and Regulations

OSHA's Construction, General Industry, Maritime, and Agriculture standards are parts under Title 29 of the Code of Federal Regulations (29 CFR). The Agriculture

standard is 29 CFR Part 1928, Occupational Safety and Health Standards for Agriculture. Each of these parts contains regulations that are broken down into subparts. The subparts are generally categorized by a topic, hazard, or exposure. For example, personal protective equipment (PPE) is primarily addressed in Subpart E of the Safety and Health Regulations for Construction (Part 1926) and in Subpart I of the Occupational Safety and Health Standards (Part 1910). Many of these regulations are performance based; that is, they require something without necessarily spelling out how compliance is to be accomplished.

The two standards applicable to the majority of the work performed by Electrical Workers are found in Part 1926, Safety and Health Regulations for Construction, and in Part 1910, Occupational Safety and Health Standards. Part 1926 applies to Construction, and Part 1910 applies to General Industry. It is interesting to note that Part 1910 contains provisions that help explain application of Part 1926, including the following:

Section 5, Duties

(a) Each employer—

    (1) shall furnish to each of his employees employment and a place of employment which are free from recognized hazards that are causing or are likely to cause death or serious physical harm to his employees;

    (2) shall comply with occupational safety and health standards promulgated under this Act.

(b) Each employee shall comply with occupational safety and health standards and all rules, regulations, and orders issued pursuant to this Act which are applicable to his own actions and conduct.

1910.12(a) Standards.

    The standards prescribed in part 1926 of this chapter are adopted as occupational safety and health standards under section 6 of the Act and shall apply, according to the provisions thereof, to every employment and place of employment of every employee engaged in construction work. Each employer shall protect the employment and places of employment of each of his employees engaged in construction work by complying with the appropriate standards prescribed in this paragraph.

1910.12(b) Definition.

For purposes of this section, "Construction work" means work for construction, alteration, and/or repair, including painting and decorating. See discussion of these terms in 1926.13 of this title.

Safety and Health Regulations for Construction

The Part 1926 Construction regulations are divided into subparts. The scope is contained in Subpart A and advises that Part 1926 sets forth the safety and health standards promulgated by the Secretary of Labor under Section 107 of the Contract Work Hours and Safety Standards Act. Currently, Part 1926 is divided into the following subparts:

1926 Subpart A:	General
1926 Subpart B:	General Interpretations
1926 Subpart C:	General Safety and Health Provisions
1926 Subpart D:	Occupational Health and Environmental Controls
1926 Subpart E:	Personal Protective and Life Saving Equipment
1926 Subpart F:	Fire Protection and Prevention
1926 Subpart G:	Signs, Signals, and Barricades
1926 Subpart H:	Materials Handling, Storage, Use, and Disposal
1926 Subpart I:	Tools—Hand and Power
1926 Subpart J:	Welding and Cutting
1926 Subpart K:	Electrical
1926 Subpart L:	Scaffolds
1926 Subpart M:	Fall Protection
1926 Subpart N:	Helicopters, Hoists, Elevators, and Conveyors
1926 Subpart O:	Motor Vehicles, Mechanized Equipment, and Marine Operations
1926 Subpart P:	Excavations
1926 Subpart Q:	Concrete and Masonry Construction
1926 Subpart R:	Steel Erection
1926 Subpart S:	Underground Construction, Caissons, Cofferdams, and Compressed Air
1926 Subpart T:	Demolition
1926 Subpart U:	Blasting and the Use of Explosives
1926 Subpart V:	Power Transmission and Distribution
1926 Subpart W:	Rollover Protective Structures; Overhead Protection
1926 Subpart X:	Ladders
1926 Subpart Y:	Commercial Diving Operations
1926 Subpart Z:	Toxic and Hazardous Substances
1926 Subpart AA:	[Reserved]
1926 Subpart BB:	[Reserved]
1926 Subpart CC:	Cranes and Derricks in Construction

For additional information, visit qr.njatcdb.org Item #1192.

Note that the construction regulations are broken down into numerous subparts, each addressing a particular topic as indicated in its title and within its scope. Some subparts cover a particular topic within its scope, such as excavations, ladders, scaffolds, and welding, whereas other subparts apply more broadly, such as the general safety and health provisions of Subpart C. Some of the Part 1926 Construction regulations will be explored here as examples of the types of provisions that could apply and require compliance.

General Safety and Health Provisions

Subpart C provides general safety and health provisions as indicated by its title. Subpart C begins by providing regulations that state, in part, "no contractor or subcontractor for any part of the contract work shall require any laborer or mechanic employed in the performance of the contract to work in surroundings or under working conditions which are unsanitary, hazardous, or dangerous to his health or safety." Additional general employer provisions from Subpart C include the following points:

1926.20(b) Accident prevention responsibilities.

1926.20(b)(1)

It shall be the responsibility of the employer to initiate and maintain such programs as may be necessary to comply with this part.

1926.20(b)(2)

Such programs shall provide for frequent and regular inspections of the job sites, materials, and equipment to be made by competent persons designated by the employers. **See Figure 3-2.**

1926.20(f)(1) Personal protective equipment.

Standards in this part requiring the employer to provide personal protective equipment (PPE), including

Figure 3-2. *Employers must provide for frequent and regular inspections of job sites by competent persons as part of their accident prevention responsibilities.*

Figure 3-3. *Employers have the responsibility to instruct workers who must enter confined or enclosed spaces about the hazards involved, necessary precautions, and the use of protective and emergency equipment required.*

respirators and other types of PPE, because of hazards to employees impose a separate compliance duty with respect to each employee covered by the requirement. The employer must provide PPE to each employee required to use the PPE, and each failure to provide PPE to an employee may be considered a separate violation.

1926.20(f)(2) Training.

Standards in this part requiring training on hazards and related matters, such as standards requiring that employees receive training or that the employer train employees, provide training to employees, or institute or implement a training program, impose a separate compliance duty with respect to each employee covered by the requirement. The employer must train each affected employee in the manner required by the standard, and each failure to train an employee may be considered a separate violation.

1926.21(b) Employer responsibility.

1926.21(b)(2)

The employer shall instruct each employee in the recognition and avoidance of unsafe conditions and the regulations applicable to his work environment to control or eliminate any hazards or other exposure to illness or injury.

1926.21(b)(6)(i)

All employees required to enter into confined or enclosed spaces shall be instructed as to the nature of the hazards involved, the necessary precautions to be taken, and in the use of protective and emergency equipment required. The employer shall comply with any specific regulations that apply to work in dangerous or potentially dangerous areas. **See Figure 3-3**.

1926.28(a)

The employer is responsible for requiring the wearing of appropriate personal protective equipment in all operations where there is an exposure to hazardous conditions or where this part indicates the need for using such equipment to reduce the hazards to the employees.

Personal Protective and Life-Saving Equipment

Subpart C states that the employer is responsible for requiring employees to wear appropriate personal protective equipment in all operations where there is an exposure to hazardous conditions or where this part indicates the need for using such equipment to reduce the hazards to the employees; the employer must provide the necessary PPE to each employee who is required to use the PPE. Some of the PPE regulations from Subpart E are shown here:

1926.95(a) Application.

Protective equipment, including personal protective equipment for eyes, face, head, and extremities, protective clothing, respiratory devices, and protective shields and barriers, shall be provided, used, and maintained in a sanitary and reliable condition wherever it is necessary by reason of hazards of processes or environment, chemical hazards, radiological hazards, or mechanical irritants encountered in a manner capable of causing injury or impairment in the function of any part of the body through absorption, inhalation, or physical contact. **See Figure 3-4.**

1926.95(b) Employee-owned equipment.

Where employees provide their own protective equipment, the employer shall be responsible to assure its adequacy, including proper maintenance and sanitation of such equipment.

Figure 3-4. *Protective equipment must be provided and used as necessary.*

Courtesy of Salisbury by Honeywell

1926.95(c) Design.

All personal protective equipment shall be of safe design and construction for the work to be performed.

1926.95(d) Payment for protective equipment.

1926.95(d)(1)

Except as provided by paragraphs (d)(2) through (d)(6) of this section, the protective equipment, including personal protective equipment (PPE), used to comply with this part, shall be provided by the employer at no cost to employees.

1926.95(d)(2)

The employer is not required to pay for non-specialty safety-toe protective footwear (including steel-toe shoes or steel-toe boots) and non-specialty prescription safety eyewear, provided that the employer permits such items to be worn off the job-site.

1926.95(d)(3)

When the employer provides metatarsal guards and allows the employee, at his or her request, to use shoes or boots with built-in metatarsal protection, the employer is not required to reimburse the employee for the shoes or boots.

1926.95(d)(4)

The employer is not required to pay for:

1926.95(d)(4)(i)

Everyday clothing, such as long-sleeve shirts, long pants, street shoes, and normal work boots; or

1926.95(d)(4)(ii)

Ordinary clothing, skin creams, or other items, used solely for protection from weather, such as winter coats, jackets, gloves, parkas, rubber boots, hats, raincoats, ordinary sunglasses, and sunscreen.

1926.95(d)(5)

The employer must pay for replacement PPE, except when the employee has lost or intentionally damaged the PPE.

1926.95(d)(6)

Where an employee provides adequate protective equipment he or she owns pursuant to paragraph (b) of this section, the employer may allow the employee to use it and is not required to reimburse the employee for that equipment. The employer shall not require an employee to provide or pay for his or her own PPE, unless the PPE is excepted by paragraphs (d)(2) through (d)(5) of this section.

1926.95(d)(7)

This section shall become effective on February 13, 2008. Employers must implement the PPE payment requirements no later than May 15, 2008.

Note to § 1926.95(d): When the provisions of another OSHA standard specify whether or not the employer must pay for specific equipment, the payment provisions of that standard shall prevail.

1926.100(a)

Employees working in areas where there is a possible danger of head injury from impact, or from falling or flying objects, or from electrical shock and burns, shall be protected by protective helmets. **See Figure 3-5.**

1926.100(b)

Helmets for the protection of employees against impact and penetration of falling and flying objects shall meet the specifications contained in American National Standards Institute, Z89.1-1969, Safety Requirements for Industrial Head Protection.

1926.100(c)

Helmets for the head protection of employees exposed to high voltage electrical shock and burns shall meet the specifications contained in American National Standards Institute, Z89.2-1971.

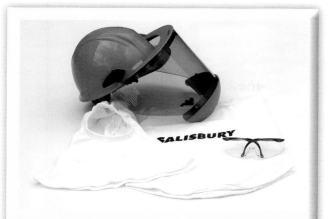

Figure 3-5. *Protective helmets must be used to protect employees where there is a potential for head injury.*

Courtesy of Salisbury by Honeywell

Eye and face protection.
1926.102(a) General.
1926.102(a)(1)
Employees shall be provided with eye and face protection equipment when machines or operations present potential eye or face injury from physical, chemical, or radiation agents. **See Figure 3-6.**
1926.102(a)(2)
Eye and face protection equipment required by this Part shall meet the requirements specified in American National Standards Institute, Z87.1-1968, Practice for Occupational and Educational Eye and Face Protection.

Figure 3-6. *Eye and face protection must be provided as necessary.*

Courtesy of Salisbury by Honeywell

Electrical

Subpart K, Electrical, indicates that it addresses electrical safety requirements that are necessary for practical safeguarding of employees involved in construction work. This part is divided into four major divisions and applicable definitions:

1926.400(a) Installation safety requirements.
Installation safety requirements are contained in 1926.402 through 1926.408. Included in this category are electric equipment and installations used to provide electric power and light on jobsites.
1926.400(b) Safety-related work practices.
Safety-related work practices are contained in 1926.416 and 1926.417. In addition to covering the hazards arising from the use of electricity at jobsites, these regulations also cover the hazards arising from the accidental contact, direct or indirect, by employees with all

energized lines, above or below ground, passing through or near the jobsite.
1926.400(c) Safety-related maintenance and environmental considerations.
Safety-related maintenance and environmental considerations are contained in 1926.431 and 1926.432.
1926.400(d) Safety requirements for special equipment. Safety requirements for special equipment are contained in 1926.441.
1926.400(e) Definitions.
Definitions applicable to this Subpart are contained in 1926.449.

An overview of a number of these requirements follows:

⦾SHA Tip

The occupational safety and health standards contained in Subpart V shall apply to the construction of electric transmission and distribution lines and equipment.

1926.404 Wiring design and protection.

1926.404(b) Branch circuits—

1926.404(b)(1) Ground-fault protection—

1926.404(b)(1)(i) General. The employer shall use either ground fault circuit interrupters as specified in paragraph (b)(1)(ii) of this section or an assured equipment grounding conductor program as specified in paragraph (b)(1)(iii) of this section to protect employees on construction sites. These requirements are in addition to any other requirements for equipment grounding conductors.

1926.416 General requirements.

1926.416(a) Protection of employees—

1926.416(a)(1)

No employer shall permit an employee to work in such proximity to any part of an electric power circuit that the employee could contact the electric power circuit in the course of work, unless the employee is protected against electric shock by deenergizing the circuit and grounding it or by guarding it effectively by insulation or other means. **See Figure 3-7.**

1926.416(a)(2)

In work areas where the exact location of underground electric powerlines is unknown, employees using jack-hammers, bars, or other hand tools which may contact a line shall be provided with insulated protective gloves.

1926.416(a)(3)

Before work is begun the employer shall ascertain by inquiry or direct observation, or by instruments, whether any part of an energized electric power circuit, exposed or concealed, is so located that the performance of the work may bring any person, tool, or machine into physical or electrical contact with the electric power circuit. The employer shall post and maintain proper warning signs where such a circuit exists. The employer shall advise employees of the location of such lines, the hazards involved, and the protective measures to be taken.

1926.416(b) Passageways and open spaces—

1926.416(b)(1)

Barriers or other means of guarding shall be provided to ensure that workspace for electrical equipment will not be used as a passageway during periods when energized parts of electrical equipment are exposed.

1926.416(b)(2)

Working spaces, walkways, and similar locations shall be kept clear of cords so as not to create a hazard to employees.

1926.417 Lockout and tagging of circuits.

1926.417(a)

Controls. Controls that are to be deactivated during the course of work on energized or deenergized equipment or circuits shall be tagged.

1926.417(b)

Equipment and circuits. Equipment or circuits that are deenergized shall be rendered inoperative and shall have tags attached at all points where such equipment or circuits can be energized. **See Figure 3-8.**

1926.417(c)

Tags. Tags shall be placed to identify plainly the equipment or circuits being worked on.

Figure 3-7. Deenergizing and grounding the circuit is one OSHA-recognized method of employee protection in Subpart K.

Photo courtesy of National Electrical Contractors Association (NECA)

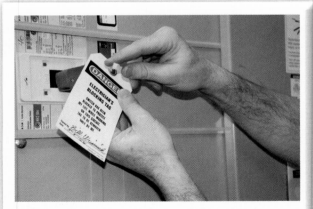

Figure 3-8. Circuits that are deenergized must be rendered inoperative and have tags attached in accordance with 1926.417(b).

Photo courtesy of National Electrical Contractors Association (NECA)

Occupational Safety and Health Standards

Currently, Part 1910 General Industry regulations are divided into 26 subparts. Note that, while a number of the subpart titles in Part 1910 are similar to those in Part 1926, they are located in a different subpart letter in each standard. For example, Fire Protection is located in Subpart F in Part 1926 and in Subpart L in Part 1910.

1910 Subpart A: General

1910 Subpart B: Adoption and Extension of Established Federal Standards

1910 Subpart C: Adoption and Extension of Established Federal Standards

1910 Subpart D: Walking–Working Surfaces

1910 Subpart E: Means of Egress

1910 Subpart F: Powered Platforms, Manlifts, and Vehicle-Mounted Work Platforms

1910 Subpart G: Occupational Health and Environmental Control

1910 Subpart H: Hazardous Materials

1910 Subpart I: Personal Protective Equipment

1910 Subpart J: General Environmental Controls

1910 Subpart K: Medical and First Aid

1910 Subpart L: Fire Protection

1910 Subpart M: Compressed Gas and Compressed Air Equipment

1910 Subpart N: Materials Handling and Storage

1910 Subpart O: Machinery and Machine Guarding

1910 Subpart P: Hand and Portable Powered Tools and Other Hand-Held Equipment

1910 Subpart Q: Welding, Cutting, and Brazing

1910 Subpart R: Special Industries

1910 Subpart S: Electrical

1910 Subpart T: Commercial Diving Operations

1910 Subpart U: [Reserved]

1910 Subpart V: [Reserved]

1910 Subpart W: Program Standard

1910 Subpart X: [Reserved]

1910 Subpart Y: [Reserved]

1910 Subpart Z: Toxic and Hazardous Substances

In addition, the requirements are generally not identical in each standard. The electrical requirements are a good example: Compare the Subpart K electrical requirements from Part 1926 covered earlier in this chapter with the Subpart S electrical requirements from Part 1910 covered later in this chapter.

For additional information, visit qr.njatcdb.org Item #1193.

General

Subpart A of Part 1910 contains the general provisions of the Occupational Safety and Health Standards. A number of these general provisions follow. Section 1910.1 addresses the purpose and scope, Section 1910.2 contains definitions, Section 1910.6 covers incorporation by reference, and Section 1910.9 addresses compliance duties owed to each employee.

1910.1(b)

This part carries out the directive to the Secretary of Labor under section 6(a) of the Act. It contains occupational safety and health standards which have been found to be national consensus standards or established Federal standards.

1910.2(c)

"Employer" means a person engaged in a business affecting commerce who has employees, but does not include the United States or any State or political subdivision of a State;

1910.2(d)

"Employee" means an employee of an employer who is employed in a business of his employer which affects commerce;

1910.6(a)(1)

The standards of agencies of the U.S. Government, and organizations which are not agencies of the U.S. Government which are incorporated by reference in this part, have the same force and effect as other standards in this part. Only the mandatory provisions (i.e., provisions containing the word "shall" or other mandatory language) of standards incorporated by reference are adopted as standards under the Occupational Safety and Health Act.

1910.9(a) Personal protective equipment.

Standards in this part requiring the employer to provide personal protective equipment (PPE), including respirators and other types of PPE, because of hazards to employees impose a separate compliance duty with respect to each employee covered by the requirement. The employer must provide PPE to each employee required to use the PPE, and each failure to provide PPE to an employee may be considered a separate violation.

1910.9(b) Training.

Standards in this part requiring training on hazards and related matters, such as standards requiring that employees receive training or that the employer train employees, provide training to employees, or institute or implement a training program, impose a separate compliance duty with respect to each employee covered by the requirement. The employer must train each affected employee in the manner required by the standard, and each failure to train an employee may be considered a separate violation. **See Figure 3-9.**

Figure 3-9. *The employer is required to train each affected employee in the manner required by the OSHA standard.*

Courtesy of Service Electric Company

Personal Protective Equipment

Personal protective equipment is covered in Subpart I of Part 1910. A number of the PPE provisions from the Occupational Safety and Health Standards are covered in this section of the text, including those related to application, design, hazard assessment and selection, training, payment, and electrical protective devices. Many of these regulations are performance based; that is, they require compliance without necessarily spelling out how to comply.

Note that the employer is required to assess the workplace to determine the need for the use of PPE. This assessment, which must be in writing, must identify the workplace evaluated, the person who performed the assessment, and the date of the assessment. PPE is also required to be provided, used, and maintained properly.

The employer is also responsible for training and retraining related to PPE, and for the adequacy of PPE, including any employee-provided PPE. Each affected employee must demonstrate an understanding of the training and the ability to use PPE properly before being allowed to perform work requiring the use of that PPE. Notice the application notes in 1910.132(g) and the fact that the employer is generally required to provide PPE at no cost to employees.

1910.132(a) Application.

Protective equipment, including personal protective equipment for eyes, face, head, and extremities, protective clothing, respiratory devices, and protective shields and barriers, shall be provided, used, and maintained in a sanitary and reliable condition wherever it is necessary by reason of hazards of processes or environment, chemical hazards, radiological hazards, or mechanical irritants encountered in a manner capable of causing injury or impairment in the function of any part of the body through absorption, inhalation, or physical contact. **See Figure 3-10.**

1910.132(b) Employee-owned equipment.

Where employees provide their own protective equipment, the employer shall be responsible to assure its adequacy, including proper maintenance and sanitation of such equipment.

1910.132(d) Hazard assessment and equipment selection.

1910.132(d)(1)

The employer shall assess the workplace to determine if hazards are present, or are likely to be present, which necessitate the use of personal protective equipment (PPE). If such hazards are present, or likely to be present, the employer shall:

1910.132(d)(1)(i)

Select, and have each affected employee use, the types of PPE that will protect the affected employee from the hazards identified in the hazard assessment;

1910.132(d)(1)(ii)

Communicate selection decisions to each affected employee; and,

1910.132(d)(1)(iii)

Figure 3-10. *Subpart I requires that necessary protective equipment be provided, used, and maintained in a sanitary and reliable condition.*

Courtesy of Salisbury by Honeywell

Select PPE that properly fits each affected employee. 1910.132(d)(2)

The employer shall verify that the required workplace hazard assessment has been performed through a written certification that identifies the workplace evaluated; the person certifying that the evaluation has been performed; the date(s) of the hazard assessment; and, which identifies the document as a certification of hazard assessment.

1910.132(e) Defective and damaged equipment.

Defective or damaged personal protective equipment shall not be used.

1910.132(f) Training.

1910.132(f)(1)

The employer shall provide training to each employee who is required by this section to use PPE. Each such employee shall be trained to know at least the following:

1910.132(f)(1)(i)

When PPE is necessary;

1910.132(f)(1)(ii)

What PPE is necessary;

1910.132(f)(1)(iii)

How to properly don, doff, adjust, and wear PPE;

1910.132(f)(1)(iv)

The limitations of the PPE; and,

1910.132(f)(1)(v)

The proper care, maintenance, useful life, and disposal of the PPE.

1910.132(f)(2)

Each affected employee shall demonstrate an understanding of the training specified in paragraph (f)(1) of this section, and the ability to use PPE properly, before being allowed to perform work requiring the use of PPE.

1910.132(f)(3)

When the employer has reason to believe that any affected employee who has already been trained does not have the understanding and skill required by paragraph (f)(2) of this section, the employer shall retrain each such employee. Circumstances where retraining is required include, but are not limited to, situations where:

1910.132(f)(3)(i)

Changes in the workplace render previous training obsolete; or

1910.132(f)(3)(ii)

Changes in the types of PPE to be used render previous training obsolete; or

1910.132(f)(3)(iii)

Inadequacies in an affected employee's knowledge or use of assigned PPE indicate that the employee has not retained the requisite understanding or skill.

1910.132(g)

Paragraphs (d) and (f) of this section apply only to 1910.133, 1910.135, 1910.136, and 1910.138. Paragraphs (d) and (f) of this section do not apply to 1910.134 and 1910.137.

1910.132(h)

Payment for protective equipment.

1910.132(h)(1)

Except as provided by paragraphs (h)(2) through (h)(6) of this section, the protective equipment, including personal protective equipment (PPE), used to comply with this part, shall be provided by the employer at no cost to employees.

1910.132(h)(2)

The employer is not required to pay for non-specialty safety-toe protective footwear (including steel-toe shoes or steel-toe boots) and non-specialty prescription safety eyewear, provided that the employer permits such items to be worn off the job-site.

1910.132(h)(3)

When the employer provides metatarsal guards and allows the employee, at his or her request, to use shoes or boots with built-in metatarsal protection, the employer is not required to reimburse the employee for the shoes or boots.

1910.132(h)(4)

The employer is not required to pay for:

1910.132(h)(4)(i)

The logging boots required by 29 CFR 1910.266(d)(1)(v);

1910.132(h)(4)(ii)

Everyday clothing, such as long-sleeve shirts, long pants, street shoes, and normal work boots; or

1910.132(h)(4)(iii)

Ordinary clothing, skin creams, or other items, used solely for protection from weather, such as winter coats, jackets, gloves, parkas, rubber boots, hats, raincoats, ordinary sunglasses, and sunscreen.

1910.132(h)(5)

The employer must pay for replacement PPE, except when the employee has lost or intentionally damaged the PPE.

1910.132(h)(6)

Where an employee provides adequate protective equipment he or she owns pursuant to paragraph (b) of this section, the employer may allow the employee to use it and is not required to reimburse the employee for that equipment. The employer shall not require an employee to provide or pay for his or her own PPE, unless the PPE is excepted by paragraphs (h)(2) through (h)(5) of this section.

1910.132(h)(7)

This paragraph (h) shall become effective on February 13, 2008. Employers must implement the PPE payment requirements no later than May 15, 2008.

Note to § 1910.132(h): When the provisions of another OSHA standard specify whether or not the employer must pay for specific equipment, the payment provisions of that standard shall prevail.

Section 1910.137 addresses electrical protective equipment. These regulations cover topics such as design requirements and in-service care and use, including daily inspection and periodic electrical tests.

A number of these regulations are covered here. Refer to Section 1910.137 to see these regulations in their entirety.

Figure 3-11. *Markings on gloves, such as type, class, size, and manufacturer information, must be nonconductive and be confined to the cuff portion of the glove.*

Courtesy of Salisbury by Honeywell

1910.137 Electrical protective devices.
　1910.137(a) Design requirements.
　1910.137(a)(1) Manufacture and marking.
　1910.137(a)(1)(i)
　Blankets, gloves, and sleeves shall be produced by a seamless process.
　1910.137(a)(1)(ii)
　Each item shall be clearly marked as follows:
　1910.137(a)(1)(ii)(A)
　Class 0 equipment shall be marked Class 0.
　1910.137(a)(1)(ii)(B)
　Class 1 equipment shall be marked Class 1.
　1910.137(a)(1)(ii)(C)
　Class 2 equipment shall be marked Class 2.
　1910.137(a)(1)(ii)(D)
　Class 3 equipment shall be marked Class 3.
　1910.137(a)(1)(ii)(E)
　Class 4 equipment shall be marked Class 4.
　1910.137(a)(1)(ii)(F)
　Non-ozone-resistant equipment other than matting shall be marked Type I.
　1910.137(a)(1)(ii)(G)
　Ozone-resistant equipment other than matting shall be marked Type II.
　1910.137(a)(1)(ii)(H)
　Other relevant markings, such as the manufacturer's identification and the size of the equipment, may also be provided.
　1910.137(a)(1)(iii)
　Markings shall be nonconducting and shall be applied in such a manner as not to impair the insulating qualities of the equipment.
　1910.137(a)(1)(iv)
　Markings on gloves shall be confined to the cuff portion of the glove. **See Figure 3-11.**
　1910.137(b) In-service care and use.
　1910.137(b)(1)
　Electrical protective equipment shall be maintained in a safe, reliable condition.
　1910.137(b)(2)
　The following specific requirements apply to insulating blankets, covers, line hose, gloves, and sleeves made of rubber:
　1910.137(b)(2)(i)
　Maximum use voltages shall conform to those listed in Table I-5.
　1910.137(b)(2)(ii) **See Figure 3-12.**

Insulating equipment shall be inspected for damage before each day's use and immediately following any incident that can reasonably be suspected of having caused damage. Insulating gloves shall be given an air test, along with the inspection. **See Figure 3-13.**
　1910.137(b)(2)(iii)
　Insulating equipment with any of the following defects may not be used:
　1910.137(b)(2)(iii)(A)
　A hole, tear, puncture, or cut;
　1910.137(b)(2)(iii)(B)
　Ozone cutting or ozone checking (the cutting action produced by ozone on rubber under mechanical stress into a series of interlacing cracks);
　1910.137(b)(2)(iii)(C)
　An embedded foreign object;
　1910.137(b)(2)(iii)(D)
　Any of the following texture changes: swelling, softening, hardening, or becoming sticky or inelastic.
　1910.137(b)(2)(iii)(E)
　Any other defect that damages the insulating properties.
　1910.137(b)(2)(iv)

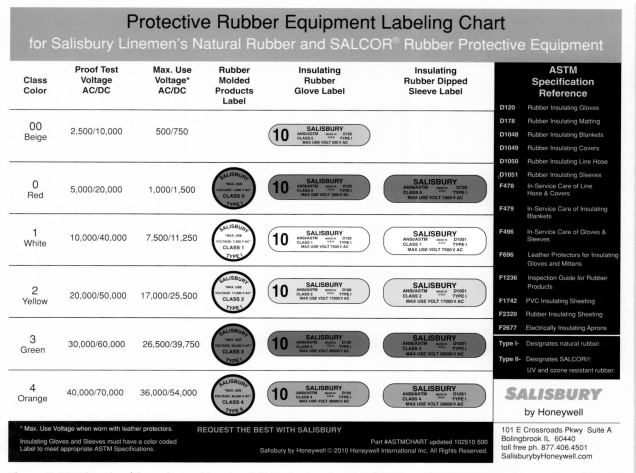

Figure 3-12. *The color of the insulating rubber glove label, rather than the color of the glove, identifies the maximum use voltage of the glove.*

Courtesy of Salisbury by Honeywell

Figure 3-13. *The required inspection of insulating gloves must include an air test.*

Courtesy of Salisbury by Honeywell

Insulating equipment found to have other defects that might affect its insulating properties shall be removed from service and returned for testing under paragraphs (b)(2)(viii) and (b)(2)(ix) of this section.

1910.137(b)(2)(v)

Insulating equipment shall be cleaned as needed to remove foreign substances.

1910.137(b)(2)(vi)

Insulating equipment shall be stored in such a location and in such a manner as to protect it from light, temperature extremes, excessive humidity, ozone, and other injurious substances and conditions.

1910.137(b)(2)(vii)

Protector gloves shall be worn over insulating gloves, except as follows:

1910.137(b)(2)(vii)(A)

Protector gloves need not be used with Class 0 gloves, under limited-use conditions, where small equipment and parts manipulation necessitate unusually high finger dexterity.

Note: Extra care is needed in the visual examination of the glove and in the avoidance of handling sharp objects.

1910.137(b)(2)(vii)(B)

Any other class of glove may be used for similar work without protector gloves if the employer can demonstrate that the possibility of physical damage to the gloves is small and if the class of glove is one class higher than that required for the voltage involved. Insulating gloves that have been used without protector gloves may not be used at a higher voltage until they have been tested under the provisions of paragraphs (b)(2)(viii) and (b)(2)(ix) of this section.

1910.137(b)(2)(viii)

Electrical protective equipment shall be subjected to periodic electrical tests. Test voltages and the maximum intervals between tests shall be in accordance with Table I-5 and Table I-6. **See Figure 3-14.**

1910.137(b)(2)(xii)

The employer shall certify that equipment has been tested in accordance with the requirements of paragraphs (b)(2)(viii), (b)(2)(ix), and (b)(2)(xi) of this section. The certification shall identify the equipment that passed the test and the date it was tested.

Note: Marking of equipment and entering the results of the tests and the dates of testing onto logs are two acceptable means of meeting this requirement.

The Control of Hazardous Energy (Lockout/Tagout)

Lockout/tagout regulations are primarily addressed in two subparts in the Part 1910, Occupational Safety and Health Standards: Subpart J, General Environmental Controls, and Subpart S, Electrical.

The Subpart J regulations are located in 1910.147, The Control of Hazardous Energy (lockout/tagout). Section 1910.147 is organized into the following subdivisions:

- Scope, application, and purpose
- Definitions
- General
- Application of control
- Lockout or tagout devices removal
- Additional requirements

Figure 3-14. Table I-5 provides the requirements for rubber insulation equipment voltage. Table I-6 details the rubber insulation equipment test intervals.

OSHA 1910.137 Table I-5 Rubber Insulating Equipment Voltage Requirements

Class of Equipment	Maximum Use Voltage* a-c-rms	Retest Voltage† a-c-rms	Retest Voltage† d-c-avg
0	1,000	5,000	20,000
1	7,500	10,000	40,000
2	17,000	20,000	50,000
3	26,500	30,000	60,000
4	36,000	40,000	70,000

*The maximum use voltage is the a-c voltage (rms) classification of the protective equipment that designates the maximum nominal design voltage of the energized system that may be safely worked. The nominal design voltage is equal to the phase-to-phase voltage on multiphase circuits. However, the phase-to-ground potential is considered to be the nominal design voltage:
1. If there is no multiphase exposure in a system area and if the voltage exposure is limited to the phase-to-ground potential; or
2. If the electrical equipment and devices are insulated or isolated or both so that the multiphase exposure on a grounded wye circuit is removed.

†The proof-test voltage shall be applied continuously for at least 1 minute, but no more than 3 minutes.

OSHA 1910.137 Table I-6 Rubber Insulating Equipment Test Intervals

Type of Equipment	When to Test
Rubber insulating line hose	Upon indication that insulating value is suspect
Rubber insulating covers	Upon indication that insulating value is suspect
Rubber insulating blankets	Before first issue and every 12 months thereafter*
Rubber insulating gloves	Before first issue and every 6 months thereafter*
Rubber insulating sleeves	Before first issue and every 12 months thereafter*

*If the insulating equipment has been electrically tested but not issued for service, it may not be placed into service unless it has been electrically tested within the previous 12 months.

The Subpart S lockout/tagout regulations are located in 1910.333(b)(2), Lockout and Tagging. One section and an accompanying note from these Subpart S requirements are included here for reference. These lockout and tagging regulations must be reviewed in their entirety for a complete look at what is required. The provisions of 1910.333(b)(2) are covered in greater detail in the chapter, "Lockout, Tagging, and Control of Hazardous Energy."

1910.333(b)(2) Lockout and Tagging.

While any employee is exposed to contact with parts of fixed electric equipment or circuits which have been deenergized, the circuits energizing the parts shall be locked out or tagged or both in accordance with the requirements of this paragraph. **See Figure 3-15.** The requirements shall be followed in the order in which they are presented.

Note 2: Lockout and tagging procedures that comply with paragraphs (c) through (f) of 1910.147 will also be deemed to comply with paragraph (b)(2) of this section provided that:

[1] The procedures address the electrical safety hazards covered by this Subpart; and

[2] The procedures also incorporate the requirements of paragraphs (b)(2)(iii)(D) and (b)(2)(iv)(B) of this section.

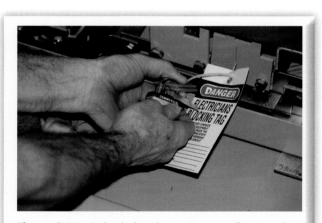

Figure 3-15. *Both a lock and a tag are generally required to comply with lockout/tagout requirements while any employee is exposed to contact with parts of fixed electric equipment or circuits that have been deenergized*

Photo courtesy of National Electrical Contractors Association (NECA)

OSHA Tip

Section 1910.269 covers the operation and maintenance of electric power generation, control, transformation, transmission, and distribution lines and equipment.

Electrical Occupational Safety and Health Standards

Subpart S details the electrical regulations of Part 1910, Occupational Safety and Health Standards. This subpart addresses electrical safety requirements that are necessary for the practical safeguarding of employees in their workplaces and is divided into four major divisions:

- Design safety standards for electrical systems
- Safety-related work practices
- Safety-related maintenance requirements
- Safety requirements for special equipment

The primary focus is on a number of the safety-related work practice regulations of Subpart S, including some from 1910.332, Training; 1910.333, Selection and Use of Work Practices; 1910.334, Use of Equipment; and 1910.335, Safeguards for Personnel Protection. In addition, a number of the definitions applicable to Subpart S located in 1910.399 are included here for review purposes.

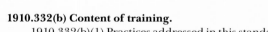

1910.332(b) Content of training.

1910.332(b)(1) Practices addressed in this standard.

Employees shall be trained in and familiar with the safety-related work practices required by 1910.331 through 1910.335 that pertain to their respective job assignments.

1910.333(a) General.

Safety-related work practices shall be employed to prevent electric shock or other injuries resulting from either direct or indirect electrical contacts, when work is performed near or on equipment or circuits which are or may be energized. The specific safety-related work practices shall be consistent with the nature and extent of the associated electrical hazards.

1910.333(a)(1) Deenergized parts.

Live parts to which an employee may be exposed shall be deenergized before the employee works on or near them, unless the employer can demonstrate that deenergizing introduces additional or increased hazards or is infeasible due to equipment design or operational limitations. Live parts that operate at less than 50 volts to ground need not be deenergized if there will be no increased exposure to electrical burns or to explosion due to electric arcs.

Note 1: Examples of increased or additional hazards include interruption of life support equipment, deactivation of emergency alarm systems, shutdown of hazardous location ventilation equipment, or removal of illumination for an area.

Note 2: Examples of work that may be performed on or near energized circuit parts because of infeasibility due to equipment design or operational limitations include testing of electric circuits that can only be performed with the circuit energized and work on circuits that form an integral part of a continuous industrial process in a chemical plant that would otherwise need to be completely shut down in order to permit work on one circuit or piece of equipment.

Note 3: Work on or near deenergized parts is covered by paragraph (b) of this section.

1910.333(a)(2) Energized parts.

If the exposed live parts are not deenergized (i.e., for reasons of increased or additional hazards or infeasibility), other safety-related work practices shall be used to protect employees who may be exposed to the electrical hazards involved. **See Figure 3-16**. Such work practices shall protect employees against contact with energized circuit parts directly with any part of their body or indirectly through some other conductive object. The work practices that are used shall be suitable for the conditions under which the work is to be performed and for the voltage level of the exposed electric conductors or circuit parts. Specific work practice requirements are detailed in paragraph (c) of this section.

Figure 3-16. *Other safety-related work practices must be used to protect employees who may be exposed to the electrical hazards involved if the exposed live parts are not deenergized.*

Courtesy of Salisbury by Honeywell

1910.333(b) Working on or near exposed deenergized parts.

1910.333(b)(1) Application.

This paragraph applies to work on exposed deenergized parts or near enough to them to expose the employee to any electrical hazard they present. Conductors and parts of electric equipment that have been deenergized but have not been locked out or tagged in accordance with paragraph (b) of this section shall be treated as energized parts, and paragraph (c) of this section applies to work on or near them.

1910.333(c) Working on or near exposed energized parts.

1910.333(c)(1)

"Application." This paragraph applies to work performed on exposed live parts (involving either direct contact or by means of tools or materials) or near enough to them for employees to be exposed to any hazard they present.

1910.333(c)(2) Work on energized equipment.

Only qualified persons may work on electric circuit parts or equipment that have not been deenergized under the procedures of paragraph (b) of this section. Such persons shall be capable of working safely on energized circuits and shall be familiar with the proper use of special precautionary techniques, personal protective equipment, insulating and shielding materials, and insulated tools.

1910.333(c)(3) Overhead lines.

If work is to be performed near overhead lines, the lines shall be deenergized and grounded, or other protective measures shall be provided before work is started. If the lines are to be deenergized, arrangements shall be made with the person or organization that operates or controls the electric circuits involved to deenergize and ground them. If protective measures, such as guarding, isolating, or insulating, are provided, these precautions shall prevent employees from contacting such lines directly with any part of their body or indirectly through conductive materials, tools, or equipment.

Note: The work practices used by qualified persons installing insulating devices on overhead power transmission or distribution lines are covered by 1910.269 of this Part, not by 1910.332 through 1910.335 of this Part. **See Figure 3-17**. Under paragraph (c)(2) of this section, unqualified persons are prohibited from performing this type of work.

1910.333(c)(3)(i) Unqualified persons.

1910.333(c)(3)(i)(A)

When an unqualified person is working in an elevated position near overhead lines, the location shall be such that the person and the longest conductive object he or she may contact cannot come closer to any unguarded, energized overhead line than the following distances:

1910.333(c)(3)(i)(A)(1)

For voltages to ground 50kV or below — 10 feet (305 cm);

Figure 3-17. *Other protective measures must be provided before work is started if that work is to be performed near overhead lines and the lines are not deenergized and grounded.*

Courtesy of Salisbury by Honeywell

1910.333(c)(3)(i)(A)(2)

For voltages to ground over 50kV — 10 feet (305 cm) plus 4 inches (10 cm) for every 10kV over 50kV.

1910.333(c)(3)(i)(B)

When an unqualified person is working on the ground in the vicinity of overhead lines, the person may not bring any conductive object closer to unguarded, energized overhead lines than the distances given in paragraph (c)(3)(i)(A) of this section.

Note: For voltages normally encountered with overhead power line, objects which do not have an insulating rating for the voltage involved are considered to be conductive.

1910.333(c)(3)(ii) Qualified persons.

When a qualified person is working in the vicinity of overhead lines, whether in an elevated position or on the ground, the person may not approach or take any conductive object without an approved insulating handle closer to exposed energized parts than shown in Table S-5 unless:

1910.333(c)(3)(ii)(A)

The person is insulated from the energized part (gloves, with sleeves if necessary, rated for the voltage involved are considered to be insulation of the person from the energized part on which work is performed), or

1910.333(c)(3)(ii)(B)

The energized part is insulated both from all other conductive objects at a different potential and from the person, or

1910.333(c)(3)(ii)(C)

The person is insulated from all conductive objects at a potential different from that of the energized part. **See Figure 3-18.**

1910.333(c)(3)(iii) Vehicular and mechanical equipment.

1910.333(c)(3)(iii)(A)

Any vehicle or mechanical equipment capable of having parts of its structure elevated near energized overhead lines shall be operated so that a clearance of 10 ft. (305 cm) is maintained. If the voltage is higher than 50kV, the clearance shall be increased 4 in. (10 cm) for every 10kV over that voltage. However, under any of the following conditions, the clearance may be reduced:

1910.333(c)(3)(iii)(A)(1)

If the vehicle is in transit with its structure lowered, the clearance may be reduced to 4 ft. (122 cm). If the

Figure 3-18. *The minimum approach distances for alternating current for qualified employees are set forth in Table S-5 in Subpart S of Part 1910.*

OSHA 1910.333 Table S-5 Approach Distances for Qualified Employees: Alternating Current

Voltage Range (Phase to Phase)	Minimum Approach Distance
300 V and less	Avoid Contact
Over 300 V, not over 750 V	1 ft. 0 in. (30.5 cm)
Over 750 V, not over 2 kV	1 ft. 6 in. (46 cm)
Over 2 kV, not over 15 kV	2 ft. 0 in. (61 cm)
Over 15 kV, not over 37 kV	3 ft. 0 in. (91 cm)
Over 37kV, not over 87.5 kV	3 ft. 6 in. (107 cm)
Over 87.5 kV, not over 121 kV	4 ft. 0 in. (122 cm)
Over 121 kV, not over 140 kV	4 ft. 6 in. (137 cm)

voltage is higher than 50kV, the clearance shall be increased 4 in. (10 cm) for every 10 kV over that voltage.

1910.333(c)(3)(iii)(A)(2)

If insulating barriers are installed to prevent contact with the lines, and if the barriers are rated for the voltage of the line being guarded and are not a part of or an attachment to the vehicle or its raised structure, the clearance may be reduced to a distance within the designed working dimensions of the insulating barrier.

1910.333(c)(3)(iii)(A)(3)

If the equipment is an aerial lift insulated for the voltage involved, and if the work is performed by a qualified person, the clearance (between the uninsulated portion of the aerial lift and the power line) may be reduced to the distance given in Table S-5.

1910.333(c)(3)(iii)(B)

Employees standing on the ground may not contact the vehicle or mechanical equipment or any of its attachments, unless:

1910.333(c)(3)(iii)(B)(1)

The employee is using protective equipment rated for the voltage; or

1910.333(c)(3)(iii)(B)(2)

The equipment is located so that no uninsulated part of its structure (that portion of the structure that provides a conductive path to employees on the ground) can come closer to the line than permitted in paragraph (c)(3)(iii) of this section.

1910.333(c)(3)(iii)(C)

If any vehicle or mechanical equipment capable of having parts of its structure elevated near energized overhead lines is intentionally grounded, employees working on the ground near the point of grounding may not stand at the grounding location whenever there is a possibility of overhead line contact. Additional precautions, such as the use of barricades or insulation, shall be taken to protect employees from hazardous ground potentials, depending on earth resistivity and fault currents, which can develop within the first few feet or more outward from the grounding point.

1910.333(c)(4) Illumination.

1910.333(c)(4)(i)

Employees may not enter spaces containing exposed energized parts, unless illumination is provided that enables the employees to perform the work safely.

1910.333(c)(4)(ii)

Where lack of illumination or an obstruction precludes observation of the work to be performed, employees may not perform tasks near exposed energized parts. Employees may not reach blindly into areas which may contain energized parts.

1910.333(c)(5) Confined or enclosed work spaces.

When an employee works in a confined or enclosed space (such as a manhole or vault) that contains exposed energized parts, the employer shall provide, and the employee shall use, protective shields, protective barriers, or insulating materials as necessary to avoid inadvertent contact with these parts. Doors, hinged panels, and the like shall be secured to prevent their swinging into an employee and causing the employee to contact exposed energized parts.

1910.333(c)(6) Conductive materials and equipment.

Conductive materials and equipment that are in contact with any part of an employee's body shall be handled in a manner that will prevent them from contacting exposed energized conductors or circuit parts. If an employee must handle long dimensional conductive objects (such as ducts and pipes) in areas with exposed live parts, the employer shall institute work practices (such as the use of insulation, guarding, and material handling techniques) which will minimize the hazard.

1910.333(c)(7) Portable ladders.

Portable ladders shall have nonconductive siderails if they are used where the employee or the ladder could contact exposed energized parts.

1910.333(c)(8) Conductive apparel.

Conductive articles of jewelry and clothing (such as watch bands, bracelets, rings, key chains, necklaces, metalized aprons, cloth with conductive thread, or metal headgear) may not be worn if they might contact exposed energized parts. However, such articles may be worn if they are rendered nonconductive by covering, wrapping, or other insulating means.

1910.333(c)(9) Housekeeping duties.

Where live parts present an electrical contact hazard, employees may not perform housekeeping duties at such close distances to the parts that there is a possibility of contact, unless adequate safeguards (such as insulating equipment or barriers) are provided. Electrically conductive cleaning materials (including conductive solids such as steel wool, metalized cloth, and silicon carbide, as well as conductive liquid solutions) may not be used in proximity to energized parts unless procedures are followed which will prevent electrical contact.

1910.333(c)(10) Interlocks.

Only a qualified person following the requirements of paragraph (c) of this section may defeat an electrical safety interlock, and then only temporarily while he or she is working on the equipment. The interlock system shall be returned to its operable condition when this work is completed.

1910.334(b)(2) Reclosing circuits after protective device operation.

After a circuit is deenergized by a circuit protective device, the circuit may not be manually reenergized until it has been determined that the equipment and circuit can be safely energized. The repetitive manual reclosing of circuit breakers or reenergizing circuits through replaced fuses is prohibited.

Note: When it can be determined from the design of the circuit and the overcurrent devices involved that the automatic operation of a device was caused by an overload rather than a fault condition, no examination of the circuit or connected equipment is needed before the circuit is reenergized.

1910.335(a) Use of protective equipment.

1910.335(a)(1) Personal protective equipment.

1910.335(a)(1)(i)

Employees working in areas where there are potential electrical hazards shall be provided with, and shall use, electrical protective equipment that is appropriate for the specific parts of the body to be protected and for the work to be performed.

Note: Personal protective equipment requirements are contained in subpart I of this part.

1910.335(a)(1)(ii)

Protective equipment shall be maintained in a safe, reliable condition and shall be periodically inspected or tested, as required by 1910.137.

1910.335(a)(1)(iii)

If the insulating capability of protective equipment may be subject to damage during use, the insulating material shall be protected. (For example, an outer covering of leather is sometimes used for the protection of rubber insulating material.)

1910.335(a)(1)(iv)

Employees shall wear nonconductive head protection wherever there is a danger of head injury from electric shock or burns due to contact with exposed energized parts.

1910.335(a)(1)(v)

Employees shall wear protective equipment for the eyes or face wherever there is danger of injury to the eyes or face from electric arcs or flashes or from flying objects resulting from electrical explosion.

1910.335(a)(2) General protective equipment and tools.

1910.335(a)(2)(i)

When working near exposed energized conductors or circuit parts, each employee shall use insulated tools or handling equipment if the tools or handling equipment might make contact with such conductors or parts. **See Figure 3-19.** If the insulating capability of insulated tools or handling equipment is subject to damage, the insulating material shall be protected.

1910.335(a)(2)(i)(A)

Fuse handling equipment, insulated for the circuit voltage, shall be used to remove or install fuses when the fuse terminals are energized.

1910.335(a)(2)(i)(B)

Ropes and handlines used near exposed energized parts shall be nonconductive.

1910.335(a)(2)(ii)

Protective shields, protective barriers, or insulating materials shall be used to protect each employee from shock, burns, or other electrically related injuries while that employee is working near exposed energized parts which might be accidentally contacted or where dangerous electric heating or arcing might occur. **See Figure 3-20.** When normally enclosed live parts are exposed for maintenance or repair, they shall be guarded to protect unqualified persons from contact with the live parts.

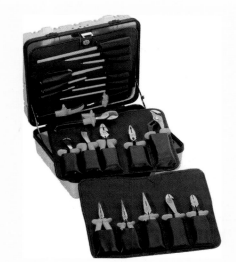

Figure 3-19. *OSHA requires that insulated tools be used if the tool could contact exposed energized conductors or circuit parts when employees are working on or near such conductors or parts.*

Courtesy of Klein Tools

1910.335(b) Alerting techniques.

The following alerting techniques shall be used to warn and protect employees from hazards which could cause injury due to electric shock, burns, or failure of electric equipment parts:

1910.335(b)(1) Safety signs and tags.

Safety signs, safety symbols, or accident prevention tags shall be used where necessary to warn employees about electrical hazards which may endanger them, as required by 1910.145.

1910.335(b)(2) Barricades.

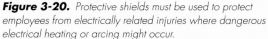

Figure 3-20. *Protective shields must be used to protect employees from electrically related injuries where dangerous electrical heating or arcing might occur.*

Courtesy of Salisbury by Honeywell

Barricades shall be used in conjunction with safety signs where it is necessary to prevent or limit employee access to work areas exposing employees to uninsulated energized conductors or circuit parts. Conductive barricades may not be used where they might cause an electrical contact hazard.

1910.335(b)(3) Attendants.

If signs and barricades do not provide sufficient warning and protection from electrical hazards, an attendant shall be stationed to warn and protect employees.

1910.399, Definitions, in part.

Deenergized. Free from any electrical connection to a source of potential difference and from electrical charge; not having a potential different from that of the earth.

Energized. Electrically connected to a source of potential difference.

Overcurrent. Any current in excess of the rated current of equipment or the ampacity of a conductor. It may result from overload, short circuit, or ground fault.

Overload. Operation of equipment in excess of normal, full-load rating, or of a conductor in excess of rated ampacity that, when it persists for a sufficient length of time, would cause damage or dangerous overheating. A fault, such as a short circuit or ground fault, is not an overload. (See Overcurrent.)

Qualified person. One who has received training in and has demonstrated skills and knowledge in the construction and operation of electric equipment and installations and the hazards involved.

Note 1 to the definition of "qualified person:" Whether an employee is considered to be a "qualified person" will depend upon various circumstances in the workplace. For example, it is possible and, in fact, likely for an individual to be considered "qualified" with regard to certain equipment in the workplace, but "unqualified" as to other equipment. (See 1910.332(b)(3) for training requirements that specifically apply to qualified persons.)

Note 2 to the definition of "qualified person:" An employee who is undergoing on-the-job training and who, in the course of such training, has demonstrated an ability to perform duties safely at his or her level of training and who is under the direct supervision of a qualified person is considered to be a qualified person for the performance of those duties.

Summary

Every day, workers are exposed to hazards associated with working on energized parts, or hazards associated with working close enough to energized parts to allow potential exposure to electrical hazards. When requirements are adhered to and appropriate procedures are in place and implemented, the likelihood of adverse incidents, injuries, and fatalities can be reduced. Many of these requirements were discussed in this chapter. However, the requirements covered here are only a few of the full set that must be considered when assessing workplace hazards and developing an effective safety program.

An understanding of the General Duty Clause; the Part 1926, Safety and Health Regulations for Construction; and the Part 1910, Occupational Safety and Health Standards, is important. Implementing and following the requirements, such as those for training, selection and use of work practices, the use of equipment, and safeguards for personnel protection, will help ensure that the workplace is free of recognized hazards.

This knowledge will also provide a foundation for what is required by OSHA and for how the requirements of NFPA 70E: Standard for Electrical Safety in the Workplace® may be an aid in accomplishing what OSHA requires. Consider this analogy: "OSHA is the *shall*, and NFPA 70E is the *how*" as you explore the requirements of NFPA 70E. Determine whether the provisions of NFPA 70E can be a compliance solution and what, if anything, NFPA 70E requires that OSHA does not already require.

REVIEW QUESTIONS

1. The Occupational Safety and Health Act (OSH Act) of 1970 states that its purpose is to assure so far as possible every working man and woman in the nation safe and __?__ working conditions and to preserve our human resources.
   - a. convenient
   - b. healthful
   - c. legal
   - d. reliable

2. The primary focus of this chapter are the regulations in Part __?__, Safety and Health Regulations for Construction, and Part __?__, General Industry Occupational Safety and Health Standards.
   - a. 1910/1926
   - b. 1926/1910
   - c. 1970/1972
   - d. 1992/2022

3. OSHA's safety and health standards have prevented countless work-related injuries, illnesses, and deaths. Unfortunately, far too many __?__ injuries and fatalities continue to occur.
   - a. obvious
   - b. preventable
   - c. unavoidable
   - d. unpreventable

4. Under the OSHA __?__, employers are responsible for providing a safe and healthful workplace for their workers.
   - a. guidelines
   - b. law
   - c. recommendations
   - d. wishes

5. OSHA's rules and regulations are __?__, they are not recommendations.
   - a. guidelines
   - b. non-mandatory
   - c. requirements
   - d. wishes

6. Many of the Occupational Safety and Health Standards regulations are __?__-based, which means that they require something without necessarily spelling out how compliance is to be accomplished.
   - a. common sense
   - b. fact
   - c. performance
   - d. prescriptive

7. The employer must provide for frequent and regular inspections of job sites by __?__ persons as part of their accident prevention responsibilities.
   - a. competent
   - b. independent
   - c. qualified
   - d. recognized

8. Subpart C requires that the __?__ is responsible for employees wearing appropriate personal protective equipment in all operations where there is an exposure to hazardous conditions.
   - a. employee
   - b. employer
   - c. union
   - d. warehouse supervisor

9. OSHA calls for personal protective equipment (PPE) to be provided by the __?__, which is required to assess the workplace to determine the need for the use of PPE.
   - a. employee
   - b. employer
   - c. union
   - d. warehouse supervisor

10. Each day workers are exposed to hazards, but when __?__ are adhered to and appropriate procedures are in place and implemented, incidents, injuries, and fatalities can be reduced.
    - a. guidelines
    - b. recommendations
    - c. requirements
    - d. wishes

4 Lockout, Tagging, and the Control of Hazardous Energy

OBJECTIVES

1. Demonstrate an understanding of the reasons why the Occupational Safety and Health Act of 1970 was enacted, the Congressional finding and purpose of the act, and the employer and employee responsibilities established in the General Duty Clause.

2. Become familiar with Safety and Health Regulations for Construction, including Subpart C, General Safety and Health Provisions, and the Subpart K provisions for lock-out and tagging of circuits.

3. Demonstrate an understanding of the Occupational Safety and Health Standards, including the Subpart S provisions related to lockout and tagging, and the six major headings of the control of hazardous energy (lockout/tagout) in Subpart J.

4. Identify the six-step process required to verify an electrically safe work condition.

REFERENCES

1. *NFPA 70E*®, 2012 Edition
2. OSHA 29 CFR Part 1910
3. OSHA 29 CFR Part 1926
4. The Occupational Safety and Health Act of 1970

CASE STUDY

A 48-year-old machine operator was killed when he was crushed inside a machine. The employer had been in business for more than 80 years and had approximately 175 employees. There were 55 employees at the facility where the incident occurred. The victim had been employed with the company for 19 years.

The employer had a written safety program with task-specific safe work procedures for all positions in the shop. Employees held informal weekly tailgate safety meetings with the supervisors, as well as formal monthly safety meetings. The company's training program was usually accomplished through on-the-job-training monitored by the supervisors.

The machine involved in the incident was completely automated. The operating portion of the machine was enclosed for safety. Whenever any regular access panel or door to the machine was opened, the machine was supposed to shut off automatically. The pedestal on which the machine sat was also enclosed, with the exception of the area where the conveyor belt exited from underneath the machine. The guarding around the pedestal had to be mechanically removed to gain access to the pedestal.

On the day of the incident, the machine stopped working, and its warning lights came on. The victim was not at his workstation. His coworkers contacted the supervisor, who discovered the victim inside the machine with his head trapped between the pedestal frame and the mobile plate frame. Coworkers called 911, and the responding fire fighters had to unbolt the guarding around the pedestal frame to gain access to the victim and extricate him. Paramedics pronounced the victim dead after they removed him from the machine.

Investigation of the incident site revealed the victim's tools lying next to the opening in the pedestal where the conveyor belt was located; in addition, several pieces of the machine product were scattered about the conveyor belt. The guard or shield on the side of the conveyor belt was unbolted on one side. These factors suggested that the victim crawled into the machine from the opening for the conveyor belt.

The cause of death, according to the death certificate, was blunt head trauma.

Source: For details of this case, see FACE Investigation #03CA006. Accessed June 4, 2012.

For additional information, visit qr.njatcdb.org Item #1194.

Introduction

The Occupational Safety and Health Administration's (OSHA's) requirements are performance oriented in many cases; that is, protection of workers is required, although the requirements do not necessarily spell out precisely how worker protection is to be accomplished. This includes requirements related to protecting workers from electrical hazards. While increasingly more employers are looking to *NFPA 70E* for guidance in an effort to understand how to comply with OSHA's performance-oriented requirements, OSHA's requirements related to lockout/tagout and working on or near exposed deenergized parts are examples where OSHA offers more specific details, including the steps that must be followed to comply with these requirements.

This chapter explores OSHA's requirements for lockout and tagging of circuits [1926.417], control of hazardous energy (lockout/tagout) [1910.147], the lockout and tagging provisions of 1910.333(b)(2), and the ways that provisions in *NFPA 70E* Section 120.1 can meet or supplement OSHA's provisions, including those in the Occupational Safety and Health Act of 1970.

Many lockout/tagout programs are based on OSHA 1910.147. OSHA notes that lockout and tagging procedures complying with 1910.147 are also deemed to comply with 1910.333(b)(2), provided that the procedures address the electrical safety hazards covered by Subpart S of 29 CFR 1910, and that those procedures incorporate the requirements of two additional paragraphs not contained in 1910.147.

In addition to developing lockout/tagout programs based on requirements from OSHA, the requirements from *NFPA 70E* Article 120, Establishing an Electrically Safe Work Condition, should be considered as part of electrical safety practices. The provisions of *NFPA 70E* Article 120 generally meet or exceed OSHA's lockout/tagout requirements and should be examined carefully for any other considerations that might lead to a more comprehensive lockout/tagout program. This includes, but is not limited to, the six-step process outlined in Section 120.1 to verify that an electrically safe work condition exists after the provisions contained in *NFPA 70E* Section 120.2 have been taken into account. The provisions of Article 120 need to be reviewed in their entirety for a full understanding of their use and application.

The Occupational Safety and Health Act of 1970

The Occupational Safety and Health Act of 1970 was enacted "to assure safe and healthful working conditions for working men and women; by authorizing enforcement of the standards developed under the Act; by assisting and encouraging the States in their efforts to assure safe and healthful working conditions; by providing for research, information, education, and training in the field of occupational safety and health; and for other purposes." This Act consists of 35 sections. Part of the content of two of those sections follows:

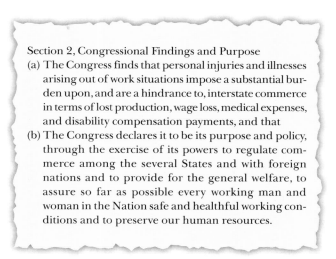

Section 2, Congressional Findings and Purpose
(a) The Congress finds that personal injuries and illnesses arising out of work situations impose a substantial burden upon, and are a hindrance to, interstate commerce in terms of lost production, wage loss, medical expenses, and disability compensation payments, and that
(b) The Congress declares it to be its purpose and policy, through the exercise of its powers to regulate commerce among the several States and with foreign nations and to provide for the general welfare, to assure so far as possible every working man and woman in the Nation safe and healthful working conditions and to preserve our human resources.

Section 2(b) includes 13 points that detail how to accomplish the purpose and satisfy the policy of assuring, so far as possible, that every working man and woman in the nation has safe and healthful working conditions and of preserving human resources.

Section 5, Duties, is commonly known as "the General Duty Clause." It identifies responsibilities for both employers and employees:

(a) Each employer –
  (1) shall furnish to each of his employees employment and a place of employment which are free from recognized hazards that are causing or are likely to cause death or serious physical harm to his employees;
  (2) shall comply with occupational safety and health standards promulgated under this Act.
(b) Each employee shall comply with occupational safety and health standards and all rules, regulations, and orders issued pursuant to this Act which are applicable to his own actions and conduct.

A fundamental understanding of the reason why the Occupational Safety and Health Act of 1970 was enacted, the Congressional finding and purpose in enacting this legislation, and employer and employee responsibilities is necessary as a foundation for understanding why safe work practices are essential "to assure safe and healthful working conditions for working men and women." Lockout and tagging of circuits, control of hazardous energy, and achieving an electrically safe work condition are among the safe work practices intended to accomplish this goal.

Safety and Health Regulations for Construction

Part 1926, Safety and Health Regulations for Construction, sets forth the safety and health standards promulgated by the Secretary of Labor under Section 107 of the Contract Work Hours and Safety Standards Act. Subpart C sets forth the general safety and health regulations of Part 1926, while Subpart K addresses electrical safety requirements within the scope of that subpart.

OSHA Tip

The standards contained in 29 CFR 1926 Subpart V apply to the construction of electric power transmission and distribution lines and equipment.

General Safety and Health Provisions

Subpart C provides general safety and health provisions for Part 1926, Safety and Health Regulations for Construction. The following definition and general provisions help clarify the application of Part 1926. Subpart C is to be referred to in its entirety for a complete understanding of these regulations.

1926.21(b)(2)
The employer shall instruct each employee in the recognition and avoidance of unsafe conditions and the regulations applicable to his work environment to control or eliminate any hazards or other exposure to illness or injury. **See Figure 4-1.**
1926.20(f)(2)
Standards in this part requiring training on hazards and related matters, such as standards requiring that employees receive training or that the employer train employees, provide training to employees, or institute or implement a training program, impose a separate compliance duty with respect to each employee covered by the requirement. The employer must train each affected employee in the manner required by the standard, and each failure to train an employee may be considered a separate violation.
1926.32(g)
For purposes of this section, "Construction work" means work for construction, alteration, and/or repair, including painting and decorating.

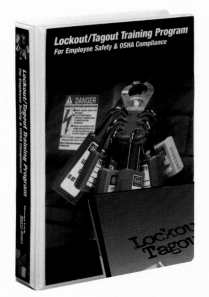

Figure 4-1. *OSHA's general safety and health provisions require that employees be instructed in any regulations applicable to their work and be able to recognize and avoid unsafe conditions.*

Courtesy of Ideal Industries, Inc.

Electrical Safety Requirements

Subpart K, Electrical, addresses electrical safety requirements that are necessary for the practical safeguarding of employees involved in construction work. Subpart K is divided into four major divisions plus an applicable definitions section. Safety-related work practices—one of those four major divisions—are contained in Sections 1926.416 and 1926.417, with Section 1926.417 addressing lockout and tagging of circuits.

1926.417 Lockout and tagging of circuits.
1926.417(a)
Controls. Controls that are to be deactivated during the course of work on energized or deenergized equipment or circuits shall be tagged.
1926.417(b)
Equipment and circuits. Equipment or circuits that are deenergized shall be rendered inoperative and shall have tags attached at all points where such equipment or circuits can be energized. **See Figure 4-2.**
1926.417(c)
Tags. Tags shall be placed to identify plainly the equipment or circuits being worked on.

Occupational Safety and Health Standards

The title of Part 1910 is the Occupational Safety and Health Standards. Per 1910.1(b), Part 1910 carries out the directive of the Secretary of Labor under section 6(a)

Figure 4-2. *Deenergized equipment or circuits shall be rendered inoperative and have tags attached at all points where such equipment or circuits can be energized. The tags must be placed so as to clearly identify the equipment or circuits being worked on.*

Courtesy of Ideal Industries, Inc.

Figure 4-3. *Both locks and tags are generally required when working on or near exposed deenergized parts.*

Courtesy of Ideal Industries, Inc.

of the Occupational Safety and Health Act of 1970. It contains occupational safety and health standards that have been found to be national consensus standards or established federal standards. A number of provisions from Subparts J and S are examined here. Subpart S, Electrical, includes 1910.333(b)(2), Lockout and Tagging. Subpart J, General Environmental Controls, includes 1910.147, The control of hazardous energy (lockout/tagout).

Lockout and Tagging

Part 1910, Subpart S, addresses electrical safety requirements that are necessary for the practical safeguarding of employees in their workplaces. It is divided into four major divisions, one of which contains safety-related work practice regulations, in Sections 1910.331 through 1910.360. Lockout and tagging within the scope of Subpart S is covered within 1910.333(b), which deals with working on or near exposed deenergized parts. **See Figure 4-3.**

1910.333(b)(1) Application.
   This paragraph applies to work on exposed deenergized parts or near enough to them to expose the employee to any electrical hazard they present. Conductors and parts of electric equipment that have been

deenergized but have not been locked out or tagged in accordance with paragraph (b) of this section shall be treated as energized parts, and paragraph (c) of this section applies to work on or near them.
   1910.333(b)(2) Lockout and Tagging.
   While any employee is exposed to contact with parts of fixed electric equipment or circuits which have been deenergized, the circuits energizing the parts shall be locked out or tagged or both in accordance with the requirements of this paragraph. **See Figure 4-4.**

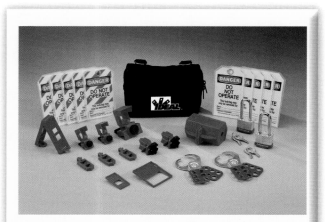

Figure 4-4. *Appropriate lockout and tagout devices must be used in compliance with any applicable requirements while any employee is exposed to contact with equipment or circuits that have been deenergized.*

Courtesy of Ideal Industries, Inc.

The requirements shall be followed in the order in which they are presented (i.e., paragraph (b)(2)(i) first, then paragraph (b)(2)(ii), etc.).

Note 1: As used in this section, fixed equipment refers to equipment fastened in place or connected by permanent wiring methods.

Note 2: Lockout and tagging procedures that comply with paragraphs (c) through (f) of 1910.147 will also be deemed to comply with paragraph (b)(2) of this section provided that:

[1] The procedures address the electrical safety hazards covered by this Subpart; and

[2] The procedures also incorporate the requirements of paragraphs (b)(2)(iii)(D) and (b)(2)(iv)(B) of this section.

Procedures

1910.333(b)(2)(i)

The employer shall maintain a written copy of the procedures outlined in paragraph (b)(2) and shall make it available for inspection by employees and by the Assistant Secretary of Labor and his or her authorized representatives. **See Figure 4-5.**

Note: The written procedures may be in the form of a copy of paragraph (b) of this section.

Deenergizing Equipment

1910.333(b)(2)(ii)
1910.333(b)(2)(ii)(A)

Safe procedures for deenergizing circuits and equipment shall be determined before circuits or equipment are deenergized.

1910.333(b)(2)(ii)(B)

The circuits and equipment to be worked on shall be disconnected from all electric energy sources. **See Figure 4-6.**

Control circuit devices, such as push buttons, selector switches, and interlocks, may not be used as the sole means for deenergizing circuits or equipment. Interlocks for electric equipment may not be used as a substitute for lockout and tagging procedures.

1910.333(b)(2)(ii)(C)

Stored electric energy which might endanger personnel shall be released. Capacitors shall be discharged and high capacitance elements shall be short-circuited and grounded, if the stored electric energy might endanger personnel.

Note: If the capacitors or associated equipment are handled in meeting this requirement, they shall be treated as energized.

1910.333(b)(2)(ii)(D)

Stored non-electrical energy in devices that could reenergize electric circuit parts shall be blocked or relieved to the extent that the circuit parts could not be accidentally energized by the device. **See Figure 4-7.**

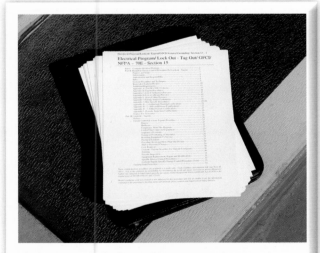

Figure 4-5. *The lockout and tagging procedures must be in writing and made available for inspection.*

Photo courtesy of National Electrical Contractors Association (NECA)

Figure 4-6. *All electrical energy sources must be accounted for.*

Courtesy of Ideal Industries, Inc.

Figure 4-7. *Stored nonelectrical energy must be blocked or relieved as necessary.*

Courtesy of Ideal Industries, Inc.

Application of Locks and Tags

1910.333(b)(2)(iii)
    1910.333(b)(2)(iii)(A)
    A lock and a tag shall be placed on each disconnecting means used to deenergize circuits and equipment on which work is to be performed, except as provided in paragraphs (b)(2)(iii)(C) and (b)(2)(iii)(E) of this section. **See Figure 4-8.**
    The lock shall be attached so as to prevent persons from operating the disconnecting means unless they resort to undue force or the use of tools.
    1910.333(b)(2)(iii)(B)
    Each tag shall contain a statement prohibiting unauthorized operation of the disconnecting means and removal of the tag. **See Figure 4-9.**

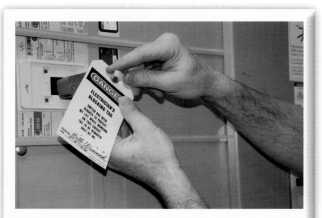

Figure 4-8. *Both a lock and a tag are generally required to be placed on a disconnecting means for circuits and equipment before work begins.*

Photo courtesy of National Electrical Contractors Association (NECA)

1910.333(b)(2)(iii)(C)
    If a lock cannot be applied, or if the employer can demonstrate that tagging procedures will provide a level of safety equivalent to that obtained by the use of a lock, a tag may be used without a lock.

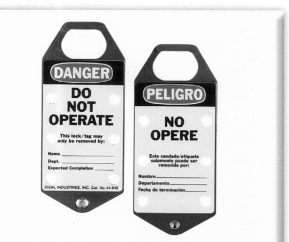

Figure 4-9. *A tag must warn against both unauthorized disconnecting means of operation and unauthorized removal of the tag.*

Courtesy of Ideal Industries, Inc.

1910.333(b)(2)(iii)(D)
    A tag used without a lock, as permitted by paragraph (b)(2)(iii)(C) of this section, shall be supplemented by at least one additional safety measure that provides a level of safety equivalent to that obtained by use of a lock. **See Figure 4-10.**
    Examples of additional safety measures include the removal of an isolating circuit element, blocking of a controlling switch, or opening of an extra disconnecting device.
    1910.333(b)(2)(iii)(E)
    A lock may be placed without a tag only under the following conditions:
    1910.333(b)(2)(iii)(E)(1)
    Only one circuit or piece of equipment is deenergized, and
    1910.333(b)(2)(iii)(E)(2)
    The lockout period does not extend beyond the work shift, and
    1910.333(b)(2)(iii)(E)(3)
    Employees exposed to the hazards associated with reenergizing the circuit or equipment are familiar with this procedure.

Figure 4-10. *Tags are permitted to be used without a lock under limited circumstances, but only where at least one additional safety measure is employed and safety equal to a lock is assured.*

Courtesy of Ideal Industries, Inc.

Figure 4-11. *A qualified person must test all circuit elements and electrical parts of equipment to verify that they are in a deenergized state.*

Verification of Deenergized Condition

1910.333(b)(2)(iv)

The requirements of this paragraph shall be met before any circuits or equipment can be considered and worked as deenergized.

1910.333(b)(2)(iv)(A)

A qualified person shall operate the equipment operating controls or otherwise verify that the equipment cannot be restarted.

1910.333(b)(2)(iv)(B)

A qualified person shall use test equipment to test the circuit elements and electrical parts of equipment to which employees will be exposed and shall verify that the circuit elements and equipment parts are deenergized. **See Figure 4-11.**

The test shall also determine if any energized condition exists as a result of inadvertently induced voltage or unrelated voltage backfeed even though specific parts of the circuit have been deenergized and presumed to be safe. If the circuit to be tested is over 600 volts, nominal, the test equipment shall be checked for proper operation immediately after this test.

Reenergizing Equipment

1910.333(b)(2)(v)

These requirements shall be met, in the order given, before circuits or equipment are reenergized, even temporarily.

1910.333(b)(2)(v)(A)

A qualified person shall conduct tests and visual inspections, as necessary, to verify that all tools, electrical jumpers, shorts, grounds, and other such devices have been removed, so that the circuits and equipment can be safely energized.

1910.333(b)(2)(v)(B)

Employees exposed to the hazards associated with reenergizing the circuit or equipment shall be warned to stay clear of circuits and equipment.

1910.333(b)(2)(v)(C)

Each lock and tag shall be removed by the employee who applied it or under his or her direct supervision. **See Figure 4-12.**

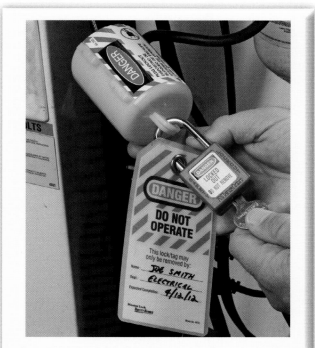

Figure 4-12. *Each lock and tag is generally required to be removed by the employee who applied it.*

Courtesy of Ideal Industries, Inc.

However, if this employee is absent from the workplace, then the lock or tag may be removed by a qualified person designated to perform this task provided that:

1910.333(b)(2)(v)(C)(1)

The employer ensures that the employee who applied the lock or tag is not available at the workplace, and

1910.333(b)(2)(v)(C)(2)

The employer ensures that the employee is aware that the lock or tag has been removed before he or she resumes work at that workplace.

1910.333(b)(2)(v)(D)

There shall be a visual determination that all employees are clear of the circuits and equipment.

The Control of Hazardous Energy (Lockout/Tagout)

Many lockout/tagout programs incorporate OSHA 29 CFR 1910.147. This standard is found in Subpart J, General Environmental Controls, of Part 1910, Occupational Safety and Health Standards, and is entitled "The control of hazardous energy (lockout/tagout)."

Recall that Note 2 to 1910.333(b)(2) clarifies that lockout and tagging procedures that comply with paragraphs (c) through (f) of 1910.147 will also be deemed to comply with paragraph (b)(2) of this section provided that two criteria are satisfied:

- The procedures address the electrical safety hazards covered by Subpart S.
- The procedures incorporate the requirements of paragraphs 1910.333(b)(2)(iii)(D) and 1910.333(b)(2)(iv)(B).

1910.333(b)(2)(iii)(D)

A tag used without a lock, as permitted by paragraph (b)(2)(iii)(C) of this section, shall be supplemented by at least one additional safety measure that provides a level of safety equivalent to that obtained by use of a lock. Examples of additional safety measures include the removal of an isolating circuit element, blocking of a controlling switch, or opening of an extra disconnecting device.

1910.333(b)(2)(iv)(B)

A qualified person shall use test equipment to test the circuit elements and electrical parts of equipment to which employees will be exposed and shall verify that the circuit elements and equipment parts are deenergized. The test shall also determine if any energized condition exists as a result of inadvertently induced voltage or unrelated voltage backfeed even though specific parts of the circuit have been deenergized and presumed to be safe. If the circuit to be tested is over 600 volts, nominal, the test equipment shall be checked for proper operation immediately after this test.

Section 1910.147 is broken down into six different major headings, each of which is indicated by a different lowercase letter:

a. Scope, application, and purpose
b. Definitions
c. General
d. Application of control
e. Release from lockout or tagout
f. Additional requirements

See the appendix in this publication on the provisions of Section 1910.147.

Typical Minimal Lockout Procedures

Appendix A to OSHA 29 CFR 1910.147 serves as a nonmandatory guideline to assist employers and employees in complying with the requirements of 1910.147 and to provide other helpful information. **See Figure 4-13.** Nothing in the appendix adds to or detracts from any of the requirements of 1910.147. Rather, a simple lockout procedure is described to assist employers in developing their own procedures so they meet the requirements of this standard. For more complex systems, more comprehensive procedures may need to be developed, documented, and utilized.

Figure 4-13. *This example of a simple lockout procedure, found in OSHA 29 CFR 1910.147 Appendix A, describes how employers may create their own procedures.*

General

The following simple lockout procedure is provided to assist employers in developing their procedures so they meet the requirements of this standard. When the energy isolating devices are not lockable, tagout may be used, provided the employer complies with the provisions of the standard which require additional training and more rigorous periodic inspections. When tagout is used and the energy isolating devices are lockable, the employer must provide full employee protection (see paragraph (c)(3)) and additional training and more rigorous periodic inspections are required. For more complex systems, more comprehensive procedures may need to be developed, documented, and utilized.

Lockout Procedure

Lockout Procedure for

_____

(Name of Company for single procedure or identification of equipment if multiple procedures are used).

Purpose

This procedure establishes the minimum requirements for the lockout of energy isolating devices whenever maintenance or servicing is done on machines or equipment. It shall be used to ensure that the machine or equipment is stopped, isolated from all potentially hazardous energy sources and locked out before employees perform any servicing or maintenance where the unexpected energization or start-up of the machine or equipment or release of stored energy could cause injury.

Compliance With This Program

All employees are required to comply with the restrictions and limitations imposed upon them during the use of lockout. The authorized employees are required to perform the lockout in accordance with this procedure. All employees, upon observing a machine or piece of equipment which is locked out to perform servicing or maintenance shall not attempt to start, energize, or use that machine or equipment.

(Type of compliance enforcement to be taken for violation of the above.)

Sequence of Lockout

(1) Notify all affected employees that servicing or maintenance is required on a machine or equipment and that the machine or equipment must be shut down and locked out to perform the servicing or maintenance.

(Name(s)/Job Title(s) of affected employees and how to notify.)

(2) The authorized employee shall refer to the company procedure to identify the type and magnitude of the energy that the machine or equipment utilizes, shall understand the hazards of the energy, and shall know the methods to control the energy.

(Type(s) and magnitude(s) of energy, its hazards and the methods to control the energy.)

(3) If the machine or equipment is operating, shut it down by the normal stopping procedure (depress the stop button, open switch, close valve, etc.).

(Type(s) and location(s) of machine or equipment operating controls.)

(4) De-activate the energy isolating device(s) so that the machine or equipment is isolated from the energy source(s).

(Type(s) and location(s) of energy isolating devices.)

(5) Lock out the energy isolating device(s) with assigned individual lock(s).

(6) Stored or residual energy (such as that in capacitors, springs, elevated machine members, rotating flywheels, hydraulic systems, and air, gas, steam, or water pressure, etc.) must be dissipated or restrained by methods such as grounding, repositioning, blocking, bleeding down, etc.

(Type(s) of stored energy—methods to dissipate or restrain.)

(7) Ensure that the equipment is disconnected from the energy source(s) by first checking that no personnel are exposed, then verify the isolation of the equipment by operating the push button or other normal operating control(s) or by testing to make certain the equipment will not operate.

Caution: Return operating control(s) to neutral or "off" position after verifying the isolation of the equipment.

(Method of verifying the isolation of the equipment.)

(8) The machine or equipment is now locked out.
"Restoring Equipment to Service." When the servicing or maintenance is completed and the machine or equipment is ready to return to normal operating condition, the following steps shall be taken.

(1) Check the machine or equipment and the immediate area around the machine to ensure that nonessential items have been removed and that the machine or equipment components are operationally intact.

(2) Check the work area to ensure that all employees have been safely positioned or removed from the area.

(3) Verify that the controls are in neutral.

(4) Remove the lockout devices and reenergize the machine or equipment. Note: The removal of some forms of blocking may require reenergization of the machine before safe removal.

(5) Notify affected employees that the servicing or maintenance is completed and the machine or equipment is ready for used.

[54 FR 36687, Sept. 1, 1989 as amended at 54 FR 42498, Oct. 17, 1989; 55 FR 38685, Sept. 20, 1990; 61 FR 5507, Feb. 13, 1996]

Definitions

OSHA defines the terms *lockout, energy-isolating device, tagout, lockout device,* and *tagout device* in 1910.147(b):

> **Lockout.** The placement of a lockout device on an energy-isolating device, in accordance with an established procedure, ensuring that the energy-isolating device and the equipment being controlled cannot be operated until the lockout device is removed.

> **Energy-isolating device.** A mechanical device that physically prevents the transmission or release of energy, including but not limited to the following: A manually operated electrical circuit breaker; a disconnect switch; a manually operated switch by which the conductors of a circuit can be disconnected from all ungrounded supply conductors, and, in addition, no pole can be operated independently; a line valve; a block; and any similar device used to block or isolate energy. Push buttons, selector switches and other control circuit type devices are not energy isolating devices.
>
> **Tagout.** The placement of a tagout device on an energy-isolating device, in accordance with an established procedure, to indicate that the energy-isolating device and the equipment being controlled may not be operated until the tagout device is removed.
>
> **Lockout device.** A device that utilizes a positive means, such as a lock, either key or combination type, to hold an energy-isolating device in the safe position and prevent the energizing of a machine or equipment. Included are blank flanges and bolted slip blinds.
>
> **Tagout device.** A prominent warning device, such as a tag and a means of attachment, which can be securely fastened to an energy isolating device in accordance with established procedure, to indicate that the energy isolating device and the equipment being controlled may not be operated until the tagout device is removed.
>
> **Affected employee.** An employee whose job requires him/her to operate or use a machine or equipment on which servicing or maintenance is being performed under lockout or tagout, or whose job requires him/her to work in an area in which such servicing or maintenance is being performed.

OSHA 1910.399 defines the terms *deenergized* and *energized* as follows; these apply to 1910 Subpart S:

> *Deenergized.* Free from any electrical connection to a source of potential difference and from electrical charge; not having a potential different from that of the earth.
>
> *Energized.* Electrically connected to a source of potential difference.

Achieving an Electrically Safe Work Condition

OSHA's lockout/tagout requirements are federal law. However, in addition to developing lockout/tagout programs and procedures for working on or near exposed deenergized parts that fulfill the requirements of 29 CFR 1910.147, 1910.333(b), and 1926.417, the process of achieving an electrically safe work condition from *NFPA 70E* Article 120 should be considered, including, but not limited to, *NFPA 70E* Section 120.1.

The Process of Achieving an Electrically Safe Work Condition

The provisions of *NFPA 70E* Article 120 generally meet or exceed OSHA's lockout/tagout requirements and should be examined carefully for considerations related to developing a more comprehensive and complete lockout/tagout program. This includes, but is not limited to, the six-step process outlined in Section 120.1 to verify that an electrically safe work condition exists after the provisions contained in *NFPA 70E* Section 120.2 have been taken into account

An electrically safe work condition is defined in *NFPA 70E* as a state in which an electrical conductor or circuit part has been

- Disconnected from energized parts
- Locked/tagged in accordance with established standards
- Tested to ensure the absence of voltage
- Grounded if determined necessary

An electrically safe work condition is achieved when the procedures outlined in *NFPA 70E* 120.2 are performed and verified by the six steps outlined in Section 120.1. Achieving an electrically safe work condition is one example of how *NFPA 70E* can supplement and enhance OSHA's requirements for lockout and tagging of circuits (1926.417), control of hazardous energy (lockout/tagout; 1910.147), and the lockout and tagging provisions of 1910.333(b)(2), among others.

Note that Article 120 is divided into three sections that provide the requirements for establishing an electrically safe work condition:

- Process of Achieving an Electrically Safe Work Condition
- Deenergized Electrical Conductors or Circuit Parts That Have Lockout/Tagout Devices Applied
- Temporary Protective Grounding Equipment

Article 120 needs to be referenced for these requirements in its entirety. Also consider *NFPA 70E* Informative Annex G, Sample Lockout/Tagout Procedure. It is provided for informational purposes only and is not a part of the requirements of *NFPA 70E*. This annex offers a sample procedure to assist employers in developing a procedure that meets the requirements of *NFPA 70E* 120.2.

Verifying an Electrically Safe Work Condition

An electrically safe work condition is achieved when the procedures outlined in *NFPA 70E* Section 120.2 are performed and verified by following the six steps outlined in Section 120.1:

Figure 4-14. *Determination of all possible sources of electrical supply is one step in verifying that an electrically safe work condition has been achieved.*

Photo courtesy of National Electrical Contractors Association (NECA)

(1) Determine all possible sources of electrical supply to the specific equipment.
Check applicable up-to-date drawings, diagrams, and identification tags. **See Figure 4-14.**

(2) After properly interrupting the load current, open the disconnecting device(s) for each source. **See Figure 4-15.**

(3) Wherever possible, visually verify that all blades of the disconnecting devices are fully open or that drawout-type circuit breakers are withdrawn to the fully disconnected position. **See Figure 4-16.**

(4) Apply lockout/tagout devices in accordance with a documented and established policy. **See Figure 4-17.**

(5) Use an adequately rated voltage detector to test each phase conductor or circuit part to verify they are deenergized. Test each phase conductor or circuit part both phase-to-phase and phase-to-ground. Before and after each test, determine that the voltage detector is operating satisfactorily. **See Figure 4-18.**

(6) Where the possibility of induced voltages or stored electrical energy exists, ground the phase conductors or circuit parts before touching them. Where it could be reasonably anticipated that the conductors or circuit parts being deenergized could contact other exposed energized conductors or circuit parts, apply ground connecting devices rated for the available fault duty. **See Figure 4-19.**

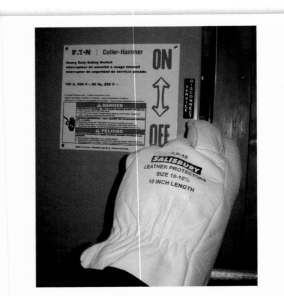

Figure 4-15. *Step 2 outlined in Section 120.1 includes opening disconnecting devices after the load current has been properly interrupted.*

Caution

Consult American National Standards Institute/Instrumentation, Systems, and Automation Society (ANSI/ISA)-61010-1 (82.02.01)/Underwriters Laboratories Inc. (UL) 61010-1 for rating and design requirements for voltage measurement and test instruments intended for use on electrical systems of 1,000 volts or less [derived from *NFPA 70E* Section 120.1(5) Informational Note].

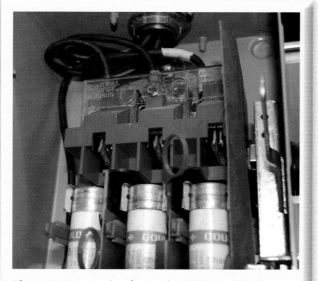

Figure 4-16. *Visual verification that all blades of the disconnecting devices are fully open is required whenever possible.*

Photo courtesy of National Electrical Contractors Association (NECA)

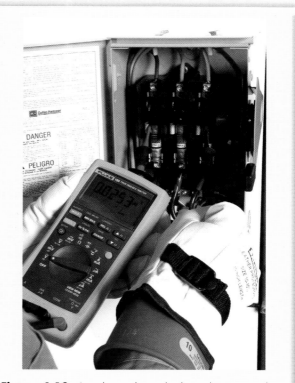

Figure. 4-18. *An adequately rated voltage detector must be used to verify the absence of both phase-to-phase and phase-to-ground voltage.*

Figure 4-17. *The documented lockout and tagout policy must be followed and devices applied in accordance with the established policy.*

Courtesy of Ideal Industries, Inc.

Figure 4-19. *Apply ground-connecting devices rated for the available fault duty as warranted.*

Photo courtesy of National Electrical Contractors Association (NECA)

Summary

OSHA's steps for lockout/tagout and detailed requirements for working on or near exposed deenergized parts must be followed. Training, planning, and preparation are critical to avoid incidents and injury.

The Occupational Safety and Health Act of 1970 was enacted "to assure safe and healthful working conditions for working men and women." The provisions of OSHA 29 CFR 1926.417 provide performance requirements for lockout and tagging of circuits within the scope of Subpart K. OSHA CFR 1910.333 (b)(2) and OSHA 29 CFR 1910.147 cover minimum steps for lockout/tagout, as well as procedures to follow when lockout or tagout devices are removed and energy is restored within the scope of their respective subparts.

Finally, consider *NFPA 70E* Article 120 to be a supplement to OSHA's requirements and procedures. Following the procedures in *NFPA 70E* Section 120.1 for verification that an electrically safe work condition has been achieved will allow for a more comprehensive lockout/tagout program.

REVIEW QUESTIONS

1. OSHA defines a __?__ as a device that utilizes a positive means such as a lock, either key or combination type, to hold an energy isolating device in the safe position and prevent the energizing of a machine or equipment.
   - a. lockout device
   - b. lockout/tagout device
   - c. tagout device
   - d. tie wrap

2. OSHA defines a __?__ as a prominent warning device, such as a tag and a means of attachment, which can be securely fastened to an energy isolating device in accordance with established procedure, to indicate that the energy isolating device and the equipment being controlled may not be operated until the device is removed.
   - a. labeling device
   - b. lockout device
   - c. lockout/tagout device
   - d. tagout device

3. Per OSHA, each __?__ shall furnish to each employee employment and a place of employment which are free from recognized hazards that are causing or are likely to cause death or serious physical harm to employees.
   - a. business manager
   - b. employer
   - c. trade
   - d. training director

4. Each __?__ shall comply with occupational safety and health standards and all rules, regulations, and orders issued pursuant to the OSHA Act which are applicable to his own actions and conduct.
   - a. craft
   - b. employee
   - c. employer
   - d. trade

5. Lockout and tagging of circuits, control of hazardous energy, and achieving an electrically safe work condition are among the safe __?__ intended to assure worker safety and health.
   - a. habits
   - b. happenings
   - c. ideas
   - d. work practices

6. Per OSHA Part 1926, Subpart C, the employer shall instruct each employee in the recognition and avoidance of __?__ conditions and the regulations applicable to his work environment to control or eliminate any hazards or other exposure to illness or injury.
   - a. abnormal
   - b. normal
   - c. safe
   - d. unsafe

7. OSHA Part 1910 states that the lock shall be attached so as to __?__ operation of the disconnecting means unless one would resort to undue force or the use of tools.
   - a. allow
   - b. discourage
   - c. encourage
   - d. prevent

8. OSHA Part 1910 states that if a lock __?__ be applied, or if the employer can demonstrate that tagging procedures will provide a level of safety equivalent to that obtained by the use of a lock, a tag may be used without a lock.
   - a. can
   - b. cannot
   - c. should not
   - d. will not

9. OSHA Part 1910 states that a tag used without a lock shall be __?__ by at least one additional safety measure that provides a level of safety equivalent to that obtained by use of a lock.
   - a. complimented
   - b. implemented
   - c. prevented
   - d. supplemented

5 Introduction to NFPA 70E

OBJECTIVES

1. Understand the history, scope, definitions, and organization of *NFPA 70E*.
2. Become familiar with the *NFPA 70E* provisions related to host and contract employer responsibilities and use of equipment.
3. Understand the concepts and requirements necessary for an effective electrical safety program and training.
4. Become familiar with the topics addressed in the three chapters and 16 annexes in *NFPA 70E*.

REFERENCES

1. *NFPA 70E, Standard for Electrical Safety in the Workplace®*, 2012 Edition
2. Appendix, *OSHA Partnership Agreement*
3. OSHA 29 CFR Part 1926
4. OSHA 29 CFR Part 1910

A 46-year-old electrical project supervisor died when he contacted an energized conductor inside a control panel. The employer was an industrial electrical contracting company that had been in operation for 10 years. It employed 20 workers, including three electrical project supervisors. The company's written safety program, which was administered by the president/CEO and the electrical project supervisors, included disciplinary procedures. The president/CEO served as safety officer on a collateral duty basis, and the supervisors held monthly safety meetings with all crew members.

The victim had worked for the company for five years and three months as an electrical project supervisor and had approximately 27 years of electrical experience. The company and victim had been working at the packaging plant for six months before the incident; the incident was the company's first fatality.

The company had been contracted to install control cabinets, conduit, wiring, and solid-state compressor motor starters for two 400-horsepower air compressors. On the day of the incident, the victim and three coworkers (one electrical worker and two helpers) arrived at the plant at 7 a.m. They were scheduled to install the last starter and to complete the wiring from the compressor motor to the starter in the control panel, and from the starter control panel to the main distribution panel. Once installation was completed, they were to check the operation of the unit.

At approximately 3:15 p.m., the starter had been installed, and all associated wiring had been completed. The victim directed a helper to turn the switch to the "on" position at the main distribution panel, approximately six feet away, to check the starter's operation. The helper turned the switch to the "on" position, energizing the components inside the starter control panel. The victim pushed the "start" button, and the starter indicator light activated, but the compressor motor did not start. When the compressor motor did not engage, the victim concluded that a problem existed inside the starter control panel. The victim directed the helper to retrieve a voltmeter so that he could check the continuity of the wiring inside the starter control panel. In the interim, the victim opened the starter control panel door without deenergizing the unit and reached inside to trace the wiring and check the integrity of the electrical leads. In doing so, he contacted the 480-volt primary lead for the motor starter with his left hand.

Current passed through the victim's left hand and body and exited through his feet to the ground. The victim yelled, and the helper immediately turned the main distribution switch to the "off" position as the victim collapsed to the floor. Emergency medical services (EMS) was called, and the helper checked the victim and immediately administered cardiopulmonary resuscitation (CPR). EMS personnel arrived in 10 to 15 minutes, continued CPR, and transported the victim to the local hospital, where he was pronounced dead one hour and 20 minutes after the incident occurred.

Source: For details of this case, see FACE Investigation #92-20. Accessed May 18, 2012.

For additional information, visit qr.njatcdb.org Item #1195.

Introduction

Both the Occupational Safety and Health Administration (OSHA) and *NFPA 70E* require an employer to implement an electrical safety program. A safety program is an organized effort to reduce injuries. It should be a subset of an overall safety program, is an integral part of an overall safety and health program, and is a vital part of establishing an electrically safe workplace. It must be understood that an effective electrical safety program cannot be duplicated from any source and implemented directly. Satisfying regulatory requirements might be possible by duplicating a program from an external source; however, for an electrical safety program to be effective, each element of the program must be discussed and analyzed on the basis of the physical environment and experience level of both supervisors and workers.

An important provision in Section 5 of the Occupational Safety and Health Act of 1970 (OSH Act) requires that workers be provided with a workplace free from recognized hazards. Yet it is not always clear how to provide a hazard-free workplace. OSHA requirements addressing electrical hazards are often written in performance language; that is, the rules define a result without providing details of how to accomplish it. Many people consider the requirements defined in *NFPA 70E* as a means to comply with the OSHA requirements related to the hazards of work involving electrical hazards.

This chapter provides an overview of the *NFPA 70E* standard, with the primary focus on the provisions of Articles 90, 100, and 110. These provisions include, but are not limited to, host and contract employer responsibilities, training requirements, electrical safety program considerations, and use of equipment.

NFPA 70E History, Introduction, and Application of Safety-Related Work Practices

NFPA 70E, *Standard for Electrical Safety in the Workplace*, consists of three chapters and 16 annexes, as well as the Foreword to *NFPA 70E* and Article 90, Introduction. Chapter 1, this chapter's main focus, is divided into five articles: 100, 105, 110, 120, and 130.

Article 100 provides definitions essential to the application of *NFPA 70E*. Article 105 addresses the application of safety-related work practices. Article 110 contains the general requirements for electrical safety-related work practices. Article 120 outlines the requirements for establishing an electrically safe work condition. Article 130 contains the provisions related to work involving electrical hazards.

History and Evolution of *NFPA 70E*

The appointment of the *NFPA 70E* Committee was announced on January 7, 1976. This committee was formed to assist OSHA in preparing electrical safety standards that would serve its requirements and that could be expeditiously promulgated through the provisions of Section 6(b) of the Occupational Safety and Health Act. A primary concern was OSHA's need for electrical regulations that addressed electrical safety-related work practices and maintenance of the electrical system considered critical to safety for employers and employees in their workplaces.

The committee found it feasible to develop a standard for electrical installations that would be compatible with OSHA requirements for safety of the employee in locations covered by the *National Electrical Code®* *(NEC)*. The new standard was named *NFPA 70E, Standard for Electrical Safety Requirements for Employee Workplaces*. The first edition was published in 1979.

The fifth edition, published in 1995, included the concepts of "limits of approach" and the establishment of an "arc." In 2000, the newly published sixth edition continued to focus on establishment of flash protection boundaries, the use of personal protective equipment, and charts to assist the user in applying appropriate protective clothing and personal protective equipment for common tasks. **See Figure 5-1.**

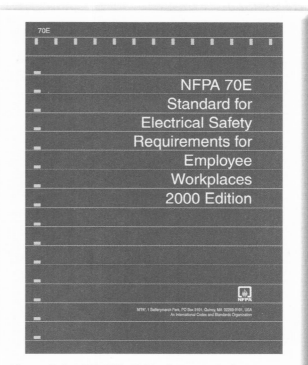

Figure 5-1. *NFPA 70E, 2000 edition, was the first version to include tables as a means to select personal and other protective equipment.*

Reprinted with permission, © Copyright 2000, National Fire Protection Association. *NFPA 70E is a registered trademark of the National Fire Protection Association, Quincy, MA.*

The seventh edition, published in 2004, reflected a name change of the document to *NFPA 70E, Standard for Electrical Safety in the Workplace*, as well as the addition of the energized electrical work permit and related requirements. **See Figure 5-2.**

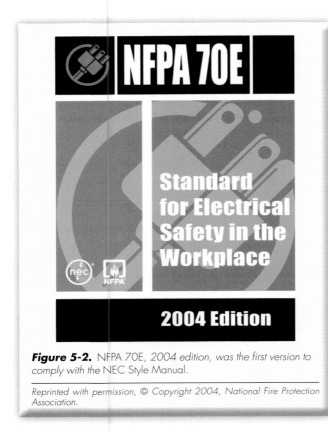

Figure 5-2. *NFPA 70E, 2004 edition, was the first version to comply with the NEC Style Manual.*

Reprinted with permission, © Copyright 2004, National Fire Protection Association.

Figure 5-3. *The 2012 edition of NFPA 70E superseded all previous editions of the standard.*

Reprinted with permission, © Copyright 2012, National Fire Protection Association.

The 2012 edition of *NFPA 70E, Standard for Electrical Safety in the Workplace*, has an effective date of August 31, 2011, and supersedes all previous editions. **See Figure 5-3.**

NFPA 70E Introductory Information

Article 90 contains the introductory information for *NFPA 70E, Standard for Electrical Safety in the Workplace*. It spells out the purpose, scope, arrangement, and organization of the 70E standard. Also covered here are the requirements related to formal interpretations as well as those related to mandatory rules, permissive rules, and explanatory material. Refer to Article 90 for these requirements in their entirety.

Purpose and Scope

The purpose of *NFPA 70E* is to provide a practical safe working area for employees relative to the hazards arising from the use of electricity. The scope in Article 90 outlines what is covered and not covered by this standard. *NFPA 70E* addresses electrical safety-related work practices for employee workplaces that are necessary for the practical safeguarding of employees relative to the hazards associated with electrical energy during activities such as the installation, inspection, operation, maintenance, and demolition of electrical conductors, electrical equipment, signaling and communications conductors and equipment, and raceways. *NFPA 70E* also includes safe work practices for employees performing other work activities that can expose them to electrical hazards as well as safe work practices as indicated in its scope.

Organization and Arrangement

The *NFPA 70E* standard is divided into the introduction, 3 chapters, and 16 annexes. Annexes are not part of the requirements of this standard but are included for informational purposes only.

Chapter 1 applies generally to safety-related work practices; Chapter 3 supplements or modifies Chapter 1 with safety requirements for special equipment. Chapter 2 applies to safety-related maintenance requirements for electrical equipment and installations in workplaces.

Rules, Explanatory Material, and Formal Interpretations

Mandatory rules of the *NFPA 70E* standard identify actions that are specifically required or prohibited.

Permissive rules, in contrast, identify actions that are allowed but not required.

Explanatory material is included in the form of informational notes, which are not enforceable as requirements. Brackets containing section references to another NFPA document are for informational purposes only and are provided as a guide to indicate the source of the extracted text.

Formal interpretation procedures have been established and are found in the NFPA Regulations Governing Committee Projects.

NFPA 70E Definitions

The scope of Article 100 indicates that only those definitions deemed essential to the proper application of *NFPA 70E* are provided there. This article is not intended to include commonly defined general terms or commonly defined technical terms. The definitions apply wherever the terms are used throughout *NFPA 70E*. A number of Article 100 definitions follow. Refer to Article 100 for a complete list of Chapter 1 definitions.

Figure 5-4. *The arc rating of a garment is typically found on a tag inside the arc-rated garment.*

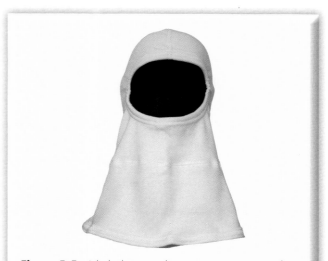

Figure 5-5. *A balaclava must be worn in conjunction with an arc-rated face shield.*

Courtesy of Salisbury by Honeywell.

Arc Flash Hazard. A dangerous condition associated with the possible release of energy caused by an electric arc.

Informational Note No. 1: An arc flash hazard may exist when energized electrical conductors or circuit parts are exposed or when they are within equipment in a guarded or enclosed condition, provided a person is interacting with the equipment in such a manner that could cause an electric arc. Under normal operating conditions, enclosed energized equipment that has been properly installed and maintained is not likely to pose an arc flash hazard.

Informational Note No. 2: See Table 130.7(C)(15)(a) and Table 130.7(C)(15)(b) for examples of activities that could pose an arc flash hazard.

Arc Flash Hazard Analysis. A study investigating a worker's potential exposure to arc flash energy, conducted for the purpose of injury prevention and the determination of safe work practices, arc flash boundary, and the appropriate levels of personal protective equipment (PPE).

Arc Rating. The value attributed to materials that describes their performance to exposure to an electrical arc discharge. The arc rating is expressed in cal/cm² and is derived from the determined value of the arc thermal performance value (ATPV) or energy of breakopen threshold (E_{BT}) (should a material system exhibit a breakopen response below the ATPV value). Arc rating is reported as either ATPV or E_{BT}, whichever is the lower value. **See Figure 5-4.**

Informational Note No. 1: Arc-rated clothing or equipment indicates that it has been tested for exposure to an electric arc. Flame-Resistant (FR) clothing without an arc rating has not been tested for exposure to an electric arc.

Informational Note No. 2: *Breakopen* is a material response evidenced by the formation of one or more holes in the innermost layer of arc-rated material that would allow flame to pass through the material.

Informational Note No. 3: ATPV is defined in ASTM F 1959–06 as the incident energy on a material or a multilayer system of materials that results in a 50 percent probability that sufficient heat transfer through the tested specimen is predicted to cause the onset of a second degree skin burn injury based on the Stoll curve, cal/cm².

Informational Note No. 4: E_{BT} is defined in ASTM F 1959–06 as the incident energy on a material or a material system that results in a 50 percent probability of breakopen. Breakopen is defined as a hole with an area of 1.6 cm² (0.5 in²) or an opening of 2.5 cm (1.0 in.) in any dimension.

Balaclava (Sock Hood). An arc-rated hood that protects the neck and head except for facial area of the eyes and nose. **See Figure 5-5.**

Boundary, Arc Flash. When an arc flash hazard exists, an approach limit at a distance from a prospective arc source within which a person could receive a second-degree burn if an electrical arc flash were to occur.

Informational Note: A second-degree burn is possible by an exposure of unprotected skin to an electric arc flash above the incident energy level of 5 J/cm² (1.2 cal/cm²).

Boundary, Limited Approach. An approach limit at a distance from an exposed energized electrical conductor or circuit part within which a shock hazard exists.

Boundary, Prohibited Approach. An approach limit at a distance from an exposed energized electrical conductor or circuit part within which work is considered the same as making contact with the electrical conductor or circuit part.

Boundary, Restricted Approach. An approach limit at a distance from an exposed energized electrical conductor or circuit part within which there is an increased risk of shock, due to electrical arc-over combined with inadvertent movement, for personnel working in close proximity to the energized electrical conductor or circuit part.

Electrical Hazard. A dangerous condition such that contact or equipment failure can result in electric shock, arc flash burn, thermal burn, or blast. **See Figure 5-6.**

Informational Note: Class 2 power supplies, listed low-voltage lighting systems, and similar sources are examples of circuits or systems that are not considered an electrical hazard.

Electrically Safe Work Condition. A state in which an electrical conductor or circuit part has been disconnected from energized parts, locked/tagged in accordance with established standards, tested to ensure the absence of voltage, and grounded if determined necessary.

Incident Energy. The amount of energy impressed on a surface, at a certain distance from the source, generated during an electrical arc event. One of the units used to measure incident energy is calories per centimeter squared (cal/cm²).

Incident Energy Analysis. A component of an arc flash hazard analysis used to predict the incident energy of an arc flash for a specified set of conditions.

Qualified Person. One who has skills and knowledge related to the construction and operation of the electrical equipment and installations and has received safety training to recognize and avoid the hazards involved.

Unqualified Person. A person who is not a qualified person.

Working On (energized electrical conductors or circuit parts). Intentionally coming in contact with energized electrical conductors or circuit parts with the hands, feet, or other body parts, with tools, probes, or with test equipment, regardless of the personal protective equipment a person is wearing. There are two categories of "working on": *Diagnostic* (testing) is taking readings or measurements of electrical equipment with approved test equipment that does not require making any physical change to the equipment; *repair* is any physical alteration of electrical equipment (such as making or tightening connections, or removing or replacing components).

Figure 5-6. *Electrical hazard is a defined term in NFPA 70E and can be the result of contact or equipment failure.*

Courtesy of Westex

Application of Safety-Related Work Practices

Chapter 1 of *NFPA 70E* covers electrical safety-related work practices and procedures for employees who are exposed to an electrical hazard in workplaces covered by the scope of this standard. These practices and procedures are intended to provide for employee safety relative to electrical hazards in the workplace. The employer must provide the safety-related work practices and train the employee, who must then implement them. Refer to Article 105 for the provisions related to the application of safety-related work practices in their entirety.

General Requirements for Electrical Safety-Related Work Practices

Article 110 contains the general requirements for electrical safety-related work practices. The following topics are addressed in this article:

- Relationships with contractors
- Training requirements
- Electrical safety program
- Use of equipment

A number of excerpts from the general requirements for electrical safety-related work practices follow. Article 110 needs to be referenced for these requirements in their entirety.

Host and Contract Employer Responsibilities

The general requirements that address relationships with contractors outline both host and contract employer responsibilities. These requirements are located in 110.1(A) and 110.1(B). A documented meeting between the host employer and the contract employer is required as part of these joint responsibilities. Among the host employer responsibilities is a requirement to provide the contract employer with the information about the installation necessary to make the assessments required by Chapter 1 of *NFPA 70E*. Among the contract employer responsibilities is a requirement that the contract employer ensure that each employee follows the work practices required by *NFPA 70E* in addition to any safety-related work rules of the host employer.

Training Requirements

The general requirements for safety training are located in 110.2. Employees are required to be trained as follows:

- To understand the specific hazards associated with electrical energy.
- In safety-related work practices and procedural requirements, as necessary, to provide protection from the electrical hazards associated with their respective job or task assignments.
- To identify and understand the relationship between electrical hazards and possible injury.

The degree of training provided depends on the risk to the employee.

Emergency Procedures

Both those employees exposed to shock hazards and those employees responsible for taking action in case of emergency are required to be trained in methods of release of victims from contact with exposed energized electrical conductors or circuit parts. Employees must also be instructed in methods of first aid and emergency procedures, if their duties warrant such training. Training of employees in approved methods of resuscitation, including CPR and automatic external defibrillator (AED) use, is required to be certified by the employer annually.

Note that Informational Note No. 1 to 110.3(F) advises that the hazard identification and risk assessment procedure may include identifying when a second person could be required and the training and equipment that the second person should have.

Training Documentation

The employer must document that each employee has received the training required by the employee training requirements of 110.2(D). This documentation must satisfy the following criteria:

- Be made when the employee demonstrates proficiency in the work practices involved.

- Be maintained for the duration of the employee's employment.
- Contain the content of the training, each employee's name, and dates of training.

Unqualified and Qualified Person Training and Retraining

An unqualified person is defined in Article 100 as a person who is not a qualified person. The provisions for unqualified person training require that these individuals be trained in, and be familiar with, any electrical safety-related practices necessary for their safety.

A qualified person is defined in Article 100 as one who has skills and knowledge related to the construction and operation of the electrical equipment and installations and has received safety training to recognize and avoid the hazards involved. The qualified person training requirements build on that definition and require that a qualified person meet the following criteria:

- Be trained and knowledgeable of the construction and operation of equipment or a specific work method, and be trained to recognize and avoid the electrical hazards that might be present with respect to that equipment or work method.
- Be familiar with the proper use of the special precautionary techniques; personal protective equipment, including the arc flash suit; insulating and shielding materials; and insulated tools and test equipment.

In addition, qualified persons permitted to work within the limited approach boundary of exposed energized electrical conductors and circuit parts operating at 50 volts or more must be trained in at least all of the following:

- The skills and techniques necessary to distinguish exposed energized electrical conductors and circuit parts from other parts of electrical equipment.
- The skills and techniques necessary to determine the nominal voltage of exposed energized electrical conductors and circuit parts.
- The approach distances specified in Table 130.4(C)(a) and Table 130.4(C)(b) and the corresponding voltages to which the qualified person will be exposed. **See Figures 5-7** and **5-8.**
- The decision-making process necessary to determine the degree and extent of the hazard and the personal protective equipment and job planning necessary to perform the task safely.

The employer must determine that each employee is complying with the safety-related work practices

Figure 5-7. NFPA 70E Table 130.4(C)(a) is shown in part. It establishes shock protection boundaries for alternating-current systems.

NFPA 70E Table 130.4(C)(a) Approach Boundaries to Energized Electrical Conductors or Circuit Parts for Shock Protection for Alternating-Current Systems (All dimensions are distance from energized electrical conductor or circuit part to employee.) (shown in part)

(1)	(2)	(3)	(4)	(5)
	Limited Approach Boundary[b]			
Nominal system voltage, range, phase to phase[a]	Exposed movable conductor[c]	Exposed fixed circuit part	Restricted approach boundary[b]; includes inadvertent movement adder	Prohibited approach boundary[b]
751 V–15 kV	3.0 m (10 ft 0 in.)	1.5 m (5 ft 0 in.)	0.7 m (2 ft 2 in.)	0.2 m (0 ft 7 in.)

[a] For single-phase systems, select the range that is equal to the system's maximum phase-to-ground voltage multiplied by 1.732.
[b] See the definition in Article 100 and text in 130.4(D) and Annex C for elaboration.
[c] This term describes a condition in which the distance between a conductor and a person is not under the control of the person. It is normally applied to overhead line conductors supported by poles.

Reprinted with permission from NFPA-70E 2012, Electrical Safety in the Workplace Copyright 2011, National Fire Protection Association, Quincy, MA 02169. This reprinted material is not the complete and official position of the NFPA on the referenced subject, which is represented only by the standard in its entirety.

Figure 5-8. Table 130.4(C)(b) is shown in part. It establishes shock protection boundaries for direct-current systems.

Table 130.4(C)(b) Approach Boundaries[a] to Energized Electrical Conductors or Circuit Parts for Shock Protection, Direct-Current Voltage Systems (shown in part)

(1)	(2)	(3)	(4)	(5)
	Limited Approach Boundary			
Nominal potential difference	Exposed moveable conductor[b]	Exposed fixed circuit part	Restricted approach boundary; includes inadvertent movement adder	Prohibited approach boundary
301 V–1 kV	3.0 m (10 ft 0 in.)	1.0 m (3 ft 6 in.)	0.3 m (1 ft 0 in.)	25 mm (0 ft 1 in.)

[a] All dimensions are distance from exposed energized electrical conductors or circuit parts to worker.
[b] This term describes a condition in which the distance between a conductor and a person is not under the control of the person. It is normally applied to overhead line conductors supported by poles.

Reprinted with permission from NFPA-70E 2012, Electrical Safety in the Workplace Copyright 2011, National Fire Protection Association, Quincy, MA 02169. This reprinted material is not the complete and official position of the NFPA on the referenced subject, which is represented only by the standard in its entirety.

required by this standard through regular supervision or inspections conducted on at least an annual basis.

The qualified employee requirements also state that employees must be trained to select an appropriate voltage detector and demonstrate how to use the correct device to verify the absence of voltage. Such training must include information that enables the employee to understand all limitations of each specific voltage detector that might be used.

Tasks performed less often than once per year require retraining before the performance of the work practices involved. Retraining must be performed at intervals not to exceed three years. An employee is required to receive additional training (or retraining) under the following conditions:

- The supervision or annual inspections indicate that the employee is not complying with the safety-related work practices.

- New technology, new types of equipment, or changes in procedures necessitate the use of safety-related work practices that are different from those that the employee would normally use.
- Safety-related work practices must be employed that are not normally used during their regular job duties.

Electrical Safety Program

As noted earlier, both OSHA and NFPA 70E require the employer to implement an electrical safety program. A safety program is an organized effort to reduce injuries. It should be a subset of an overall safety program, an integral part of an overall safety and health program, and a vital part of establishing an electrically safe workplace. An effective electrical safety program cannot be duplicated from any source and implemented directly. Satisfying regulatory requirements might be possible by duplicating a program from an external source; however, for an electrical safety program to be effective, each element of the program must be discussed and analyzed on the basis of the physical environment and experience level of both supervisors and workers. This section examines the electrical safety program considerations and requirements set forth in *NFPA 70E*.

Establishing an Effective Electrical Safety Program

The Electrical Safety Program Guide, by Jones and Jones, makes some excellent points about establishing an effective electrical safety program. The prose and figures in this section are reprinted with permission from that text.

For most organizations, the initial program setup is difficult. Thinking about the entire program is somewhat overwhelming. However, any single piece of the program helps to get it started. One method of looking at the program is to envision the discrete elements as spokes or segments of a wheel. **See Figure 5-9.** A comprehensive program will consider the segments as discrete parts of the program. To keep the wheel rolling, all segments must be present.

It is possible for a single segment to exist, but all of these segments must be present and adequately addressed for an electrical safety program to be complete. Components of an electrical safety program may include the following:

- Policies and procedures
- Site assessment
- Task assessment
- Personal protective equipment (PPE) requirements
- Hazardous boundaries and hazard/risk analysis
- Administration
- Lockout/tagout
- Training
- Auditing and recordkeeping
- Budgeting

This and other chapters address a number of these considerations in the manner in which they are addressed in *NFPA 70E*.

The electrical safety program will always be a work in progress. Many variables and outside influences will interfere with getting it started or with its forward progress. Most highly successful companies and programs have a history of numerous setbacks and, in some cases, situations where the entire company or project appeared to be destined for disaster. Overcoming negative issues with a positive approach and a dedicated focus on the program goals can achieve success. A critical component of planning a program is to justify the need for one in the minds of company management.

Electrical Safety Program Objectives

H. W. Heinrich was a psychologist in the 1930s who changed the way the world considered safety fundamentals. He developed a theory that states for every 300 recordable injuries, approximately 30 lost-time injuries and one fatality will occur. Over the years, these relationships have proved to be relatively accurate. Some people feel that if the energy source is electrical, then a zero can be taken from the numbers—in other words, three lost-time injuries and one fatality per every 30 recordable injuries. This relationship might be used to help justify funding for an electrical safety program.

The safety triangle demonstrating Heinrich's relationships shows the relationship between behaviors and incidents. **See Figure 5-10.** Developing an understanding of how these segments of the triangle relate to one another brings clarity to the strategies required for a comprehensive program. The triangle shows that, by not having a comprehensive safety program with a strong safety culture, the company will feel the negative effects of these behaviors in a matter of time.

If employees do not have adequate training for the electrical hazards they face, the company is inviting disaster. If a company measures recordable injuries and establishes a program to deal with those injuries, they are, by definition, addressing lost-time injuries as well. The same relationship exists between incidents and recordable

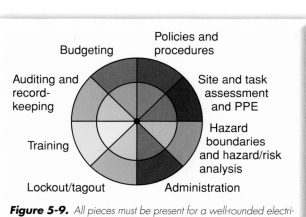

Figure 5-9. *All pieces must be present for a well-rounded electrical safety program.*

Figure 5-10. *Heinrich's triangle illustrates the relationship of behaviors to injuries.*

injuries. The next step—obviously—if a company program is based on monitoring and improving behaviors, is that incidents, recordable injuries, and lost-time injuries are all addressed.

Heinrich's relationship suggests that a program that is focused on eliminating fatalities has a chance to be successful. However, if the program is focused on eliminating lost-time injuries, fatalities also will be eliminated. Likewise, if recordable injuries are eliminated, both lost-time injuries and fatalities will be eliminated. Efforts that are focused lower in the triangle have more data stand a greater chance of success. Note that underpinning Heinrich's triangle are both near misses and unsafe acts. An effective electrical safety program will tend to place emphasis on unsafe acts.

The first priority of the planning process should be to generate an outline that defines the electrical safety goals for the organization. This outline should be consistent and align with the company philosophy on safety to enhance wide acceptance of the program. Electrical safety has obstacles that do not exist in other disciplines. These obstacles include hazards that are not readily visible, along with the stereotypes, misinformation, and lack of knowledge about electrical hazards that exist in the workplace. Many organizations rely on the electrical workers' training and experience as a substitute for safe work practices and proper protective equipment. Typically, a company would not let its employees handle harmful chemicals without proper procedures and protective equipment, but the same company might consistently expect workers to be exposed to energized electrical circuits without safe work practices and protective equipment.

Many managers believe that all incidents and injuries can be prevented and have the goal of "zero incidents in the workplace." This is a realistic goal that can drive the processes of accident/incident reporting and accident investigation and analysis. Why should an incident in the workplace be an acceptable occurrence as part of a job task? Developing an electrical safety program with a proactive vision can result in lower operating costs and strong employee support. In addition to federal regulations, the company should incorporate into its program best work practices and techniques, both internally and within peer groups and organizations. Improving work practices is a critical process to ensure that the program is effective and up-to-date with new techniques and procedures.

Goals must be attainable and embraced by senior management for the program to have credibility. Upper management must take an active support role for the program to be successful. The program plan must be attainable within the framework of the organization, and the mission statement should set the tone for the scope of the program and its components. The scope of the program is directly related to the budget and should spell out what steps, procedures, and policies will be required to attain the goal of zero incidents. The philosophy of management toward the electrical safety program should be spelled out in a procedure.

Program Cost Considerations
Planning must consider both direct and indirect costs and benefits. Direct benefits such as lower insurance cost through safe operations are quantifiable. Measurable benefits are essential to justify expenditure. Indirect benefits can be substantial for a company. Companies want to be perceived as a great place to work. A safe workplace has a far-reaching effect for a company. The concept of a good place to work helps with community relations, hiring and retaining a talented workforce, and municipality issues. Many companies are proactive in promoting good health for employees. Equating safety with good health under one umbrella delivers the message that the company cares about its employees. A well-trained employee performs his or her work with a higher level of confidence and increased productivity. The electrical safety program affects all of the employees, not just the people working on the systems.

Source: The Electrical Safety Program Guide, Second Edition, by Jones and Jones. Sudbury, MA: Jones & Bartlett Learning, 2011; pp. 15–19.

NFPA 70E and the Electrical Safety Program

An electrical safety program will always be a work in progress. Such a program cannot be effective unless it is complete and fully implemented in the workplace. Many who open the pages of *NFPA 70E* expect to find a comprehensive, ready-to-go electrical safety program laid out for them. What *NFPA 70E* actually provides is not a complete electrical safety program, but rather a number of minimum considerations that must be evaluated and integrated into the framework of an existing electrical safety program to make it more comprehensive. Specific requirements are provided in 110.3 as to what must be included in an *NFPA 70E* compliant electrical safety program. These are divided into eight major categories. Information is also provided in *NFPA 70E* Informative Annex E.

General Safety Program Requirements

The first of eight considerations is the general requirement for the employer to implement and document an electrical safety program that provides work practices and procedures that will protect workers for the electrical hazards, voltage, energy level, and circuit conditions to which they will be exposed. Recall that an electrical hazard is a defined term. Therefore, the degree of an electrical hazard generally includes electric shock, arc flash, or blast from either contact or equipment failure.

Awareness and Self-Discipline

The second of eight factors that *NFPA 70E* requires to be included in the electrical safety program is awareness and self-discipline. It is required that the electrical safety program meet the following criteria:
- Be designed to provide an awareness of the potential electrical hazards to employees who work in an environment with the presence of electrical hazards.

- Be developed to provide the required self-discipline for all employees who must perform work that may involve electrical hazards.
- Instill safety principles and controls.

Electrical Safety Program Principles, Controls, and Procedures

The third, fourth, and fifth of the eight factors that *NFPA 70E* requires to be included in the electrical safety program are program principles, controls, and procedures. The electrical safety program must identify the principles upon which it is based, the controls by which it is measured and monitored, and the procedures that are to be used for work inside the limited approach and arc flash boundaries. No specific requirements indicate what those program principles, controls, and procedures must be. While reference is made to the information in Informative Annex E.1, E.2, and E.3 through informational notes, these notes and annex information are nonmandatory. Informative Annexes E.1, E.2, and E.3 include examples of considerations as part of the required principles, controls, and procedures. **See Figure 5-11.**

Hazard Identification and Risk Assessment Procedure

The sixth element required by *NFPA 70E* to be included in an electrical safety program is a hazard identification evaluation and risk assessment procedure. This procedure must be used before work is started within the limited approach boundary or within the arc flash boundary of energized electrical conductors and circuit parts operating at 50 volts or more or where an electrical hazard exists. It must specify the process to be used by the employee before work is started to identify hazards and assess risks, including potential risk mitigation strategies.

NFPA 70E Annex F offers nonmandatory insight and tools to assist in the understanding of these mandatory hazard identification evaluation and risk assessment procedure requirements. Figure F.1(a) offers an example nonmandatory flow chart illustrating steps and decisions for consideration when performing an electrical work risk assessment. **See Figure 5-12.**

Figure 5-11. *NFPA 70E Informative Annexes E.1, E.2, and E.3 offer nonmandatory examples of the electrical safety program principles, controls, and procedures required by 70E 110.3(C), (D), and (E).*

Informative Annex E Electrical Safety Program

This informative annex is not a part of the requirements of this NFPA document but is included for informational purposes only.

(See 110.3, Electrical Safety Program.)

E.1 Typical Electrical Safety Program Principles. Electrical safety program principles include, but are not limited to, the following:

1. Inspect/evaluate the electrical equipment
2. Maintain the electrical equipment's insulation and enclosure integrity
3. Plan every job and document first-time procedures
4. De-energize, if possible (*see 120.1*)
5. Anticipate unexpected events
6. Identify and minimize the hazard
7. Protect the employee from shock, burn, blast, and other hazards due to the working environment
8. Use the right tools for the job
9. Assess people's abilities
10. Audit the principles

E.2 Typical Electrical Safety Program Controls. Electrical safety program controls can include, but are not limited to, the following:

1. Every electrical conductor or circuit part is considered energized until proven otherwise.
2. No bare-hand contact is to be made with exposed energized electrical conductors or circuit parts operating at 50 volts or more, unless the bare-hand method is properly used.
3. De-energizing and electrical conductor or circuit part and making it safe to work on is, in itself, a potentially hazardous task.

4. The employer develops programs, including training, and the employees apply them.
5. Procedures are to be used as tools to identify the hazards and to develop plans to eliminate/control the hazards.
6. Employees are to be trained to qualify them for working in an environment influenced by the presence of electrical energy.
7. Tasks to be performed on or near exposed energized electrical conductors and circuit parts are to be identified/categorized.
8. A logical approach is to be used to determine the potential hazard of task.
9. Precautions appropriate to the working environment are to be identified and used.

E.3 Typical Electrical Safety Program Procedures. Electrical safety program procedures can include, but are not limited to, the following:

1. Purpose of task
2. Qualifications and number of employees to be involved
3. Hazardous nature and extent of task
4. Limits of approach
5. Safe work practices to be used
6. Personal protective equipment involved
7. Insulating materials and tools involved
8. Special precautionary techniques
9. Electrical diagrams
10. Equipment details
11. Sketches/pictures of unique features
12. Reference data

Reprinted with permission from NFPA-70E 2012, Electrical Safety in the Workplace Copyright 2011, National Fire Protection Association, Quincy, MA 02169. This reprinted material is not the complete and official position of the NFPA on the referenced subject, which is represented only by the standard in its entirety.

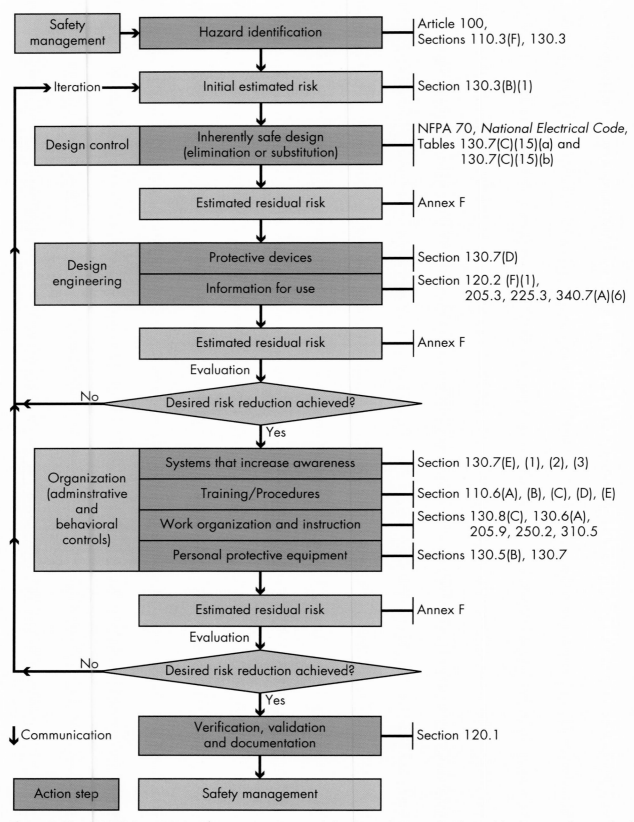

Figure 5-12. NFPA 70E *Figure F.1(a) in Informative Annex F is intended to illustrate the steps to be taken and the decisions to be considered when performing an electrical work risk assessment.*

Additional guidance is offered in an informational note to these requirements. This note recommends that consideration be given to when a second person might be required, and which training and equipment the second person would need. This informational note reminder also offers insight into the portion of the training requirements where employees who are exposed to shock hazards and those responsible for taking action in case of emergency are required to be trained in methods of release of victims from contact with exposed energized equipment. Among other things, a second person should be trained in how to safely release a worker who is "hung up." The insulated rescue hook is an example of equipment that could be used to release a victim from contact with energized parts. **See Figure 5-13.**

Job Briefing

The seventh of the eight topics that *NFPA 70E* requires to be included in an electrical safety program is the

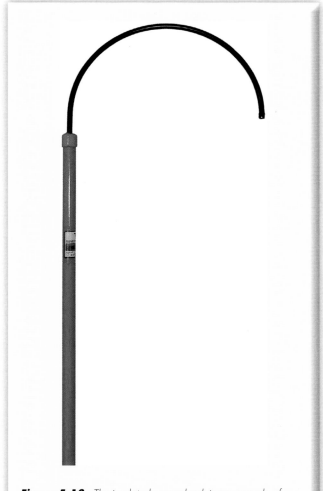

Figure 5-13. *The insulated rescue hook is an example of equipment that could be used to release a victim from contact with energized parts.*

Courtesy of Salisbury by Honeywell.

job briefing. These provisions are broken down into three categories:

- General job briefing requirements
- Requirements for repetitive and similar tasks
- Requirements for routine work

General Job Briefing Requirements

The general job briefing requirements mandate that job briefings be conducted by the employee in charge before commencement of work and include everyone who will be involved. Additional job briefings are required if changes that might affect the safety of employees occur during the course of the work.

Informative Annex I offers an example of what job briefing documentation could contain and look like, using a checklist format. **See Figure 5-14.**

Keep in mind that Informative Annex I is nonmandatory in both format and content. At a minimum, the following subjects must be covered as part of a job briefing by the employee in charge in accordance with the general job briefing requirements:

- Hazards associated with the job
- Work procedures involved
- Special precautions
- Energy source controls
- Personal protective equipment requirements
- Information on the energized electrical work permit

The provisions addressing repetitive and similar tasks require that at least one job briefing be conducted before the start of the first job of the day or shift if the work or operations to be performed during the work day or shift are repetitive and similar.

The requirements addressing routine work state that, prior to starting work, a brief discussion is satisfactory if the work involved is routine and if the employee is qualified for the task. In contrast, a more extensive discussion must be conducted if the work is complicated or particularly hazardous, or if the employee cannot be expected to recognize and avoid the hazards involved in the job.

Electrical Safety Auditing

The last of the eight topics that *NFPA 70E* requires to be included in an electrical safety program is documented program auditing. The electrical safety program must be audited to verify that the principles and procedures of the electrical safety program are in compliance at a frequency of not more than three years.

Field work is required to be audited to verify that the requirements contained in the procedures of the electrical safety program are being followed. Appropriate revisions to the training program or revisions to the procedures must be made when such

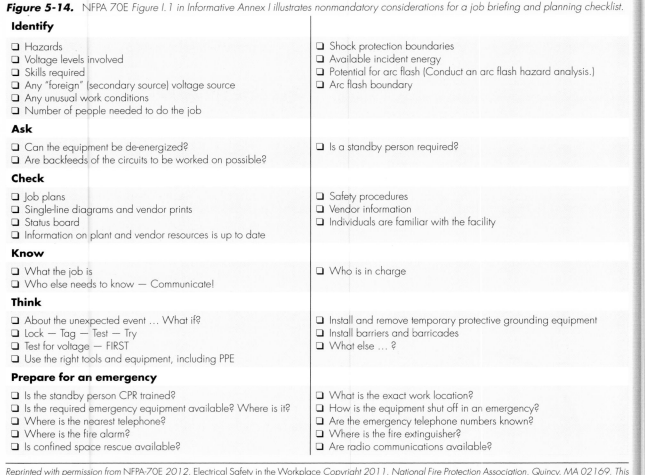

Figure 5-14. NFPA 70E *Figure I.1 in Informative Annex I illustrates nonmandatory considerations for a job briefing and planning checklist.*

Identify

- ❑ Hazards
- ❑ Voltage levels involved
- ❑ Skills required
- ❑ Any "foreign" (secondary source) voltage source
- ❑ Any unusual work conditions
- ❑ Number of people needed to do the job

- ❑ Shock protection boundaries
- ❑ Available incident energy
- ❑ Potential for arc flash (Conduct an arc flash hazard analysis.)
- ❑ Arc flash boundary

Ask

- ❑ Can the equipment be de-energized?
- ❑ Are backfeeds of the circuits to be worked on possible?

- ❑ Is a standby person required?

Check

- ❑ Job plans
- ❑ Single-line diagrams and vendor prints
- ❑ Status board
- ❑ Information on plant and vendor resources is up to date

- ❑ Safety procedures
- ❑ Vendor information
- ❑ Individuals are familiar with the facility

Know

- ❑ What the job is
- ❑ Who else needs to know — Communicate!

- ❑ Who is in charge

Think

- ❑ About the unexpected event … What if?
- ❑ Lock — Tag — Test — Try
- ❑ Test for voltage — FIRST
- ❑ Use the right tools and equipment, including PPE

- ❑ Install and remove temporary protective grounding equipment
- ❑ Install barriers and barricades
- ❑ What else … ?

Prepare for an emergency

- ❑ Is the standby person CPR trained?
- ❑ Is the required emergency equipment available? Where is it?
- ❑ Where is the nearest telephone?
- ❑ Where is the fire alarm?
- ❑ Is confined space rescue available?

- ❑ What is the exact work location?
- ❑ How is the equipment shut off in an emergency?
- ❑ Are the emergency telephone numbers known?
- ❑ Where is the fire extinguisher?
- ❑ Are radio communications available?

Reprinted with permission from NFPA-70E 2012, Electrical Safety in the Workplace Copyright 2011, National Fire Protection Association, Quincy, MA 02169. This reprinted material is not the complete and official position of the NFPA on the referenced subject, which is represented only by the standard in its entirety.

auditing determines that the principles and procedures of the electrical safety program are not being followed.

Groundbreaking Development

By now, it should be obvious that *NFPA 70E* Section 110.3 and *NFPA 70E* Annex E offer a number of considerations that should be evaluated and integrated into the framework of an existing electrical safety program to make it more comprehensive and complete. Is it effectively becoming electrical safety culture? The answer is clear: absolutely. The following is one example of this evolution, provided by Stephen M. Lipster, Training Director of the Electrical Trades Center, in Columbus, Ohio.

In late 2001, representatives from the Columbus offices of OSHA, the National Electrical Contractors Association (NECA), and the International Brotherhood of Electrical Workers (IBEW) met in the boardroom of the IBEW/NECA training facility in Columbus, Ohio. The purpose of this meeting was to explore common ground in an effort to create a safer workplace for

Electrical Workers. Using compiled data, the representatives identified several particular areas to target: fall protection, worker education, and electrical safe work practices. This was the beginning of what was to become the Central Ohio IBEW/NECA/OSHA Partnership.

Once target areas were identified, combining the resources of all the organizations at the table to sharpen the focus made sense. While this made sense philosophically, some substantial barriers had to be overcome before the Partnership could come into being. The historical relationship between OSHA and NECA was, not without reason, somewhat confrontational. The real and perceived wounds and wrongs of the past had to be set aside, and a new environment of cooperation fostered. Much to the parties' credit, by small steps, this new environment developed and flourished. The barrier faced by the IBEW was entirely singular. While supporting safety and safety education from its inception, the Columbus area local union had real reservations about becoming a party to the Partnership.

By law, the safety and well-being of a worker are solely the employer's responsibility. Does it make sense for the local union to become a legal partner to a safety program? Happily, the IBEW chose the high road. The local union took the position that no one had a more vested right in the safety and health of its members than the local union itself and signed on as a partner.

Once the formidable barriers had been surmounted, a steering committee was formed to define the partnership and craft a document defining the goals, measures, and outcomes of this yet-to-be named enterprise. Using similar OSHA partnership documents as a template, the steering committee produced a total of 16 drafts before crafting a final document acceptable to all parties. The document and enterprise are called the IBEW/NECA/OSHA Partnership. The IBEW participation in the project was groundbreaking and demanded appropriate recognition. The Partnership adopted strong language concerning fall protection and required, at a minimum, an OSHA 10-hour course for all workers and an OSHA 30-hour course for all persons trained at the level of foreman or higher (according to the collective bargaining agreement). Perhaps the most remarkable section of the Partnership was the adoption of *NFPA 70E, Standard for Electrical Safety in the Workplace*, as the de facto standard for working on energized circuits. This was the first time this particular standard was voluntarily adopted in a collaborative effort. Finally, the Partnership was designed to be a living document that would grow and change with the evolving workplace.

Since the birth of the Partnership in 2002, more than 900 Electrical Workers have completed an OSHA 10- or 30-hour course, the *NFPA 70E* standard has undergone three revisions, and the Partnership has also undergone two substantial revisions. After 10 years, the Partnership continues to be a viable enterprise that provides a safer, smarter workplace for central Ohio Electrical Workers.

The Central Ohio IBEW/NECA/OSHA Partnership has served as a model for other organizations in a number of different areas of the United States to develop and implement similar agreements. A copy of the second generation of that unique and groundbreaking Partnership is included as an appendix, *OSHA Partnership Agreement*, to this text.

Use of Equipment

Familiarity with the *NFPA 70E* requirements related to the use of equipment is important because tasks where these requirements might apply are performed with some frequency. 70E 110.4 addresses five major topics:

- Test instruments and equipment
- Portable electric equipment

- Ground-fault circuit interrupter (GFCI) protection
- GFCI protection devices
- Overcurrent protection modification

Test Instruments and Equipment

This section stipulates that only qualified persons are permitted to perform tasks such as testing, troubleshooting, and voltage measuring within the limited approach boundary of energized electrical conductors or circuit parts operating at 50 volts or more or where an electrical hazard exists. Test instruments, equipment, and their accessories are required to be rated for circuits and equipment to which they will be connected. Informational note guidance references ANSI/ISA-61010-1 (82.02.01)/UL 61010-1, *Safety Requirements for Electrical Equipment for Measurement, Control, and Laboratory Use—Part 1: General Requirements*, for rating and design requirements for voltage measurement and test instruments intended for use on electrical systems of 1,000 volts or less. **See Figure 5-15.**

Additional requirements in Section 110.4 related to test instruments and equipment pertain to the following topics:

- Design
- Visual inspection
- Operation verification

Portable Electric Equipment

This section addresses the use of cord- and plug-connected equipment, including cord sets (extension cords) and covers four main topics:

- Handling
- Grounding-type equipment
- Visual inspection of portable cord- and plug-connected equipment and flexible cord sets
- Connecting attachment plugs

Figure 5-15. *The Overvoltage Installation Category is one of the ratings to be considered for test equipment.*

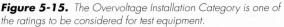
Courtesy of Fluke Corporation

GFCI Protection and GFCI Protection Devices

These provisions require that employees be provided with ground-fault circuit-interrupter protection where required by applicable state, federal, or local codes and standards, and recognize listed cord sets or devices incorporating listed GFCI protection for personnel identified for portable use as accomplishing this protection.

GFCI protection is also required to be provided when an employee is outdoors and operating or using cord- and plug-connected equipment supplied by 125-volt or 15-, 20-, or 30-ampere circuits. Where employees working outdoors operate or use equipment supplied by other than 125-volt or 15-, 20-, or 30-ampere circuits, an assured equipment grounding conductor program is to be implemented.

GFCI protection devices must be tested in accordance with the manufacturer's instructions.

Overcurrent Protection Modification

NFPA 70E requirements related to use of equipment conclude by addressing modification of overcurrent protection. Overcurrent protection of circuits and conductors is not permitted to be modified, even on a temporary basis, beyond that permitted by applicable portions of electrical codes and standards dealing with overcurrent protection.

Underground Electrical Lines and Equipment

NFPA 70E includes provisions that introduce a number of concepts from OSHA's 29 CFR Part 1926 requirements for underground installations.

NFPA 70E addresses this topic in Section 110.5 and requires that, before excavation starts, and where there exists a reasonable possibility of contacting electrical lines or equipment, the employer must take the necessary steps to contact the appropriate owners or authorities to identify and mark the location of the electrical lines or equipment. When it has been determined that a reasonable possibility for contacting electrical lines or equipment exists, a hazard analysis must be performed to identify the appropriate safe work practices to be used during the excavation. **See Figure 5-16.**

Figure 5-16. *The employer must contact the appropriate owners or authorities to identify and mark the location of the electrical lines or equipment prior to excavation.*

Establishing an Electrically Safe Work Condition

An electrically safe work condition is defined in *NFPA 70E* as a state in which an electrical conductor or circuit part has been

- Disconnected from energized parts.
- Locked/tagged in accordance with established standards.
- Tested to ensure the absence of voltage.
- Grounded if determined necessary.

An electrically safe work condition is achieved when these actions are performed in accordance with the procedures of *NFPA 70E* 120.2 and verified by the six steps outlined in Section 120.1. Achieving an electrically safe work condition is one example of how *NFPA 70E* can supplement and enhance OSHA's requirements for lockout and tagging of circuits [1926.417], control of hazardous energy (lockout/tagout) [1910.147], and the lockout and tagging provisions of 1910.333(b)(2), among others.

Article 120 is divided into the following three sections, which provide the requirements for establishing an electrically safe work condition:

- Process of Achieving an Electrically Safe Work Condition
- De-energized Electrical Conductors or Circuit Parts That Have Lockout/Tagout Devices Applied
- Temporary Protective Grounding Equipment

Article 120 needs to be referenced for these requirements in their entirety. Also consider Informative Annex G, Sample Lockout/Tagout Procedure. The procedure in this annex is provided for informational purposes only and is not a part of the requirements of *NFPA 70E*. The sample procedure in Annex G is intended to assist employers in developing a procedure that meets the requirements of *NFPA 70E* 120.2.

Work Involving Electrical Hazards

Article 130 contains the provisions related to work involving electrical hazards as defined in *NFPA 70E*. Electrical hazard is defined as a dangerous condition such that contact or equipment failure can result in electric shock, arc flash burn, thermal burn, or blast. Therefore, by definition, Article 130 covers work involving these four hazards caused by either contact or equipment failure.

Note that Article 130 is divided into the following eight sections, which provide the requirements for work involving electrical hazards:

- General
- Electrically Safe Working Conditions

- Working While Exposed to Electrical Hazards
- Approach Boundaries to Energized Electrical Conductors or Circuit Parts
- Arc Flash Hazard Analysis
- Other Precautions for Personnel Activities
- Personal and Other Protective Equipment
- Work Within the Limited Approach Boundary or Arc Flash Boundary of Overhead Lines

While the chapter "Justification, Assessment, and Documentation of Energized Work" provides an expanded look at these and other topics, *NFPA 70E* needs to be referenced for the Article 130 provisions in their entirety.

Additional Considerations

It is important to remember that Section 90.3 provides the following arrangement of the standard:

- Chapter 1 applies generally for safety-related work practices.
- Chapter 2 applies to safety-related maintenance requirements for electrical equipment and installations in workplaces.
- Chapter 3 supplements or modifies Chapter 1 with safety requirements for special equipment.

As an overview, the following is a list of the three chapters, their articles, and the topics addressed in each:

Chapter 1: Safety-Related Work Practices
 Article 100: Definitions
 Article 105: Application of Safety-Related Work Practices
 Article 110: General Requirements for Electrical Safety-Related Work Practices
 Article 120: Establishing an Electrically Safe Work Condition
 Article 130: Work Involving Electrical Hazards
Chapter 2: Safety-Related Maintenance Requirements
 Article 200: Introduction
 Article 205: General Maintenance Requirements
 Article 210: Substations, Switchgear Assemblies, Switchboards, Panelboards, Motor Control Centers, and Disconnect Switches
 Article 215: Premises Wiring
 Article 220: Controller Equipment
 Article 225: Fuses and Circuit Breakers
 Article 230: Rotating Equipment
 Article 235: Hazardous (Classified) Locations
 Article 240: Batteries and Battery Rooms
 Article 245: Portable Electric Tools and Equipment
 Article 250: Personal Safety and Protective Equipment

Chapter 3: Safety Requirements for Special Equipment
    Article 300: Introduction
    Article 310: Safety-Related Work Practices for Electrolytic Cells
    Article 320: Safety Requirements Related to Batteries and Battery Rooms
    Article 330: Safety-Related Work Practices for Use of Lasers
    Article 340: Safety-Related Work Practices: Power Electronic Equipment
    Article 350: Safety-Related Work Requirements: Research and Development Laboratories

While *NFPA 70E* contains 16 annexes, these are not part of the requirements of the standard but rather are included for informational purposes only.

The following list identifies the 16 annexes and their topics:
    Annex A: Referenced Publications
    Informative Annex B: Informational References
    Informative Annex C: Limits of Approach
    Informative Annex D: Incident Energy and Arc Flash Boundary Calculation Methods
    Informative Annex E: Electrical Safety Program
    Informative Annex F: Hazard Analysis, Risk Estimation, and Risk Evaluation Procedure
    Informative Annex G: Sample Lockout/Tagout Procedure
    Informative Annex H: Guidance on Selection of Protective Clothing and Other Personal Protective Equipment
    Informative Annex I: Job Briefing and Planning Checklist
    Informative Annex J: Energized Electrical Work Permit
    Informative Annex K: General Categories of Electrical Hazards
    Informative Annex L: Typical Application of Safeguards in the Cell Line Working Zone
    Informative Annex M: Layering of Protective Clothing and Total System Arc Rating
    Informative Annex N: Example Industrial Procedures and Policies for Working Near Overhead Electrical Lines and Equipment
    Informative Annex O: Safety-Related Design Requirements
    Informative Annex P: Aligning Implementation of This Standard with Occupational Health and Safety Management Standards

Summary

This chapter provides an overview of the *NFPA 70E* standard, with a primary focus on the provisions of Articles 90, 100, and 110. These provisions include, but are not limited to, host and contract employer responsibilities, training requirements, electrical safety program considerations, and use of equipment. Understanding the definitions from Article 100 and the requirements of Articles 110, 120, and 130 is essential in any effort to comply with these requirements. Recognize that Chapters 2 and 3 of *NFPA 70E* also contain numerous additional electrical safety requirements. The entire *NFPA 70E* standard and all applicable OSHA requirements need to be reviewed and considered. An effective electrical safety program, training, and work practices will adopt and implement appropriate requirements from all sources of information.

An electrical safety program and effective training provide a foundation for a workplace free from recognized hazards. Achieving an electrically safe work condition is the primary safety-related work practice and must be honored unless the employer demonstrates infeasibility or a greater hazard. Additional safety-related work practices must be determined through both a shock hazard analysis and an arc flash hazard analysis before work is performed. Energized work may include tasks such as verifying the absence of voltage to even place equipment in an electrically safe work condition. Only qualified persons are permitted to work on electrical conductors or circuit parts that have not been put into an electrically safe work condition, and an Energized Electrical Work Permit is generally required if live parts are not placed in an electrically safe work condition.

The importance of establishing an electrical safety program and appropriate training cannot be overstated. An electrical safety program is essential as part of an effort to train and protect workers from electrical hazards and a fundamental component of providing an effective overall safety and health program for worker safety. It is important to review and understand the eight elements that *NFPA 70E* requires to be included in an electrical safety program as minimum considerations for making improvements to a new or existing electrical safety program.

REVIEW QUESTIONS

1. *NFPA 70E* requires the employer to implement a(n) _?_ program. This program is an organized effort to reduce injuries and should be a part of an overall safety program that is a vital part of establishing an electrically safe workplace.
   a. cost benefit analysis
   b. electrical safety
   c. loss prevention
   d. research and development

2. The *NFPA 70E* Committee was created in 1976. The first edition was published in _?_.
   a. 1970
   b. 1976
   c. 1979
   d. 1995

3. _?_ covers electrical safety-related work practices and procedures for employees who are exposed to an electrical hazard in workplaces covered within the scope of this standard.
   a. Annex A
   b. Chapter 1
   c. Chapter 2
   d. Chapter 3

4. Training of employees in approved methods of resuscitation, including cardiopulmonary resuscitation (CPR) and automatic external defibrillator (AED) use, is required to be certified by the employer _?_.
   a. annually
   b. biannually
   c. quarterly
   d. semiannually

5. A(n) _?_ is defined in Article 100 as one who has skills and knowledge related to the construction and operation of the electrical equipment and installations and has received safety training to recognize and avoid the hazards involved.
   a. Electrical Worker
   b. manager
   c. qualified person
   d. unqualified person

6. An electrical safety program will always be a work in progress. Overcoming negative issues with a _?_ and a dedicated focus on the program goals can achieve success.
   a. complex approach
   b. negative approach
   c. positive approach
   d. simple approach

7. *NFPA 70E* requires that an electrical safety program be designed to provide awareness of the potential electrical hazards to employees who work in an environment with the presence of electrical hazards and that the program be developed to provide the required _?_ for all employees who must perform work that may involve electrical hazards.
   a. equipment
   b. regulations
   c. self-awareness
   d. self-discipline

8. General job briefing requirements mandate that job briefings be conducted by the employee in charge before commencement of work and include everyone who will be involved.
   a. true
   b. false

9. Article _?_ contains the provisions related to work involving electrical hazards.
   a. 100
   b. 130
   c. 200
   d. 330

10. This chapter provided an overview of the *NFPA 70E* standard, with the primary focus on the provisions of Articles 90, 100, and 110. These provisions include, but are not limited to _?_.
    a. electrical safety program considerations
    b. host and contract employer responsibilities
    c. training requirements
    d. all of the above

6 Justification, Assessment, and Implementation of Energized Work

OBJECTIVES

1. Understand when energized work is permitted and the elements of an energized electrical work permit.

2. Understand the requirements related to shock hazard analysis and arc flash hazard analysis.

3. Understand the requirements for personal and other protective equipment.

4. Know the requirements related to precautions for personnel activities, alerting techniques, and work within the limited approach or arc flash boundary of overhead lines.

REFERENCES

1. *NFPA 70E,*® 2012 Edition

A 60-year-old electrical worker was electrocuted when he contacted an energized 277-volt circuit. The incident occurred in an office building that was being renovated.

The employer had no electrical training or safety program, explaining that the workers were hired from the union hall and were certified by the union as trained Journeymen or apprentices. As part of the contract with the site owner, the employer was required to follow all of the site owner's safety rules, including the enforcement of a lockout/tagout procedure.

Three employees of the electrical contractor were at the site on the morning of the incident: the foreman, a Journeyman electrician (the victim), and an apprentice. At 7:15 a.m., the foreman gave the victim a set of blueprints and explained the job to him. He was told to install new two-by-two-foot fluorescent lighting fixtures in two offices that were being expanded. He was instructed to install a fixture tail to one fixture and then to connect it to the existing lights. The Journeyman electrician was also told to reconnect the power on a second bank of lights in an adjacent area that had been disconnected a week earlier. The foreman did not instruct the victim to deenergize the circuit breakers.

The Journeyman and the apprentice went to the site. At some point, the victim asked the apprentice, "What's up there?" The apprentice replied, "There are a hot switch leg and another fixture tail to be tied into the new fixtures," to which the victim said, "Okay." The switch leg was an energized 277-volt electrical cable that had been previously connected to a wall switch in a partition wall. The fixture tail was an electrical cable from another fixture that had not yet been connected and was deenergized.

At approximately 9:30 a.m., after a coffee break, the victim resumed work on wiring the fixtures. He contacted the 277 volts apparently while stripping the insulation from the switch leg. The electric shock burned the victim's left hand and knocked him from the ladder. The apprentice heard the Journeyman fall and went to his aid. He checked the victim for a pulse and found none. An office employee then started cardiopulmonary resuscitation (CPR). The paramedics arrived soon after, and the victim was transported to the local hospital, where he was pronounced dead at 10:36 a.m.

It is not known why the victim was working on the energized switch leg. The apprentice thought that the victim grabbed the wrong cable and failed to test both the black and white wires. The foreman later stated that there was no reason for the circuits to be energized and that the crew always tested every circuit beforehand. Although the contractor had access to the breaker, no one had deenergized the circuits in that area. The Occupational Safety and Health Administration (OSHA) noted that the second bank of lights adjacent to the accident site was deenergized by taping off the wall switch.

The county medical examiner attributed the cause of death to electrocution.

Source: For details of this case, see New Jersey Case Report 92NJ007. Accessed July 19, 2012.

For additional information, visit qr.njatcdb.org item #1196.

Introduction

Article 130 contains the provisions related to work involving electrical hazards. As described earlier, an *electrical hazard* is defined in *NFPA 70E* as follows:

Electrical Hazard. A dangerous condition such that contact or equipment failure can result in electric shock, arc flash burn, thermal burn, or blast.

Reprinted with permission from NFPA-70E *2012,* Electrical Safety in the Workplace *Copyright 2011, National Fire Protection Association, Quincy, MA 02169. This reprinted material is not the complete and official position of the NFPA on the referenced subject, which is represented only by the standard in its entirety.*

Therefore, by definition, Article 130 covers work involving these four hazards caused by either contact or equipment failure. **See Figure 6-1**.

Figure 6-1. *An electrical hazard is a dangerous condition that may be caused by contact or equipment failure.*

Courtesy of Westex

Article 130 is divided into eight sections that provide the requirements for work involving electrical hazards:

- General
- Electrically Safe Working Conditions
- Working While Exposed to Electrical Hazards
- Approach Boundaries to Energized Electrical Conductors or Circuit Parts
- Arc Flash Hazard Analysis
- Other Precautions for Personnel Activities
- Personal and Other Protective Equipment
- Work Within the Limited Approach Boundary or Arc Flash Boundary of Overhead Lines

An overview, excerpts, and abbreviated content from the requirements for work involving electrical hazards follow. Article 130 needs to be referenced for a complete understanding of these requirements in their entirety.

Application, Justification, and Documentation

The first part of Article 130 addresses application, justification, and documentation through general application requirements, general requirement that energized conductors and circuit parts be put into an electrically safe work condition before work is performed, justification for energized work, and the documentation requirements of the energized electrical work permit.

General

Article 130 begins with the general requirement stating that all requirements of Article 130 apply whether an incident energy analysis is completed or whether Table 130.7(C)(15)(a), Table 130.7(C)(15)(b), and Table 130.7(C)(16) are used in lieu of an incident energy analysis in accordance with 130.5, Exception. The significance of this rule may not be immediately apparent, as these referenced requirements are covered somewhat later in Article 130. As an overview, recognize that incident energy analysis or Table 130.7(C)(15)(a), Table 130.7(C)(15)(b), and Table 130.7(C)(16) are the two ways recognized for protective clothing and other personal protective equipment application with an arc flash hazard analysis in Article 130. This is indicated in 130.5(B), which states that one of the following is to be used for the selection of PPE for work inside the arc flash boundary:

1. Incident Energy Analysis
2. Hazard/Risk Categories

The definition of incident energy analysis is reviewed here to aid in the understanding and application of this rule:

Incident Energy Analysis. A component of an arc flash hazard analysis used to predict the incident energy of an arc flash for a specified set of conditions.

Reprinted with permission from NFPA-70E *2012,* Electrical Safety in the Workplace *Copyright 2011, National Fire Protection Association, Quincy, MA 02169. This reprinted material is not the complete and official position of the NFPA on the referenced subject, which is represented only by the standard in its entirety.*

This rule might appear to apply only to an incident energy analysis and the referenced tables. In reality, all of the Article 130 provisions—such as those related to the energized electrical work permit, electrical hazard analysis, approach boundaries, other precautions for personnel activities, personal and other protective equipment, and overhead lines—apply no matter whether an incident energy analysis is completed or the referenced tables are used.

Examples of the application of this rule are provided later in this chapter when the requirements for incident energy analysis and the referenced tables are covered in detail.

Electrically Safe Working Conditions

The second of the eight sections in Article 130, titled Electrically Safe Working Conditions, covers rules for energized work and the energized electrical work permit. It generally requires that energized electrical conductors and circuit parts be put into an electrically safe work condition before an employee performs work if either of two conditions exist:

- The employee is within the limited approach boundary.
- The employee interacts with equipment where conductors or circuit parts are not exposed, but an increased risk of injury from an exposure to an arc flash hazard exists.

There is an exception to this rule that applies as long as all of the conditions spelled out in the exception are met, including a risk assessment that does not identify unacceptable risks for the task.

Energized Work

Energized work is genrally prohibited. Such work is permitted only under three conditions:

- Where the employer can demonstrate that deenergizing introduces additional hazards or increased risk.
- Where the employer can demonstrate that the task to be performed is infeasible in a deenergized state due to equipment design or operational limitations.
- For energized electrical conductors and circuit parts that operate at less than 50 volts, where the capacity of the source and any overcurrent protection between the energy source and the worker are considered and it is determined that there will be no increased exposure to electrical burns or to explosion due to electric arcs.

Two informational notes provide application examples for two of these three conditions. **See Figure 6-2**.

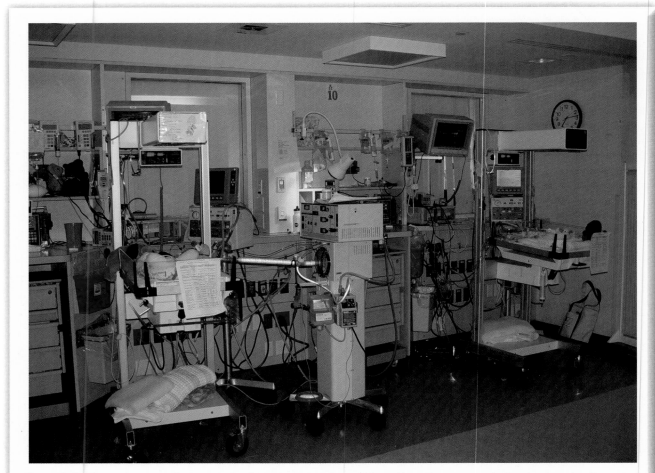

Figure 6-2. *The employer must demonstrate that deenergizing introduces additional hazards or increased risk or that the task to be performed is infeasible in a deenergized state due to equipment design or operational limitations before energized work is permitted.*

Courtesy of Michael J. Johnston-NECA

Examples provided in these informational notes cover the following:

- Additional hazards or increased risk, including but not limited to, interruption of life-support equipment, deactivation of emergency alarm systems, and shutdown of hazardous location ventilation equipment.
- Work that might be performed within the limited approach boundary of exposed energized electrical conductors or circuit parts because of infeasibility due to equipment design or operational limitations, including performing diagnostics and testing (for example, start-up or troubleshooting) electric circuits that can only be done with the circuit energized, and work on circuits that form an integral part of a continuous process that would otherwise need to be completely shut down in order to permit work on one circuit or piece of equipment.

Energized Electrical Work Permit

An energized electrical work permit is required when working within the limited approach boundary or the arc flash boundary of exposed energized electrical conductors or circuit parts that are not placed in an electrically safe work condition where energized work is permitted due to increased or additional hazards or infeasibility. This work is considered energized electrical work and is to be performed by written permit only. **See Figure 6-3**.

Figure 6-3. *The information on the energized electrical work permit is required to be covered during a job briefing.*

Courtesy of Service Electric Company

The energized electrical work permit shall include, but not be limited to, the following:

1. Description of the circuit and equipment to be worked on and their location
2. Justification for why the work must be performed in an energized condition
3. Description of the safe work practices to be employed
4. Results of the shock hazard analysis, including the following elements:
   - Limited approach boundary
   - Restricted approach boundary
   - Prohibited approach boundary
   - Necessary shock personal and other protective equipment to safely perform the assigned task
5. Results of the arc flash hazard analysis, including the following elements:
   - Available incident energy or hazard/risk category
   - Necessary personal protective equipment to safely perform the assigned task
   - Arc flash boundary
6. Means employed to restrict the access of unqualified persons from the work area
7. Evidence of completion of a job briefing, including a discussion of any job-specific hazards
8. Energized work approval (for example, authorizing or responsible management, safety officer, or owner) signature(s)

An exception to these energized electrical work permit requirements is also specified for work performed within the limited approach boundary of energized electrical conductors or circuit parts by qualified persons related to certain tasks in accordance with the conditions of that exception.

Examples include:

- Tasks such as testing, troubleshooting, and voltage measuring
- Visual inspection where the restricted approach boundary will not be crossed

Note that it is not an exception to the need for appropriate safe work practices and personal protective equipment, however.

A nonmandatory example of an energized electrical work permit is found in Figure J.1 in Informative Annex J. **See Figure 6-4**. This example is provided in Informational Annex J for informational purposes only and is not a part of the requirements of *NFPA 70E*.

ENERGIZED ELECTRICAL WORK PERMIT

PART I: TO BE COMPLETED BY THE REQUESTER:

Job/Work Order Number _____

(1) Description of circuit/equipment/job location: _____

_____

(2) Description of work to be done: _____

(3) Justification of why the circuit/equipment cannot be de-energized or the work deferred until the next scheduled outage: _____

_____

_____

_____   _____

Requester/Title   Date

PART II: TO BE COMPLETED BY THE ELECTRICALLY QUALIFIED PERSONS DOING THE WORK:

Check when Complete

(1) Detailed job description procedure to be used in performing the above detailed work: ☐

_____

_____

(2) Description of the safe work practices to be employed: _____ ☐

_____

(3) Results of the shock hazard analysis: _____
    (a) Limited approach boundary ☐
    (b) Restricted approach boundary ☐
    (c) Prohibited approach boundary ☐
    (d) Necessary shock personal and other protective equipment to safely perform assigned task ☐

(4) Results of the arc flash hazard analysis: _____
    (a) Available incident energy or hazard/risk category ☐
    (b) Necessary arc flash personal and other protective equipment to safely perform the assigned task ☐
    (c) Arc flash boundary ☐

(5) Means employed to restrict the access of unqualified persons from the work area: _____ ☐

_____ ☐

(6) Evidence of completion of a job briefing, including discussion of any job-related hazards: _____

(7) Do you agree the above-described work can be done safely? ☐ Yes ☐ No   (If no, return to requester.)

_____   _____

Electrically Qualified Persons(s)   Date

_____   _____

Electrically Qualified Persons(s)   Date

PART III: APPROVAL(S) TO PERFORM THE WORK WHILE ELECTRICALLY ENERGIZED:

_____   _____

Manufacturing Manager   Maintenance/Engineering Manager

_____   _____

Safety Manager   Electrically Knowledgeable Person

_____   _____

General Manager   Date

Note: Once the work is complete, forward this form to the site Safety Department for review and retention.

Reprinted with permission from NFPA-70E 2012, Electrical Safety in the Workplace Copyright 2011, National Fire Protection Association, Quincy, MA 02169. This reprinted material is not the complete and official position of the NFPA on the referenced subject, which is represented only by the standard in its entirety.   NFPA 70E

Figure 6-4. NFPA 70E *Figure J.1 from Informative Annex J illustrates a nonmandatory example of considerations for an energized electrical work permit.*

Working While Exposed to Electrical Hazards

Two categories of requirements are subsumed within the "working while exposed to electrical hazards" heading, with two subheadings under each category.

The first category includes the general requirement that safety-related work practices be used to safeguard employees from injury while they are exposed to electrical hazards from conductors or circuit parts that are or can become energized. The specific safety-related work practices must be consistent with the nature and extent of the associated electrical hazards. This general requirement then addresses requirements for both a safe work condition and an unsafe work condition in 130.3(A). Energized electrical conductors and circuit parts must be put into an electrically safe work condition before work commences inside the limited approach boundary unless the energized work is justified per 130.2(A). Only qualified persons are permitted to work on that which has not been placed into an electrically safe work condition.

The second category of requirements under the "working while exposed to electrical hazards" heading addresses working within the limited approach boundary of exposed electrical conductors or circuit parts that are or might become energized. The limited approach boundary is primarily addressed in 130.4, Approach Boundaries to Energized Electrical Conductors or Circuit Parts.

These requirements are further broken down under two headings: Electrical Hazard Analysis and Safety Interlocks. The electrical hazard analysis rules require, in part, that appropriate safety-related work practices be determined before any person is exposed to the electrical hazards involved by using both shock hazard analysis and arc flash hazard analysis. **See Figure 6-5**.

The requirements for shock hazard analysis and arc flash hazard analysis are covered in 130.4 and 130.5, respectively. The safety interlocks rules require, in part, that only qualified persons be able to defeat or bypass an electrical safety interlock while working on the equipment and that the safety interlock system be returned to its operable condition when the work is completed.

Approach Boundaries

This section almost exclusively addresses shock protection. An informational note points out that, in certain instances, the arc flash boundary may be a greater distance than the limited approach boundary and that these boundaries are independent of each other. The distance associated with each boundary is determined by the nominal phase-to-phase system voltage range as established in Table 130.4(C)(a) and Table 130.4(C)(b). Conversely, the arc flash boundary distance is not a fixed distance and is not based solely on voltage.

Requirements related to approach boundaries for energized electrical conductors or circuit parts are segmented into four major headings:

- Shock Hazard Analysis
- Shock Protection Boundaries
- Approach to Exposed Energized Electrical Conductors or Circuit Parts Operating at 50 Volts or More
- Approach by Unqualified Persons

Shock Hazard Analysis

The first of the four topics related to approach boundaries to energized electrical conductors or circuit parts addresses shock hazard analysis. As covered earlier, requirements addressing working within the limited approach boundary of exposed electrical conductors or circuit parts that are or might become energized mandate that an electrical hazard analysis be performed [130.3(B)(1)]. One component of an electrical hazard analysis is a shock hazard analysis. **See Figure 6-6**.

Figure 6-5. *An electrical hazard analysis consists of both shock and arc flash hazard analyses. Appropriate safety-related work practices must be determined before exposure to any electrical hazard.*

Courtesy of Westex

ALWAYS WEAR YOUR GLOVES

Figure 6-6. *Determination of the personal protective equipment necessary to minimize the possibility of electric shock is one reason why a shock hazard analysis is required.*

Courtesy of Salisbury by Honeywell

A shock hazard analysis is required to make the following determinations:

- The voltage to which personnel will be exposed
- The boundary requirements
- The personal protective equipment necessary to minimize the possibility of electric shock to personnel

Shock Protection Boundaries

The second of the four topics related to approach boundaries to energized electrical conductors or circuit parts addresses shock protection boundaries. *NFPA 70E* establishes three shock protection boundaries: limited approach, restricted approach, and prohibited approach. Review the following four definitions from Article 100 to better understand these three boundaries and their application. Pay attention to how each of the three approach boundaries for shock differs from the others. By definition, a shock hazard is possible by either contact or approach.

Shock Hazard. A dangerous condition associated with the possible release of energy caused by contact or approach to energized electrical conductors or circuit parts.

Boundary, Limited Approach. An approach limit at a distance from an exposed energized electrical conductor or circuit part within which a shock hazard exists.

Boundary, Restricted Approach. An approach limit at a distance from an exposed energized electrical conductor or circuit part within which there is an increased risk of shock, due to electrical arc-over combined with inadvertent movement, for personnel working in close proximity to the energized electrical conductor or circuit part.

Boundary, Prohibited Approach. An approach limit at a distance from an exposed energized electrical conductor or circuit part within which work is considered the same as making contact with the electrical conductor or circuit part.

Reprinted with permission from NFPA-70E *2012,* Electrical Safety in the Workplace *Copyright 2011, National Fire Protection Association, Quincy, MA 02169. This reprinted material is not the complete and official position of the NFPA on the referenced subject, which is represented only by the standard in its entirety.*

The three shock boundaries are applicable where approaching personnel are exposed to energized electrical conductors or circuit parts.

Approach to Exposed Energized Electrical Conductors or Circuit Parts Operating at 50 Volts or More

The third of the four topics related to approach boundaries to energized electrical conductors or circuit parts addresses the approach to exposed energized electrical conductors or circuit parts operating

at 50 volts or more. These provisions are found in 130.4(C) and are reproduced here for reference purposes:

No qualified person shall approach or take any conductive object closer to exposed energized electrical conductors or circuit parts operating at 50 volts or more than the restricted approach boundary set forth in Table 130.4(C)(a) and Table 130.4(C)(b), unless any of the following apply:

(1) The qualified person is insulated or guarded from the energized electrical conductors or circuit parts operating at 50 volts or more. Insulating gloves or insulating gloves and sleeves are considered insulation only with regard to the energized parts upon which work is being performed. If there is a need for an uninsulated part of the qualified person's body to cross the prohibited approach boundary, a combination of 130.4(C)(1), 130.4(C)(2), and 130.4(C)(3) shall be used to protect the uninsulated body parts.

(2) The energized electrical conductors or circuit part operating at 50 volts or more are insulated from the qualified person and from any other conductive object at a different potential.

(3) The qualified person is insulated from any other conductive object as during live-line bare-hand work.

Reprinted with permission from NFPA-70E *2012,* Electrical Safety in the Workplace *Copyright 2011, National Fire Protection Association, Quincy, MA 02169. This reprinted material is not the complete and official position of the NFPA on the referenced subject, which is represented only by the standard in its entirety.*

The requirements of 130.4(C) reference Table 130.4(C)(a) and Table 130.4(C)(b). Table 130.4(C)(a) is intended for use with AC systems. **See Figure 6-7**. Table 130.4(C)(b) is intended for use with DC voltage systems. **See Figure 6-8**. Note that both tables have five columns numbered (1) to (5) from left to right. Column (1) addresses nominal system voltage and nominal potential difference, respectively. Columns (2) through (5) address boundary distances. Also pay attention to the notes attached to the tables. For example, Note c to Table 130.4(C)(a) and Note b to Table 130.4(C)(b) clarify the application of columns (2) and (3) and indicate which of these columns would be used to determine the limited approach boundary for a particular application.

Approach by Unqualified Persons

The fourth of the four topics related to approach boundaries to energized electrical conductors or circuit parts addresses approach by unqualified persons. This topic is further divided into two additional parts. The first addresses the situation in which unqualified persons are working at or close to the limited approach boundary.

Figure 6-7. NFPA 70E *Table 130.4(C)(a) is illustrated in part and provides the approach boundaries for shock protection up to 750 V for AC systems.*

NFPA 70E Table 130.4(C)(a) Approach Boundaries to Energized Electrical Conductors or Circuit Parts for Shock Protection for Alternating-Current Systems (All dimensions are distance from energized electrical conductor or circuit part to employee.) (shown in part)

(1)	(2)	(3)	(4)	(5)
	Limited Approach Boundary[b]			
Nominal System Voltage, Range, Phase to Phase[a]	Exposed Moveable Conductor[c]	Exposed Fixed Circuit Part	Restricted Approach Boundary[b]; Includes Inadvertent Movement Adder	Prohibited Approach Boundary[b]
301 V–750 V	3.0 m (10 ft 0 in.)	1.0 m (3 ft 6 in.)	0.3 m (1 ft 0 in.)	25 mm (0 ft 1 in.)

Note: For arc flash boundary, see 130.5(A).
[a] *For single-phase systems, select the range that is equal to the system's maximum phase-to-ground voltage.*
[b] *See definition in Article 100 and text in 130.4(D)(2) and Annex C for elaboration.*
[c] *This term describes a condition in which the distance between the conductor and a person is not under the control of the person. The term is normally applied to overhead line conductors supported by poles.*

Reprinted with permission from NFPA-70E 2012, Electrical Safety in the Workplace Copyright 2011, National Fire Protection Association, Quincy, MA 02169. This reprinted material is not the complete and official position of the NFPA on the referenced subject, which is represented only by the standard in its entirety.

Figure 6-8. NFPA 70E *Table 130.4(C)(b) is illustrated in part and provides the approach boundaries for shock protection up to 1 kV for DC systems.*

NFPA 70E Table 130.4(C)(b) Approach Boundaries[a] to Energized Electrical Conductors or Circuit Parts for Shock Protection, Direct-Current Voltage Systems (shown in part)

(1)	(2)	(3)	(4)	(5)
	Limited Approach Boundary			
Nominal Potential Difference	Exposed Moveable Conductor[b]	Exposed Fixed Circuit Part	Restricted Approach Boundary; Includes Inadvertent Movement Adder	Prohibited Approach Boundary
301 V–1 kV	3.0 m (10 ft 0 in.)	1.0 m (3 ft 6 in.)	0.3 m (1 ft 0 in.)	25 mm (0 ft 1 in.)

[a] *All dimensions are distance from exposed energized conductors or circuit parts to worker.*
[b] *This term describes a condition in which the distance between the conductor and a person is not under the control of the person. The term is normally applied to overhead line conductors supported by poles.*

Reprinted with permission from NFPA-70E 2012, Electrical Safety in the Workplace Copyright 2011, National Fire Protection Association, Quincy, MA 02169. This reprinted material is not the complete and official position of the NFPA on the referenced subject, which is represented only by the standard in its entirety.

The second provides requirements related to when unqualified persons enter the limited approach boundary. Unqualified persons are generally not permitted to approach or cross the limited approach boundary. There are conditions that must be met for an unqualified person to either cross or work at or close to the limited approach boundary. An unqualified person is never permitted to cross the restricted approach boundary.

Arc Flash Hazard Analysis

Arc flash hazard analysis is defined in Article 100, and related requirements are primarily located in Article 130.

Arc Flash Hazard Analysis. A study investigating a worker's potential exposure to arc flash energy, conducted for the purpose of injury prevention and the determination of safe work practices, arc flash boundary, and the appropriate levels of personal protective equipment (PPE).

Reprinted with permission from NFPA-70E 2012, Electrical Safety in the Workplace Copyright 2011, National Fire Protection Association, Quincy, MA 02169. This reprinted material is not the complete and official position of the NFPA on the referenced subject, which is represented only by the standard in its entirety.

The requirements of 130.5 begin by stating that an arc flash hazard analysis is to make the following determinations:

- The arc flash boundary
- The incident energy at the working distance
- The personal protective equipment that people within the arc flash boundary must use

In addition, the arc flash hazard analysis must meet the following criteria:

- Be updated when a major modification or renovation takes place
- Be reviewed periodically, within a span not to exceed 5 years, to account for changes in the electrical distribution system that could affect the results of the arc flash hazard analysis
- Take into consideration the design of the overcurrent protective device and its opening time, including its condition of maintenance

An exception to these provisions allows the requirements of 130.7(C)(15) and 130.7(C)(16) to be used in lieu of determining the incident energy at the working distance.

Five informational notes to 130.5 provide insight and additional information related to these requirements including, but not limited to, the importance of proper and adequate maintenance.

The requirements of 130.5 include three subdivisions:

- Arc Flash Boundary
- Protective Clothing and Other Personal Protective Equipment (PPE) for Application with an Arc Flash Hazard Analysis
- Equipment Labeling

Arc Flash Boundary

The first of three subdivisions provides the requirements for determining the arc flash boundary for systems 50 volts and greater in 130.5(A). This boundary is the distance at which the incident energy equals 1.2 calories per square centimeter. This distance is generally required to be calculated. Although the requirements of *NFPA 70E* do not indicate which method is to be used to determine the arc flash boundary, an informational note references *NFPA 70E* Informative Annex D regarding information on estimating the arc flash boundary.

An arc flash boundary distance is provided in Tables 130.7(C)(15)(a) and 130.7(C)(15)(b) based on the parameters of these tables. Among other things, the arc flash boundary determination must take into consideration the design of the overcurrent protective device (OCPD) and its opening time, including its condition of maintenance.

Protective Clothing and Other Personal Protective Equipment

The second of the three subdivisions of 130.5 addresses the requirements related to protective clothing and other personal protective equipment for application with an arc flash hazard analysis. **See Figure 6-9.**

There are two recognized methods for the selection of protective clothing and other PPE: incident energy analysis and hazard/risk categories. The 130.5(B) requirements state that, where it has been determined that work will be performed within the arc flash boundary, one of the following methods must be used for the selection of protective clothing and other PPE:

1. Incident Energy Analysis
2. Hazard/Risk Categories

The incident energy analysis must:

- Be documented by the employer.
- Determine the incident energy exposure of the worker.
- Be based on the working distance of the employee's face and chest areas from a prospective arc source for the specific task to be performed.

Figure 6-9. *Determination of protective clothing and other personal protective equipment is one reason why an arc flash hazard analysis is required.*

Courtesy of Salisbury by Honeywell

- Be used to determine the arc-rated clothing and other PPE to be used by the employee.

Since incident energy increases as the distance from the arc flash decreases, additional PPE must be used for any parts of the body that are closer than the distance at which the incident energy was determined.

Similar to the case for the arc flash boundary, *NFPA 70E* does not specify how the incident energy is to be obtained. An informational note advises referencing Informative Annex D for information on estimating the incident energy and Table H.3(b) in Informative Annex H for information on selecting arc-rated clothing and other PPE. Again, among other things, the arc flash hazard analysis must take into consideration the design of the OCPD and its opening time, including its condition of maintenance.

Equipment Labeling

The third of the three subdivisions of 130.5 addresses the requirements related to equipment labeling. **See Figure 6-10**.

The provisions of 130.5(C) state that electrical equipment such as switchboards, panelboards, industrial control panels, meter socket enclosures, and motor control centers that are in other than dwelling units, and that are likely to require examination, adjustment, servicing, or maintenance while energized, must be field marked with a label containing all the following information:

- The nominal system voltage
- The arc flash boundary

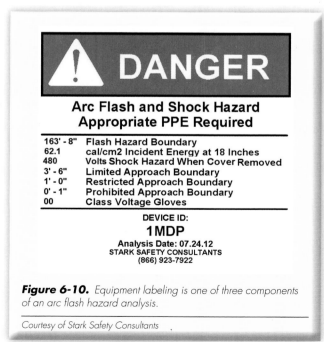

Figure 6-10. *Equipment labeling is one of three components of an arc flash hazard analysis.*

Courtesy of Stark Safety Consultants

- At least one of the following:
  - Available incident energy and the corresponding working distance
  - Minimum arc rating of clothing
  - Required level of PPE
  - Highest Hazard/Risk Category for the equipment

An exception notes that labels applied prior to September 30, 2011, are acceptable if they contain the available incident energy or required level of PPE.

An additional consideration in 130.5(C) requires the method of calculation and the data supporting the information for the label to be documented.

Other Precautions for Personnel Activities

The next category of requirements related to work involving electrical hazards addresses "other precautions for personnel activities." These provisions address the following precautions for personnel activities:

- Alertness
- Blind Reaching
- Illumination
- Conductive Articles Being Worn
- Conductive Materials, Tools, and Equipment Being Handled
- Confined or Enclosed Work Spaces
- Doors and Hinged Panels
- Housekeeping Duties
- Occasional Use of Flammable Materials
- Anticipating Failure
- Routine Opening and Closing of Circuits
- Reclosing Circuits After Protective Device Operation

A number of these topics are similar to what OSHA requires in Subpart S of 29 CFR Part 1910. Each of these topics is an important consideration that must be carefully reviewed in their entirety, with most—if not all—of these issues likely to be included in an electrical safety program or procedures that are based on *NFPA 70E*.

An abbreviated look at a number of these 130.6 provisions follows. Again, these, like all *NFPA 70E* requirements, must be reviewed in their entirety for a full and complete understanding:

- Instruction must generally be provided to be alert at all times for work inside the limited approach boundary.
- Working inside the limited approach boundary is generally prohibited while alertness is recognizably impaired.

- Illumination must be provided that enables work to be performed safely prior to employee entry into spaces containing electrical hazards.
- Conductive articles cannot be worn where they present an electrical contact hazard.
- Conductive materials, tools, and equipment must be handled in a manner that prevents accidental contact with energized electrical conductors or circuit parts.
- Conductive materials must not approach exposed energized electrical conductors or circuit parts closer than that permitted by 130.2.
- Protective shields, protective barriers, or insulating materials must generally be provided and used to avoid the effects of electrical hazards and inadvertent contact with exposed energized parts before entry into a confined or enclosed space.

Personal and Other Protective Equipment

The requirements related to personal and other protective equipment are broken down into two broad categories: personal protective equipment and other protective equipment. PPE, such as arc flash protective equipment, shock protection, and head, face, eye, and hearing protection, is covered in 130.7(C). **See Figure 6-11**. Other protective equipment, such as insulated tools, protective shields, and rubber insulating equipment used for protection from accidental contact, is covered in 130.7(D). **See Figure 6-12**.

Keep the rules of 130.1 in mind as personal and other protective equipment requirements are considered. All requirements in Article 130 are applicable whether an incident energy analysis is completed or if Table 130.7(C)(15)(a), Table 130.7(C)(15)(b), and Table 130.7(C)(16) are used.

The requirements of 130.7 begin with two topic areas that apply to both the personal protective and other protective equipment requirements. The two topic areas include general requirements and provisions addressing care of equipment. The general requirement in 130.7(A) requiring that PPE be provided and used based on the potential hazard is almost identical to what OSHA requires in 1910.335(a)(1)(i). There are three informational notes associated with the general requirements, and one such note associated with care of equipment. Informational Note No. 1 to 130.7(A) states that, even when protection is selected in accordance

Figure 6-11. *Personal protective equipment includes arc flash protective equipment.*

Courtesy of Salisbury by Honeywell

Figure 6-12. *Other protective equipment includes rubber insulating equipment.*

Courtesy of Salisbury by Honeywell

with the PPE requirements of 130.7, some situations could result in burns to the skin, although burn injury should be reduced and survivable.

Section 130.7(B) addresses the need for protective equipment to be:

- Maintained in a safe, reliable condition.
- Visually inspected before each use.
- Stored in a manner to prevent damage.

ⓞSHA Tip

1910.335(a)(1)(i)

Employees working in areas where there are potential electrical hazards shall be provided with, and shall use, electrical protective equipment that is appropriate for the specific parts of the body to be protected and for the work to be performed.

Personal Protective Equipment (PPE)

The Article 130 requirements for PPE are primarily covered in 130.7(C). These PPE requirements are broken down into the following 16 categories:

- General
- Movement and Visibility
- Head, Face, Neck, and Chin (Head Area) Protection
- Eye Protection
- Hearing Protection
- Body Protection
- Hand and Arm Protection
- Foot Protection
- Factors in Selection of Protective Clothing
- Arc Flash Protective Equipment
- Clothing Material Characteristics
- Clothing and Other Apparel Not Permitted
- Care and Maintenance of Arc-Rated Clothing and Arc-Rated Arc Flash Suits
- Standards for Personal Protective Equipment (PPE)
- Selection of Personal Protective Equipment When Required for Various Tasks
- Protective Clothing and Personal Protective Equipment

In addition to the general requirements of 130.7(A) that cover both personal and other protective equipment, certain general requirements apply specifically to personal protective equipment. These requirements in 130.7(C)(1) address both shock and arc flash protective equipment. **See Figure 6-13**.

When an employee is working within the restricted approach boundary, he or she must wear PPE in accordance with 130.4. When an employee is working within the arc flash boundary, he or she is required to wear protective clothing and other personal protective equipment in accordance with 130.5.

All parts of the body inside the arc flash boundary must be protected. When arc-rated clothing is worn as protection, it is required to cover all ignitible clothing. The arc-rated clothing worn must also allow for movement and visibility.

Protection

The PPE requirements of 130.7(C) provide a number of protection requirements for both general and specific parts of the body. Keep in mind the provisions of 130.1 as the following protection requirements are explored. These and other requirements may be applicable even where Table 130.7(C)(15)(a), Table 130.7(C)(15)(b), and Table 130.7(C)(16) are used in lieu of an incident energy analysis in accordance with 130.5, Exception.

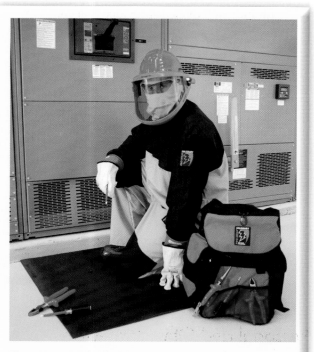

Figure 6-13. *Shock protection must be worn within the restricted approach boundary, and arc flash protection must be worn within the arc flash boundary.*

Courtesy of Salisbury by Honeywell

Note that the arc flash protective equipment requirements of 130.7(C)(10) may modify or supplement one or more of these requirements as indicated by the informational note to the following head, face, neck, chin, and eye protection requirements, for example. **See Figure 6-14**.

Section 130.7(C) is divided into 16 subdivisions: 130.7(C)(1) through (14) are used to select PPE to fulfill shock and arc flash PPE needs. 130.7(C)(15) addresses the selection of PPE when required for various tasks in lieu of the incident energy analysis of 130.5(B)(1). 130.7(C)(16) addresses protective clothing and PPE once the hazard/risk category has been identified from Table 130.7(C)(15)(a) and Table 130.7(C)(15)(b). Recall that, per 130.1, 130.7(C)(1) through (14) are also applicable to the provisions of 130.7(C)(15), Table 130.7(C)(15)(a), Table 130.7(C)(15)(b), 130.7(C)(16), and Table 130.7(C)(16).

An overview of these PPE requirements follows. Again, these, like all requirements in *NFPA 70E*, need to be reviewed and applied in their entirety. Note that Informative Annex H provides additional non-mandatory information that may help in the understanding and application of these and other requirements.

These PPE requirements begin with those related to head, face, neck, and chin (head area) protection.

Figure 6-14. *Specific arc flash protective equipment requirements may modify or supplement other PPE requirements for specific parts of the body.*

Courtesy of Salisbury by Honeywell

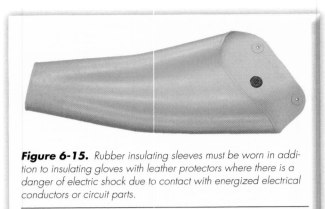

Figure 6-15. *Rubber insulating sleeves must be worn in addition to insulating gloves with leather protectors where there is a danger of electric shock due to contact with energized electrical conductors or circuit parts.*

Courtesy of Salisbury by Honeywell

that recommendation with a unanimous 24-0 vote with the following response:

Committee Statement: "Removing the word 'body' and replacing it with the word 'torso' could mislead the user of the standard by implying that other parts of the body need not be protected. The committee concludes that this section is intended to address all parts of the body."

Hand and arm protection requirements are set forth in 130.7(C)(7). **See Figure 6-15.** These hand and arm protection provisions are divided into the following categories:

- Shock protection
- Arc flash protection
- Maintenance and use

In addition to the need for nonconductive head protection and protective equipment for the face, neck, and chin, as necessary, hairnets and beard nets, where used, must be arc rated when working within the arc flash boundary. Eye protection is generally required in addition to face protection and must be worn whenever there is danger of injury from electric arcs, flashes, or from flying objects resulting from electrical explosion.

Hearing protection is required in every hazard/risk category in Table 130.7(C)(16). Hearing protection is also addressed in 130.7(C)(5) and is required when working inside the arc flash boundary.

The "body protection" provisions of 130.7(C) (6) require arc-rated clothing wherever there is possible exposure to an electric arc flash above 1.2 cal/cm². Body protection includes all parts of the body as clarified in the committee statement on Proposal 70E-271 for the 2009 edition of *NFPA 70E*. That proposal sought to replace the word "body" with the word "torso." The *NFPA 70E* Technical Committee rejected

(7) **Hand and Arm Protection.** Hand and arm protection shall be provided in accordance with 130.7(C) (7)(a), (b), and (c).

(a) **Shock Protection.** Employees shall wear rubber insulating gloves with leather protectors where there is a danger of hand injury from electric shock due to contact with energized electrical conductors or circuit parts. Employees shall wear rubber insulating gloves with leather protectors and rubber insulating sleeves where there is a danger of hand and arm injury from electric shock due to contact with energized electrical conductors or circuit parts. Rubber insulating gloves shall be rated for the voltage for which the gloves will be exposed.

Exception: Where it is necessary to use rubber insulating gloves without leather protectors, the requirements of ASTM F 496, Standard Specification for In-Service Care of Insulating Gloves and Sleeves, shall be met.

Informational Note: Table 130.7(C)(15)(a) and Table 130.7(C)(15)(b) provide further information on tasks where rubber insulating gloves are required.

(b) Arc Flash Protection. Hand and arm protection shall be worn where there is possible exposure to arc flash burn. The apparel described in 130.7(C)(10)(d) shall be required for protection of hands from burns. Arm protection shall be accomplished by the apparel described in 130.7(C)(6).

(c) Maintenance and Use. Electrical protective equipment shall be maintained in a safe, reliable condition. Insulating equipment shall be inspected for damage before each day's use and immediately following any incident that can reasonably be suspected of having caused damage. Insulating gloves shall be given an air test, along with the inspection. Electrical protective equipment shall be subjected to periodic electrical tests. Test voltages and the maximum intervals between tests shall be in accordance with Table 130.7(C)(7)(c). **See Figure 6-16.**

Informational Note: See OSHA 1910.137 and ASTM F 496, Standard Specification for In-Service Care of Insulating Gloves and Sleeves.

Reprinted with permission from NFPA-70E 2012, Electrical Safety in the Workplace Copyright 2011, National Fire Protection Association, Quincy, MA 02169. This reprinted material is not the complete and official position of the NFPA on the referenced subject, which is represented only by the standard in its entirety.

Note that, while foot protection is covered in 130.7(C)(8), these requirements address protection against step and touch potential by requiring dielectric overshoes. This requirement does not address foot protection as it relates to protection against arc flash. See Table 130.7(C)(16) and Section 130.7(C)(10)(e), which specify that foot protection is required in all exposures greater than four calories per square centimeter.

Factors in Selection of Protective Clothing

A number of factors must be considered in the selection of protective clothing. It is important to note that protective clothing may include both flammable and arc-rated garments. Refer to Table 130.7(C)(16) for an example of where Hazard/Risk Category 0 recognizes flammable material as protective clothing, such as non-melting or untreated natural fiber with a fabric weight of at least 4.5 ounces per square yard. Exception No. 1 to 130.7(C)(12) allows nonmelting, flammable clothing to be used for Hazard/Risk Category 0 as described in Table 130.7(C)(16).

The 130.7(C)(9) requirements address factors in the selection of protective clothing. Flammable clothing, as well as all associated parts of the body, must be covered when arc-rated clothing is required.

Requirements in this section address the use of nonmelting flammable fiber garments, the need for arc-rated garments worn as outer layers, and the prohibition of meltable fibers used as underlayers next to the skin. **See Figure 6-17.** A number of these considerations include:

- If nonmelting, flammable fiber garments are used as underlayers, the system arc rating must be sufficient to prevent breakopen of the innermost arc-rated layer at the expected arc exposure incident energy level to prevent ignition of flammable underlayers.
- Garments that are not arc rated cannot increase the arc rating of a garment or of a clothing system.
- Garments worn as outer layers over arc-rated clothing, such as jackets or rainwear, must also be made from arc-rated material.
- Meltable fibers are generally not permitted in fabric underlayers (underwear) next to the skin.

Figure 6-16. NFPA 70E Table 130.7(C)(7)(c) provides voltages and the maximum intervals between tests for rubber insulating equipment.

NFPA 70E Table 130.7(C)(7)(c) Rubber Insulating Equipment, Maximum Test Intervals

Rubber Insulating Equipment	When to Test	Governing Standard for Test Voltage*
Blankets	Before first issue; every 12 months thereafter†	ASTM F 479
Covers	If insulating value is suspect	ASTM F 478
Gloves	Before first issue; every 6 months thereafter†	ASTM F 496
Line hose	If insulating value is suspect	ASTM F 478
Sleeves	Before first issue; every 12 months thereafter†	ASTM F 496

*ASTM F 478, Standard Specification for In-Service Care of Insulating Line Hose and Covers; ASTM F 479, Standard Specification for In-Service Care of Insulating Blankets; ASTM F 496, Standard Specification for In-Service Care of Insulating Gloves and Sleeves.

†If insulating equipment has been electrically tested but not issued for service, it is not permitted to be placed into service unless it has been electrically tested within the previous 12 months.

Reprinted with permission from NFPA-70E 2012, Electrical Safety in the Workplace Copyright 2011, National Fire Protection Association, Quincy, MA 02169. This reprinted material is not the complete and official position of the NFPA on the referenced subject, which is represented only by the standard in its entirety.

Figure 6-17. *Garments worn as outer layers over arc-rated clothing must be made from arc-rated material.*

Courtesy of Salisbury by Honeywell

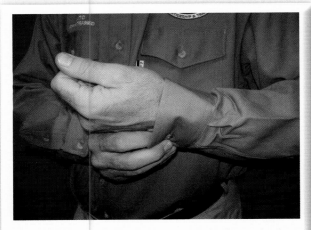

Figure 6-18. *Shirt sleeves are required to be fastened at the wrists and shirts and jackets closed at the neck to cover potentially exposed areas as completely as possible.*

- An incidental amount of elastic used on non-melting fabric underwear or socks shall be permitted.

Finally, coverage, fit, and interference are additional topics that are included in the requirements to be considered as factors in the selection of protective clothing. **See Figure 6-18.** A number of these considerations include:

- Tight-fitting clothing must be avoided.
- Arc-rated apparel must fit properly so as not to interfere with the task.
- Clothing must cover potentially exposed areas as completely as possible.

- Shirt sleeves must be fastened at the wrists and shirts and jackets must be closed at the neck.
- The garment selected shall result in the least interference with the task but still provide the necessary protection.

Clothing Material Characteristics

NFPA 70E Article 130 also provides requirements addressing the clothing material characteristics; these are primarily located in 130.7(C)(11). If the protective clothing is arc rated, it must meet the requirements of 130.7(C)(12) and 130.7(C)(14). Table 130.7(C)(14) identifies the standards that PPE, such as arc-rated apparel, rubber insulating gloves, leather protectors, and arc-rated face protective products, must conform to. Three informational notes provide information related to the application of this provision.

130.7(C)(11) further requires that clothing incorporating fabrics, zipper tapes, and findings made from flammable synthetic materials that melt at temperatures below 315°C (600°F), such as acetate, acrylic, nylon, polyester, polyethylene, polypropylene, and spandex, either alone or in blends, must not be used. An informational note to this requirement provides insight into this provision.

Types of Clothing and Other Apparel Not Permitted

The PPE requirements given in 130.7(C)(12) address the types of clothing and other apparel that are not permitted. Note that these requirements reference 130.7(C)(11), which in turn refers back to 130.7(C)(12) as well as to Table 130.7(C)(14). Table 130.7(C)(14) identifies the standards that PPE must conform to.

Clothing and other apparel (such as hard hat liners and hair nets) made from materials that do not meet the requirements of 130.7(C)(11) regarding melting, or made from materials that do not meet the flammability requirements, are generally not be permitted to be worn.

An informational note and two exceptions to the requirements of 130.7(C)(12) are provided. Exception No. 1 to 130.7(C)(12) conditionally allows nonmelting, flammable clothing to be used as underlayers when it is worn under arc-rated clothing as well as when it is used for Hazard/Risk Category 0 per Table 130.7(C)(16).

Arc Flash Protective Equipment

The arc flash protective equipment requirements are divided into five categories. In addition to detailing unique and specific arc flash protective equipment requirements, these rules supplement or modify other requirements in 130.7(C), such as indicated in the informational note to 130.7(C)(3). The five major topics addressed by 130.7(C)(10) are as follows:

- Arc flash suits
- Head protection
- Face protection
- Hand protection
- Foot protection

Arc Flash Suits

The first of the five major topics addressed in 130.7(C)(10) is arc flash suits. **See Figure 6-19.**

By definition, an arc flash suit must cover the entire body except for the hands and feet. A number of arc flash suit considerations include:

- Design must permit easy and rapid removal.
- The entire arc flash suit must have an arc rating suitable for the arc flash exposure.
- Air hoses and pump housing must be either covered by arc-rated materials or constructed of nonmelting and nonflammable materials, if used.

Head Protection

The second of the five major topics addressed in 130.7(C)(10) is head protection. **See Figure 6-20.**

In addition to meeting other head protection requirements, such as those in 130.7(C)(3), 130.7(C)(14), and Table 130.7(C)(14), head protection must meet head protection requirements for arc flash protective equipment. These considerations include:

- An arc-rated balaclava is required to be used with an arc-rated faceshield when the back of the head is within the arc flash boundary.

- An arc-rated hood shall be permitted to be used instead of an arc-rated faceshield and balaclava combination unless the anticipated incident energy exposure exceeds 12 cal/cm^2.

Recall that 130.1 states that all requirements of Article 130 apply whether an incident energy analysis is completed or Table 130.7(C)(15)(a), Table 130.7(C)(15)(b), and Table 130.7(C)(16) are used in lieu of an incident energy analysis in accordance with 130.5, Exception. Apply the head protection requirements for arc flash protective equipment, as applicable, as protective clothing and equipment is determined by Table 130.7(C)(16), including the protection required for Hazard/Risk Category 0.

Face Protection

The third of the five major topics addressed in 130.7(C)(10) is face protection. In addition to meeting other face protection requirements, such as those in 130.7(C)(3), 130.7(C)(14), and Table 130.7(C)(14), face protection must meet the specific face protection requirement for arc flash protective equipment provided in 130.7(C)(10)(c). In particular, face shields with an adequate arc rating must be worn, and eye protection must be worn even when face protection is worn. **See Figure 6-21.** Face shields must also incorporate wrap-around guarding that protects the face, chin, forehead, ears, and neck area. An informational note to these requirements provides insight into this provision, including the potential need for additional illumination, as these shields are tinted and can reduce visual acuity and color perception.

Figure 6-19. *The entire arc flash suit, including the hood's face shield, must have an arc rating that is suitable for the arc flash exposure.*

Courtesy of Salisbury by Honeywell

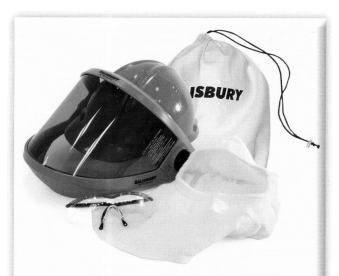

Figure 6-20. *An arc-rated hood or an arc-rated balaclava in conjunction with an arc-rated face shield is required when the back of the head is within the arc flash boundary.*

Courtesy of Salisbury by Honeywell

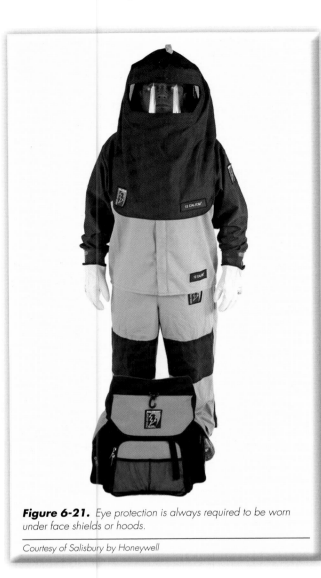

Figure 6-21. *Eye protection is always required to be worn under face shields or hoods.*

Courtesy of Salisbury by Honeywell

Hand Protection

The fourth of the five major topics addressed in 130.7(C)(10) is hand protection. In addition to meeting other hand protection requirements, such as those in 130.7(C)(7), 130.7(C)(14), and Table 130.7(C)(14), hand protection must meet the requirements for arc flash protective equipment. Heavy-duty leather gloves or arc-rated gloves must be worn when required for arc flash protection. Both arc-rated gloves and heavy-duty leather gloves are conditionally recognized as arc flash protection for the hands. Where insulating rubber gloves are used for shock protection, leather protectors must be worn over the rubber gloves. Additional information is provided in the two informational notes. This includes the thickness of leather that would qualify leather gloves as heavy duty and insight into the protection provided by leather protectors worn over rubber insulating gloves for arc flash protection.

Foot Protection

The fifth of the five major topics addressed in 130.7(C)(10) is foot protection. In addition to meeting other foot protection requirements, such as those in 130.7(C)(8), 130.7(C)(14), and Table 130.7(C)(14), foot protection must be in the form of heavy-duty leather work shoes where arc flash exposures exceed 4 cal/cm².

Compare this with the foot protection required by Table 130.7(C)(16); leather work shoes are required for Hazard/Risk Categories 1, 2, 3, and 4 per this table.

Care and Maintenance of Arc-Rated Clothing and Arc-Rated Arc Flash Suits

Several requirements cover the care and maintenance of arc-rated clothing and arc flash suits. **See Figure 6-22.** These rules are primarily located in 130.7(C)(13) and are divided into four categories:
- Inspection
- Manufacturer's Instructions
- Storage
- Cleaning, Repairing, and Affixing Items

Compliance with ASTM F 1506 as provided for by Table 130.7(C)(14), and the following summary of provisions are among what is required by these rules:
- Apparel must be inspected before each use.
- Garments that are contaminated or damaged to the extent that their protective qualities are impaired cannot be used.
- Manufacturers' instructions for care, maintenance, and cleaning must be followed.
- The same arc-rated materials used to manufacture the clothing must be used for repair.
- Guidance in ASTM F 1506 must be followed when trim, name tags, or logos are affixed.

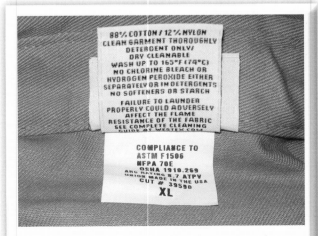

Figure 6-22. *The garment manufacturer's instructions for care and maintenance of arc-rated apparel must be followed to avoid loss of protection.*

Refer to *NFPA 70E* to explore the requirements in the four categories of topics that were introduced in the list above. These and all related requirements must be referred to in their entirety for a full and complete understanding.

Standards for Personal Protective Equipment (PPE)

Not only does *NFPA 70E* require protective equipment, but it also requires that much of this protective equipment meet specific standards. Notably, two tables specify these protective equipment standards. Other protective equipment standards, such as those for arc protective blankets, insulated hand tools, ladders, line hose, safety signs and tags, and temporary protective grounds, are identified in Table 130.7(F). Personal protective equipment must conform to the standards specified in Table 130.7(C)(14). This includes PPE such as arc-rated apparel, eye and face protection, arc-rated face protection, fall protection, leather protectors, rubber insulating gloves, hard hats, and arc-rated rainwear, for example. **See Figure 6-23**.

Note that non–arc-rated or flammable fabrics are not covered by a standard in Table 130.7(C)(14), even though they are permitted to be used in some cases. The informational note to 130.7(C)(14) points to 130.7(C)(11) and 130.7(C)(12) for reference.

Selection of Personal Protective Equipment Using Hazard/Risk Categories

One of the required outcomes of an arc flash hazard analysis in accordance with 130.5 is determining the personal protective equipment that people within the arc flash boundary must use. Two methods may be used to make this determination: an incident energy analysis per 130.5(B)(1) or use of hazard/risk categories per 130.5(B)(2).

The requirement and informational notes in 130.7(C)(15) address protective equipment

Figure 6-23. NFPA 70E Table 130.7(C)(14) identifies 16 standards on protective equipment that must be followed for personal protective equipment.

NFPA 70E Table 130.7(C)(14) Standards on Protective Equipment

Subject	Document Title	Document Number and Revision
Apparel—Arc Rated	Standard Performance Specification for Flame Resistant and Arc Rated Textile Materials for Wearing Apparel for Use by Electrical Workers Exposed to Momentary Electric Arc and Related Thermal Hazards	ASTM F 1506–10a
Aprons—Insulating	Standard Specification for Electrically Insulating Aprons	ASTM F 2677–08a
Eye and Face Protection—General	Practice for Occupational and Educational Eye and Face Protection	ANSI/ASSE Z87.1-2003
Face—Arc Rated	Standard Test Method for Determining the Arc Rating and Standard Specification for Face Protective Products	ASTM F 2178–08
Fall Protection	Standard Specifications for Personal Climbing Equipment	ASTM F 887–10
Footwear—Dielectric Specification	Standard Specification for Dielectric Footwear	ASTM F 1117–03(2008)
Footwear—Dielectric Test Method	Standard Test Method for Determining Dielectric Strength of Dielectric Footwear	ASTM F 1116–03(2008)
Footwear—Standard Performance Specification	Standard Specification for Performance Requirements for Foot Protection	ASTM F 2413–05
Footwear—Standard Test Method	Standard Test Methods for Foot Protection	ASTM F 2412–05
Gloves—Leather Protectors	Standard Specification for Leather Protectors for Rubber Insulating Gloves and Mittens	ASTM F 696–06
Gloves—Rubber Insulating	Standard Specification for Rubber Insulating Gloves	ASTM D 120–09
Gloves and Sleeves—In-Service Care	Standard Specification for In-Service Care of Insulating Gloves and Sleeves	ASTM F 496–08
Head Protection—Hard Hats	Personal Protection—Protective Headwear for Industrial Workers	ANSI/ISEA Z89.1-2009
Rainwear—Arc Rated	Standard Specification for Arc and Flame Resistant Rainwear	ASTM F 1891–06
Rubber Protective Products—Visual Inspection	Standard Guide for Visual Inspection of Electrical Protective Rubber Products	ASTM F 1236–96(2007)
Sleeves—Insulating	Standard Specification for Rubber Insulating Sleeves	ASTM D 1051–08

determination where hazard/risk categories, rather than an incident energy analysis, constitute the method used to select protective equipment. It is important to note the parameters that must be met, such as short-circuit current capacities and maximum fault-clearing times, to determine the appropriate protective equipment via hazard/risk categories rather than an incident energy analysis. The assumed maximum short-circuit current capacities and maximum fault clearing times for various tasks are listed in Table 130.7(C)(15)(a). For tasks not listed, or for power systems with greater than the assumed maximum short-circuit current capacity or with longer than the assumed maximum fault clearing times, an incident energy analysis shall be required in accordance with 130.5. The informational notes illustrate details of this process, such as how the hazard/risk category, work tasks, and protective equipment identified in Table 130.7(C)(15)(a) were determined, including why the hazard/risk category number may have been reduced.

In accordance with the requirements in 130.7(C)(15), Table 130.7(C)(15)(a) and Table 130.7(C)(15)(b) are to be used to determine the hazard/risk category and requirements for use of rubber insulating gloves and insulated and insulating hand tools for a task for AC and DC equipment, respectively. **See Figures 6-24** and **6-25.**

Figure 6-24. NFPA 70E Table 130.7(C)(15)(a) is used to determine the hazard/risk category and requirements for use of rubber insulating gloves and insulated and insulating hand tools for a task performed on AC equipment. Table 130.7(C)(16) is used to select additional protective equipment based on the determined hazard/risk category.

NFPA 70E Table 130.7(C)(15)(a) Hazard/Risk Category Classification and Use of Rubber Insulating Gloves and Insulated and Insulating Hand Tools-Alternating Current Equipment (Formerly TABLE 130.7(C)(9) (shown in part)

Tasks Performed on Energized Equipment	Hazard/Risk Category	Rubber Insulating Gloves	Insulated and Insulating Hand Tools
Panelboards or other equipment rated > 240 V and up to 600 V			
Parameters: Maximum of 25 kA short circuit current available; maximum of 0.03 sec (2 cycle) fault clearing time; minimum 18 in. working distance			
Potential arc flash boundary with exposed energized conductors or circuit parts using above parameters: 30 in.			
Work on energized electrical conductors and circuit parts, including voltage testing	2	Y	Y

Y = Yes (required). N: No (not required)

Notes:

(1) Rubber insulating gloves are gloves rated for the maximum line-to-line voltage upon which work will be done.

(2) Insulated and insulating hand tools are tools rated and tested for the maximum line-to-line voltage upon which work will be done, and are manufactured and tested in accordance with ASTM F 1505, Standard Specification for Insulated and Insulating Hand Tools.

(3) The use of "N" does not indicate that rubber insulating gloves and insulated and insulating hand tools are not required in all cases. Rubber insulating gloves and insulated and insulating hand tools may be required by 130.4, 130.8 (C) (7), and 130.8(D).

(4) For equipment protected by upstream current limiting fuses with arcing fault current in their current limiting range (1/2 cycle fault clearing time or less), the hazard/risk category required may be reduced by one number.

(5) For power systems up to 600 V the arc flash boundary was determined by using the following information: When 0.03 second trip time was used, that indicated MCC or panelboard equipment protected by a molded-case circuit breaker. Working distance used was 18 in. (455 mm). Arc gap used was 32 mm for switchgear and 25 mm for MCC and protective device type 0 for all. When 0.33 or 0.5 second trip time was used, that indicated a LVPCB (drawout circuit breaker) in switchgear. Working distance used was 24 in. (610 mm). Arc gap used was 32 mm and protective device type 0 for all. All numbers were rounded up or down depending on closest multiple of 5.

(6) For power systems from 1 kV to 38 kV the arc flash boundary was determined by using the following information: No maximum values were given in the 2009 edition of NFPA 70E for short-circuit current or operating time. Two sets of equations were performed: 35 kA AIC and 0.2 second operating time and 26 kA AIC and 0.2 second operating time. 0.2 seconds as used by adding the typical maximum total clearing time of the circuit breaker to an estimated value for relay operation. This coincides with the IEEE 1584 values of 0.18 second operating time and 0.08 tripping time rounded off. A short-circuit current of 35 kA was used as a maximum (HRC-4 @ ~ 40 cal/cm2) and 26 kA was used to compare the effects of lowering the short circuit current (HRC-4 @ ~ 30 cal/cm2). Working distance used was 36 in. (909 mm), arc gap was 6 in. (455 mm), and protective device type 0 for all.

Figure 6-25. NFPA 70E Table 130.7(C)(15)(b) is used to determine the hazard/risk category and requirements for use of rubber insulating gloves and insulated and insulating hand tools for a task performed on energized DC equipment. Table 130.7(C)(16) is used to select additional protective equipment based on the determined hazard/risk category.

NFPA 70E Table 130.7(C)(15)(b) Hazard/Risk Category Classifications and Use of Rubber Insulating Gloves and Insulated and Insulating Hand Tools—Direct Current Equipment (shown in part)

Tasks Performed on Energized Equipment	Hazard/Risk Category[a]	Rubber Insulating Gloves[b]	Insulated and Insulating Hand Tools
Storage batteries, direct-current switchboards and other direct-current supply sources ≥ 250 V ≤ 600 V			
Parameters: Voltage: 600 V Maximum arc duration and working distance: 2 sec @ 18 in.			
Work on energized electrical conductors and circuit parts, including voltage testing where arcing current is ≥ 1 kA and < 1.5 kA	1	Y	Y
Potential arc flash boundary using above parameters at 1.5 kA: 36 in.			

Y: Yes (required).

[a] If acid exposure is possible, the clothing is required to be protected from acid and arc rated to the hazard according to ASTM F 1891 or equivalent and evaluated by ASTM F 1296 for acid protection.

[b] In clean rooms or other electrical installations, that do not permit leather protectors for arc flash exposure, ASTM F 496 is required to be followed for use of rubber insulating gloves without leather protectors, and the rubber gloves chosen are required to be arc rated to the potential exposure level of the hazard/risk category.

Reprinted with permission from NFPA-70E 2012, Electrical Safety in the Workplace Copyright 2011, National Fire Protection Association, Quincy, MA 02169. This reprinted material is not the complete and official position of the NFPA on the referenced subject, which is represented only by the standard in its entirety.

Protective Clothing and PPE Determination Using Hazard/Risk Category

Table 130.7(C)(16) can only be used to determine the required PPE for the task based on a hazard/risk category determined from Table 130.7(C)(15)(a) and Table 130.7(C)(15)(b). A hazard/risk category cannot be determined through an incident energy analysis. This protective clothing and equipment determination from Table 130.7(C)(16) is an adjunct to the protective equipment identified in Table 130.7(C)(15)(a) and Table 130.7(C)(15)(b). Once the hazard/risk category has been identified from Table 130.7(C)(15)(a) and 130.7(C)(15)(b), and the requirements of 130.7(C)(15), Table 130.7(C)(16) is to be used to determine the required PPE for the task. This clothing and equipment must be used when working within the arc flash boundary. **See Figure 6-26.**

The requirements and informational notes of 130.7(C)(16) detail the application of Table 130.7(C)(16). **See Figure 6-27.** Informational Note No. 2 states that, while some situations could result in burns to the skin, even with the protection described in Table 130.7(C)(16), burn injury should be reduced and survivable. Due to the explosive effect of some arc events, physical trauma injuries could occur. The PPE require-

ments of this section do not address protection against physical trauma other than exposure to the thermal effects of an arc flash.

Other Protective Equipment

Personal protective equipment is covered in 130.7(C) and includes equipment such as arc flash

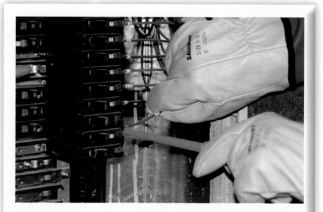

Figure 6-26. Tables 130.7(C)(15)(a) and 130.7(C)(15)(b) indicate the need for rubber insulating gloves and insulated and insulating hand tools in addition to the hazard/risk category for use in conjunction with Table 130.7(C)(16) after the justification for energized work is demonstrated by the employer.

Figure 6-27. *NFPA 70E Table 130.7(C)(16) is used to determine protective clothing and other PPE to be used when working within the arc flash boundary based on the hazard/risk category determined from Table 130.7(C)(15)(a) and Table 130.7(C)(15)(b).*

NFPA 70E Table 130.7(C)(16) Protective Clothing and Personal Protective Equipment (PPE) (shown in part)

Hazard/Risk Category	Protective Clothing and PPE
2	**Arc-Rated Clothing, Minimum Arc Rating of 8 cal/cm².** *(See Note 3.)* Arc-rated long-sleeve shirt and pants or arc-rated coverall Arc-rated flash suit hood or arc-rated face shield *(See Note 2)* and arc-rated balaclava Arc-rated jacket, parka, rainwear, or hard hat liner (AN) **Protective Equipment** Hard hat Safety glasses or safety goggles (SR) Hearing protection (ear canal inserts) Heavy duty leather gloves *(See Note 1.)* Leather work shoes

AN: as needed (optional). AR: as required. SR: selection required.

Notes:

(1) If rubber insulating gloves with leather protectors are required by Table 130.7(C)(9), additional leather or arc-rated gloves are not required. The combination of rubber insulating gloves with leather protectors satisfies the arc flash protection requirement.

(2) Face shields are to have wrap-around guarding to protect not only the face but also the forehead, ears, and neck, or alternatively, an arc-rated arc flash suit hood is required to be worn.

(3) Arc rating is defined in Article 100 and can be either the arc thermal performance value (ATPV) or energy of break open threshold (E_{BT}). ATPV is defined in ASTM F 1959, Standard Test Method for Determining the Arc Thermal Performance Value of Materials for Clothing, as the incident energy on a material, or a multilayer system of materials, that results in a 50 percent probability that sufficient heat transfer through the tested specimen is predicted to cause the onset of a second-degree skin burn injury based on the Stoll curve, in cal/cm². E_{BT} is defined in ASTM F 1959 as the incident energy on a material or material system that results in a 50 percent probability of breakopen. Arc rating is reported as either ATPV or EBT, whichever is the lower value.

protective equipment, shock protection, and head, face, eye, and hearing protection. Other protective equipment, such as insulated tools, protective shields, and rubber insulating equipment used for protection from accidental contact, is covered in 130.7(D). The other protective equipment required by 130.7(D) must comply with those requirements as well as the standards provided in Table 130.7(F). **See Figure 6-28**.

The requirements of 130.7(D) are broken down into nine categories of insulated tools and equipment:

- Requirements for Insulated Tools
- Fuse or Fuse Holding Equipment
- Ropes and Handlines
- Fiberglass-Reinforced Plastic Rods
- Portable Ladders
- Protective Shields
- Rubber Insulating Equipment
- Voltage-Rated Plastic Guard Equipment
- Physical or Mechanical Barriers

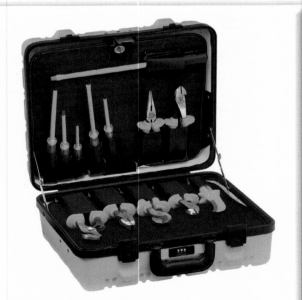

Figure 6-28. *Other protective equipment, such as insulated tools, must comply with the standards provided in Table 130.7(F).*

Courtesy of Klein Tools

The requirements and informational note addressing insulated tools and equipment are located in 130.7(D)(1). The other protective equipment requirements of 130.7(D) begin with general requirements that apply to the nine categories of insulated tools and equipment. These general provisions of 130.7(D)(1) state:

- Employees must use insulated tools or handling equipment, or both, when working inside the limited approach boundary of exposed energized electrical conductors or circuit parts where tools or handling equipment might make accidental contact.
- Table 130.7(C)(15)(a) and Table 130.7(C)(15)(b) provide further information for tasks that require insulated and insulating hand tools.
- Insulated tools must be protected from damage to the insulating material.

The specific requirements for each of the nine categories of insulated tools and equipment are contained in 130.7(D)(1)(a) through (i). Be sure to refer to *NFPA 70E* to explore the requirements in the nine categories of topics of 130.7(D)(1) that were introduced in the list above. These and all related requirements must be referred to in their entirety for a full and complete understanding.

Standards for Other Protective Equipment

NFPA 70E requires protective equipment and also mandates that much of this protective equipment meet specific standards. **See Figure 6-29**. In particular, personal protective equipment must conform to the standards specified in Table 130.7(C)(14); other protective equipment required in 130.7(D) must conform to the standards given in Table 130.7(F). Other protective equipment standards, such as those for arc protective blankets, insulated hand tools, ladders, line hose, safety signs and tags, and temporary protective grounds, are identified in Table 130.7(F). **See Figure 6-30**.

Alerting Techniques

Alerting techniques are located in 130.7 and are recognized as a form of protective equipment. These protective equipment requirements are broken down into the following categories:

- Safety Signs and Tags
- Barricades

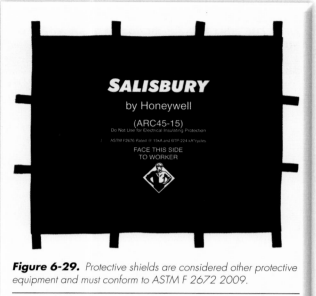

Figure 6-29. *Protective shields are considered other protective equipment and must conform to ASTM F 2672 2009.*

Courtesy of Salisbury by Honeywell

- Attendants
- Look-Alike Equipment

Safety signs and tags are covered by one of the standards in Table 130.7(F). One of the required elements of the energized work permit is "means employed to restrict access of unqualified persons from the work area." The following alerting techniques identified in 130.7(E) may be among the ways to meet this requirement. Refer to *NFPA 70E* to explore the requirements in the four categories of alerting techniques of 130.7(E) that were introduced in the list above. These and all related requirements must be referred to in their entirety for a full and complete understanding.

Work on Overhead Lines

The final section of Article 130 deals with overhead line work—specifically, requirements for work within the limited approach boundary or arc flash boundary of overhead lines. The requirements addressing this work are located in Section 130.8. This section is divided into six categories. These and all requirements related to this work must be referred to in their entirety for a full and complete understanding. These six categories are:

- Uninsulated and Energized
- Determination of Insulation Rating

Figure 6-30. NFPA 70E *The standards that other protective equipment must meet are provided in Table 130.7(F).*

NFPA 70E Table 130.7(F) Standards on Other Protective Equipment

Subject	Document	Document Number and Revision
Arc Protective Blankets	Standard Test Method for Determining the Protective Performance of an Arc Protective Blanket for Electric Arc Hazards	ASTM F 2676 2009
Blankets	Standard Specification for Rubber Insulating Blankets	ASTM D 1048 – 05
Blankets—In-Service Care	Standard Specification for In-Service Care of Insulating Blankets	ASTM F 479 – 06
Covers	Standard Specification for Rubber Insulating Gloves	ASTM D 1049 – 98(2002)e1
Fiberglass Rods—Live-Line Tools	Standard Specification for Fiberglass-Reinforced Plastic (FRP) Rod and Tube Used in Live Line Tools	ASTM F 711 – 02(2007)
Insulated Hand Tools	Standard Specification for Insulated and Insulating Hand Tools	ASTM F 1505 - 07
Ladders	American National Standard for Ladders—Wood Safety Requirements	ANSI A14.1-2007
	American National Standard for Ladders—Fixed—Safety Requirements	ANSI A14.3-2008
	American National Standard Safety Requirements for Job Made Wooden Ladders	ANSI A14.4-2009
	American National Standard for Ladders—Portable Reinforced Plastic—Safety Requirements	ANSI A14.5-2007
Line Hose	Standard Specification for Rubber Insulating Line Hose	ASTM D 1050 – 05e1
Line Hose and Covers— In-Service Care	Standard Specification for In-Service Care of Insulating Line Hose and Covers	ASTM F 478 – 09
Plastic Guard	Standard Test Methods and Specifications for Electrically Insulating Plastic Guard Equipment for Protection of Workers	ASTM F 712 – 06
PVC Sheeting	Standard Specification for PVC Insulating Sheeting	ASTM F 1742 – 03e1
Safety Signs and Tags	Series of Standards for Safety Signs and Tags	ANSI Z535 Series
Shield Performance on Live-Line Tools	Standard Test Method for Determining the Protective Performance of a Shield Attached on Live Line Tools or on Racking Rods for Electric Arc Hazards	ASTM F 2522 - 05
Temporary Protective Grounds— In-Service Testing	Standard Specification for In-Service Test Methods for Temporary Grounding Jumper Assemblies Used on De-Energized Electric Power Lines and Equipment	ASTM F 2249 – 03(2009)
Temporary Protective Grounds— Test Specification	Standard Specifications for Temporary Protective Grounds to Be Used on De-energized Electric Power Lines and Equipment	ASTM F 855 – 09

Reprinted with permission from NFPA-70E 2012, Electrical Safety in the Workplace Copyright 2011, National Fire Protection Association, Quincy, MA 02169. This reprinted material is not the complete and official position of the NFPA on the referenced subject, which is represented only by the standard in its entirety.

- Employer and Employee Responsibility **See Figure 6-31**
- Deenergizing or Guarding **See Figure 6-32**
- Approach Distances for Unqualified Persons
- Vehicular and Mechanical Equipment

Notice that the only mention of where the requirements of 130.8 apply is in the title—that is, Work Within the Limited Approach Boundary or Arc Flash Boundary of Overhead Lines. The definition of each of those boundaries is as follows per *NFPA 70E:*

Boundary, Arc Flash. When an arc flash hazard exists, an approach limit at a distance from a prospective arc source within which a person could receive a second degree burn if an electrical arc flash were to occur.

Figure 6-31. *The requirements for work within the limited approach boundary or arc flash boundary of overhead lines within the scope of NFPA 70E address topics including employer and employee responsibility.*

Courtesy of Service Electric Company

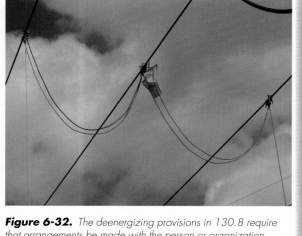

Figure 6-32. *The deenergizing provisions in 130.8 require that arrangements be made with the person or organization that operates or controls the lines to deenergize them and visibly ground them at the point of work if the lines are to be deenergized.*

Courtesy of Service Electric Company

Boundary, Limited Approach. An approach limit at a distance from an exposed energized electrical conductor or circuit part within which a shock hazard exists.

Reprinted with permission from NFPA-70E 2012, Electrical Safety in the Workplace Copyright 2011, National Fire Protection Association, Quincy, MA 02169. This reprinted material is not the complete and official position of the NFPA on the referenced subject, which is represented only by the standard in its entirety.

Recall that the limited approach boundary distance is determined from Table 130.4(C)(a) for AC systems and Table 130.4(C)(b) for DC systems. Note c to Table 130.4(C)(a) and Note b to Table 130.4(C)(b) advise, in part, that the limited approach boundary for an exposed movable conductor normally applies to overhead line conductors supported by poles. Recall, too, that the arc flash boundary for systems 50 volts and greater is the distance at which the incident energy equals 1.2 cal/cm^2 per 130.5(A).

The requirements of 130.8 reference both qualified persons and unqualified persons. Recall that training requirements for both qualified and unqualified persons are located in 110.2(D). Qualified and unqualified persons are also defined in Article 100. The *NFPA 70E* definitions are provided here for review and reference purposes:

Qualified Person. One who has skills and knowledge related to the construction and operation of the electrical equipment and installations and has received safety training to recognize and avoid the hazards involved. [70, 2011]

Unqualified Person. A person who is not a qualified person.

Reprinted with permission from NFPA-70E 2012, Electrical Safety in the Workplace Copyright 2011, National Fire Protection Association, Quincy, MA 02169. This reprinted material is not the complete and official position of the NFPA on the referenced subject, which is represented only by the standard in its entirety.

Be sure to refer to *NFPA 70E* to explore the requirements in the six categories of 130.8 that were introduced here.

Recognizing that the 130.8 provisions—like all of Article 130—are not independent of the rest of the requirements of *NFPA 70E*, be sure to reference these and apply these and all applicable *NFPA 70E* rules and definitions, including those in Articles 90, 100, 110, 120, and 130.

Summary

This chapter covered requirements related to work involving electrical hazards through overview, abbreviated content, and excerpts. Article 130 needs to be referenced to look at these requirements in their entirety.

The Article 130 requirements are broken down into eight sections covering topics including, but not limited to, those related to demonstration of justification for energized work, the energized electrical work permit, a shock hazard analysis, an arc flash hazard analysis (including equipment labeling), personal protective equipment, and other protective equipment.

These requirements, like all *NFPA 70E* requirements, apply within the scope of the standard spelled out in Article 90. Remember to review and apply all pertinent definitions in Article 100, the general requirements of Article 110, and the requirements for establishing an electrically safe work condition in Article 120. In addition, review and apply any relevant *NFPA 70E* Chapter 2 and 3 requirements and any of the related information in the Informative Annexes.

1. Incident energy analysis is defined as a component of an arc flash hazard analysis used to predict the incident energy of an arc flash for a(n) _?_ set of conditions.
   a. general
   b. nominal
   c. specified
   d. unspecified

2. An energized electrical work permit is _?_ when working within the limited approach boundary or the arc flash boundary where energized work is permitted due to increased or additional hazards or infeasibility.
   a. expected
   b. optional
   c. required
   d. unnecessary

3. An electrical hazard analysis consists of _?_ shock hazard analysis and arc flash hazard analysis.
   a. both
   b. calculated
   c. either
   d. neither

4. One reason why a shock hazard analysis is required is to determine the personal protective equipment necessary to _?_ the possibility of electric shock.
   a. eliminate
   b. maximize
   c. minimize
   d. remove

5. Under no circumstance is an unqualified person permitted to cross the _?_ approach boundary.
   a. arc flash
   b. limited
   c. restricted
   d. secured

6. The arc flash hazard analysis must take into consideration the design of the overcurrent protective device and its opening time, _?_ its condition of maintenance.
   a. eliminating
   b. excluding
   c. ignoring
   d. including

7. Which one of the following is NOT among the category of requirements related to "other precautions for personnel activities" in 130.6?
   a. Alertness, blind reaching, and confined or enclosed work spaces
   b. Conductive articles being worn, and doors and hinged panels
   c. Electrical hazard, arc blast, and cost benefit analysis
   d. Illumination, housekeeping, and anticipating failure

8. Shock protection must be worn within the _?_ approach boundary and arc flash protection must be worn within the _?_ boundary.
   a. limited, arc blast
   b. prohibited, arc blast
   c. prohibited, limited
   d. restricted, arc flash

9. When using hazard/risk categories to select personal protective equipment, it is important to note the parameters that must be met, such as short-circuit current capacities and _?_ fault clearing times, to determine the appropriate protective equipment.
   a. anticipated
   b. approximated
   c. maximum
   d. minimum

10. A hazard/risk category _?_ be determined through an incident energy analysis.
    a. can
    b. cannot
    c. should
    d. should not

7 Incident Energy Varies by Fault Current Magnitude and Duration

OBJECTIVES

1. Understand two types of short-circuit (fault) current—bolted and arcing—and the relationship between the two.
2. Understand how to determine fault-clearing times from OCPD time–current curves.
3. Understand OCPD generalized incident energy profiles for circuit breakers and fuses.

REFERENCE

1. *NFPA 70E®*, 2012 Edition

CASE STUDY

A 33-year-old male electrical engineer died from injuries sustained in an electrical fire while he was trying to get the serial number from an electrical panel. The victim received second- and third-degree burns to 90% of his body, due to an explosion in an electrical panel. The victim lived for six days after the incident occurred.

The victim had come in for his final paycheck and was no longer employed by the company at the time of the incident. He went to investigate a problem concerning an electrical panel that had caused a power outage and an employee injury one hour earlier that morning. During the incident, coworkers heard screams and went to the location where the victim was located. No one witnessed the actual incident. When the coworkers arrived at the site, they found the victim on fire and screaming that he had not touched anything.

Interviews were conducted during the following week, including an interview with the employee who was injured in the earlier incident on the day of the fatality. The building owner refused to be interviewed and would not allow an on-site investigation to take place. The employer had been working at this location for approximately 10 weeks, and was in the process of moving when the incident occurred. Its industry uses many locations for short periods of time and then moves on. The building had a prior history of electrical problems, according to both the Los Angeles Building and Safety office and the owner of a car dealership located downstairs below the incident site.

The Medical Examiner's report stated that the cause of death was sepsis, due to multi-organ failure, secondary to massive thermal burns and inhalation injury.

Source: For details of this case, see California Case Report 92CA00201. Accessed October 23, 2012.

For additional information, visit qr.njatcdb.org Item #1197.

Introduction

In this chapter, the concepts of bolted short-circuit (fault) current, arcing fault current, and overcurrent protective device (OCPD) opening time or clearing time are presented. In addition, the chapter provides a brief overview of how to read OCPD time–current curves as well as describes the relationship between an OCPD time–current curve, fault current, and resulting arcing fault incident energy.

This background information is necessary to prepare for the chapter "Arc-Flash Hazard Analysis Methods." The depth of knowledge an individual must master from this chapter depends on the role the person has in electrical safe work practices related to 130.5 Arc-Flash Hazard Analysis. This material is important for those workers who must determine arc-flash boundaries (AFB) (130.5(A)), arc-flash incident energies (130.5(B)(1)), and hazard risk categories (130.5(B)(2)). When performing an arc-flash hazard analysis to determine the AFB and arcing fault incident energy (for the appropriate PPE arc rating), it is important to note that the available short-circuit (fault) current and the OCPD clearing time are parameters that affect the results. For arcing fault incident energy calculations, the OCPD clearing time depends on the OCPD characteristics and the arcing fault current magnitude. The magnitude of the possible arcing fault current depends on the available short-circuit (bolted) current.

When performing an arc-flash hazard analysis with the hazard risk category (HRC) method, the maximum short-circuit current available and the maximum fault clearing time for the type or class of OCPD must be within the parameter values specified in Table 130.7(C)(15)(a); meeting these criteria is a condition of use of the table.

Short-Circuit Current

Overcurrent is defined as either an overload current or a short-circuit current, which often is referred to as a fault current.

Overload current is an excessive current relative to normal operating current, but one that is confined to the normal conductive path provided by the conductors, circuit components, and loads of the distribution system. Harmless overloads are routinely caused by temporary inrush or surge currents that occur when motors are started or transformers are energized. Such temporary overload currents are normal occurrences. In contrast, potentially harmful overloads can result from improperly designed or operated equipment, worn equipment, or too many loads on one circuit.

A sustained overload current results in overheating of conductors and other components and can cause deterioration of insulation, which may eventually result in severe damage, fires, and short circuits, if the overload current is not interrupted in sufficient time to prevent damage.

As the name implies, a *short-circuit current* is a current that flows outside the normal conducting path. One generally accepted definition of *short circuit* is when a phase or ungrounded conductor comes in contact with, or arcing current flows between, another phase conductor, neutral, or ground.

Whereas overloads are modest multiples of the normal current, a short circuit can be many hundreds or thousands of times larger than the normal operating current. The magnitude of available bolted short-circuit currents in commercial and industrial facilities varies dramatically, from roughly 1,000 amperes to more than 200,000 amperes.

Two categories of short-circuit (fault) currents are distinguished: bolted and arcing. In addition, a bolted or arcing short circuit can occur between various parts of a circuit, such as line-line (L-L), line-line-line (L-L-L), line-neutral (L-N), line-ground (L-G), or any combination of L, N, and G.

Bolted Short-Circuit (Fault) Currents

A bolted fault condition represents a "solid" (bolted or welded) connection of relatively low (or assumed "zero") impedance. Because of the low-impedance path, a 3-phase bolted fault condition is typically assumed to be the highest level of fault current. The exceptions to this assumption are not covered. **See Figure 7-1**. A line-to-line bolted fault current will be approximately 87% of the 3-phase bolted fault current. **See Figure 7-2**. Bolted line-to-ground fault currents can range from 25%

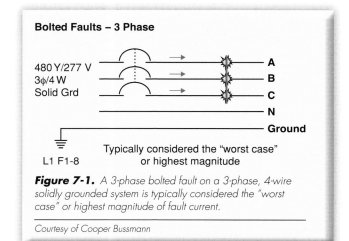

Bolted Faults – 3 Phase

480 Y/277 V
3φ/4 W
Solid Grd

A
B
C
N
Ground

L1 F1-8

Typically considered the "worst case" or highest magnitude

Figure 7-1. *A 3-phase bolted fault on a 3-phase, 4-wire solidly grounded system is typically considered the "worst case" or highest magnitude of fault current.*

Courtesy of Cooper Bussmann

to 125% of the 3-phase value, depending on the distance from the source.

The magnitude of a bolted short-circuit current is a function of the system voltage and the impedance for the electrical source to the point of the fault. In an electrical system, the bolted short-circuit current varies depending on the point in the system at which the fault occurs. Calculating the bolted short-circuit current that would flow if a bolted fault occurred in a system is a well-established procedure that is commonly performed. The available (bolted) short-circuit current is needed to ensure that OCPD interrupting ratings (*NEC* 110.9) and equipment short-circuit current ratings (*NEC* 110.10) are greater than the available short-circuit current where they are applied. Also, performing coordination studies and ensuring the choice of OCPDs are selectively coordinated (*NEC* 700.27 and other sections) require knowing the available short-circuit currents. As will be established in this chapter, the arcing fault current magnitude for arc-flash hazard analysis is directly related to the available short-circuit current.

A simple 3-phase electrical system is depicted as a one-line diagram. **See Figure 7-3**. The highest available 3-phase short-circuit current is typically found at the service transformer's secondary terminals and service equipment. The available 3-phase short-circuit currents at various points in the system are shown with

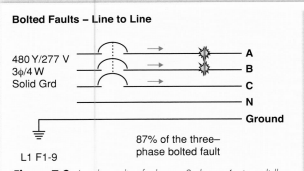

Bolted Faults – Line to Line

480 Y/277 V
3φ/4 W
Solid Grd

A
B
C
N
Ground

L1 F1-9

87% of the three–phase bolted fault

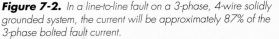

Figure 7-2. *In a line-to-line fault on a 3-phase, 4-wire solidly grounded system, the current will be approximately 87% of the 3-phase bolted fault current.*

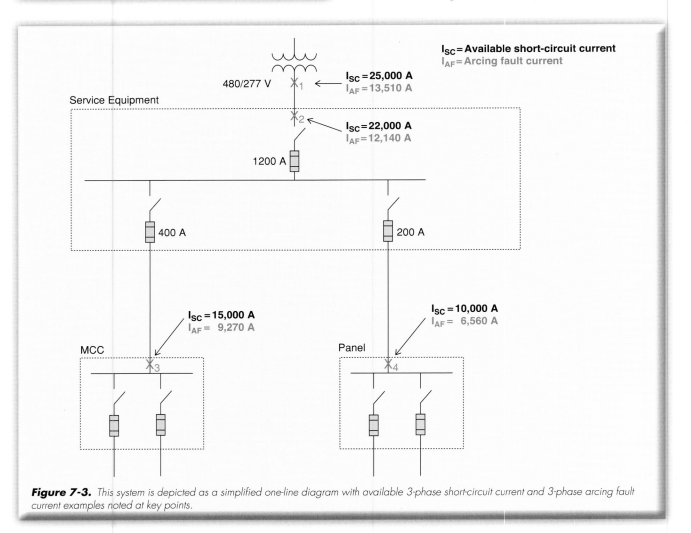

Figure 7-3. *This system is depicted as a simplified one-line diagram with available 3-phase short-circuit current and 3-phase arcing fault current examples noted at key points.*

"X" symbols, with the value for the symmetrical root mean square (rms) amperes denoted by I_{SC} (such as X_3 with 15,000 amperes available short-circuit current). In addition, the value for the 3-phase arcing fault current, expressed in rms amperes, at each point in the system is shown with "X" symbols (such as 9,270 amperes arcing fault current at X_3).

The chapter "Fundamentals of 3-Phase Bolted Fault Currents" covers available short-circuit current in more detail and provides a calculation method and examples.

Arcing Faults

An arcing fault condition does not have a solid bolted connection between conductors or buses; instead, an arcing current path through the air between conductive parts with an associated arc resistance is present. **See Figure 7-4**. Arcing faults can be 3-phase, phase to phase, phase to ground, or phase to neutral, or can include one or more phases with ground and/or neutral.

An arcing fault current is always less than the available bolted fault current value at a specific location due to the additional resistance of the arc. Its magnitude depends on many variables. In some cases, an arcing current is not sustainable, meaning that it will self-extinguish. A self-sustaining arcing fault, in contrast, will persist for a relatively long time (several cycles or longer). The ability of an arc to be initiated and sustained reflects many variables, including whether the fault is on a single-phase or 3-phase circuit, the system voltage, the length of the arc (arc gap), the available bolted short-circuit current, and whether the arcing source is within an enclosure (for instance, switchgear cubicle) or in the open (for instance, utility distribution conductors on an outdoor pole). The size of the enclosure also makes a difference: The smaller the enclosure, the easier it is for the arc to continue. For example, an arc that self-extinguishes in a Size 5 motor starter may be self-sustaining in a Size 1 motor starter.

Generally, for low-voltage systems that experience high-thermal-intensity arcing fault events, if an arcing fault is initiated on a 1-phase circuit (L-L, L-N, or L-G on a 1-phase system) or an L-L, L-N, or L-G circuit (on a 3-phase system without the third phase being present), the arcing current may very well self-extinguish (that is, not sustain itself). By comparison, an arcing fault current that is initiated on a 3-phase circuit as a L-L arcing fault, L-N arcing fault, or L-G arcing fault can quickly (in less than one-thousandth of a second) escalate into a 3-phase arcing fault. **See Figure 7-5**. However, a 3-phase arcing fault current may or may not sustain itself based on the parameters of the electrical circuit and variables of the fault conditions, such as arc-gap spacing, the current's containment within an enclosure, the size of enclosure, the available bolted short-circuit current, and other factors.

Determining the 3-Phase Arcing Fault Current

Whereas calculating the 3-phase bolted short-circuit current available is rather straightforward and several methods and industry tools are available to assist with this process, calculating 3-phase arcing fault currents is not as straightforward and the results are not as certain. This ambiguity arises because of the nature of arcing faults as well as the many variables pertaining to the immediate surroundings that can affect an arcing fault event.

A direct relationship exists between the magnitude of the available bolted short-circuit current and the arcing fault current. For a given point in the electrical system, the arcing fault current is less than the available bolted short-circuit current. The greater the available short-circuit current, the greater the arcing fault current. However, the relationship between the magnitude of the available bolted short-circuit current and the magnitude of the arcing fault is not linear. In other words, the arcing fault current is not a fixed percentage of the available short-circuit current over the range of low to high available short-circuit currents.

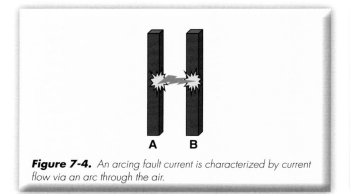

Figure 7-4. An arcing fault current is characterized by current flow via an arc through the air.

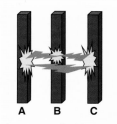

Figure 7-5. A 3-phase arcing fault has arcing current from phase A to phase B, from phase B to phase C, and from phase C to phase A (in addition, the arcing current can go from any of the phases to ground or neutral).

Two different methods can be used to determine the arcing current. Both methods are based on knowing the available short-circuit current (bolted short-circuit current) and the point where the arcing fault current is to be determined. The minimum and maximum arcing current can be determined using percentages based on older industry research and experience. As shown in D.5 of *NFPA 70E* Informative Annex D, for 480-volt, 3-phase systems, the typical minimum sustainable arcing fault (at 480 volts) is 38% of the 3-phase bolted short-circuit current and the maximum 3-phase (line-to-line-to-line) arcing current is 89% of the 3-phase bolted short-circuit current. In theory, the arcing current magnitude could vary from 38% to 89% of the 3-phase short-circuit current. Years ago this method of calculation provided some information about the arcing current, but in today's environment its results may not be sufficient.

A more accurate calculation method, based on extensive arc-flash testing, is now available in the 2002 Institute of Electrical and Electronics Engineers (IEEE) 1584, *Guide for Performing Arc-Flash Hazard Calculations*. (The IEEE 1584 equations shown in this chapter are from IEEE 1584 Guide. © 2002, by IEEE. All rights reserved.) The arcing current can be calculated by the formulas found in *NFPA 70E* Informative Annex D.7.2, taken from IEEE 1584 as follows (note that the term "arc-in-a-box" refers to an arc that occurs within a confined space, such as when working on enclosed equipment):

For systems under 1 kV [per equation D.7.2(a)]:
$$\lg I_a = K + 0.662\ \lg I_{bf} + 0.0966\ V + 0.000526\ G + 0.5588\ V(\lg I_{bf})\ 0.00304\ G\ (\lg I_{bf})$$

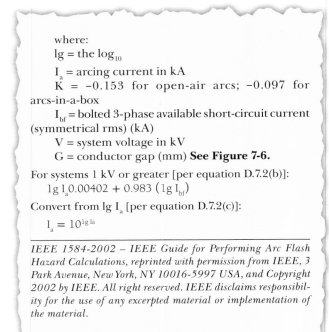

where:
$\lg$ = the $\log_{10}$
I_a = arcing current in kA
K = –0.153 for open-air arcs; –0.097 for arcs-in-a-box
I_{bf} = bolted 3-phase available short-circuit current (symmetrical rms) (kA)
V = system voltage in kV
G = conductor gap (mm) **See Figure 7-6.**

For systems 1 kV or greater [per equation D.7.2(b)]:
$$\lg I_a 0.00402 + 0.983\ (\lg I_{bf})$$

Convert from $\lg I_a$ [per equation D.7.2(c)]:
$$I_a = 10^{\lg I_a}$$

IEEE 1584-2002 – IEEE Guide for Performing Arc Flash Hazard Calculations, reprinted with permission from IEEE, 3 Park Avenue, New York, NY 10016-5997 USA, and Copyright 2002 by IEEE. All right reserved. IEEE disclaims responsibility for the use of any excerpted material or implementation of the material.

The arcing fault currents for X_1 to X_4 were determined by using the previous 2002 IEEE 1584 equation with the respective available short-circuit currents and other appropriate parameters.

Using the formula for systems less than 1 kilovolt, the 3-phase arcing current values can be calculated and compared to the available 3-phase short-circuit current. **See Figure 7-7**. This table is valid only for arcs-in-a-box, 480-volt systems, and applies to switchgear, motor control centers (MCCs), or panelboards. Note that the arcing current in this table varies from 43% to 63% of the 3-phase short-circuit current value for switchgear at

Figure 7-6. *NFPA 70E Table D.7.2, Factors for Equipment and Voltage Classes, provides the conductor gap for various types of electrical equipment that are used in IEEE 1584.*

TABLE D.7.2 Factors for Equipment and Voltage Classes

System Voltage (kV)	Type of Equipment	Typical Conductor Gap (mm)	Distance Exponent Factor X
0.208–1	Open air	10–40	2.000
	Switchgear	32	1.473
	MCCs and panels	25	1.641
	Cables	13	2.000
>1–5	Open air	102	2.000
	Switchgear	13–102	0.973
	Cables	13	2.000
>5–15	Open air	13–153	2.000
	Switchgear	153	0.973
	Cables	13	2.000

Reprinted with permission from NFPA-70E 2012, Electrical Safety in the Workplace Copyright 2011, National Fire Protection Association, Quincy, MA 02169. This reprinted material is not the complete and official position of the NFPA on the referenced subject, which is represented only by the standard in its entirety.

Figure 7-7. *A comparison of the arcing fault current-in-a-box calculated using IEEE 1584 equations for various levels of available short-circuit currents for two different equipment types: 480-volt switchgear (1¼″ conductor gap) and 480-volt MCC/panel (1″ conductor gap).*

480-Volt Switchgear		480-Volt MCC/Panel	
3-Phase Short-Circuit Current (kA)	3-Phase Arcing Current (kA)	3-Phase Short-Circuit Current (kA)	3-Phase Arcing Current (kA)
10	6.30	10	6.56
20	11.22	20	11.85
30	15.72	30	16.76
40	19.98	40	21.43
50	20.46	50	25.93
60	28.01	60	30.30
70	31.85	70	34.57
80	35.59	80	38.74
90	39.26	90	42.84
100	42.86	100	46.88

480 volts or from 47% to 66% for MCC or panelboards at 480 volts. Also, note that the arcing current as a percentage of the 3-phase short-circuit current (bolted) increases as the short-circuit current magnitude decreases.

OCPD Clearing Time

The OCPD clearing time is an important factor in the resultant level of arc-flash hazard. Determining an OCPD clearing time typically involves reading the OCPD time–current curves.

Reading OCPD Time–Current Curves

This section describes how OCPDs respond to various levels of overcurrent and explains how to read time–current curves for an OCPD. It represents a simplified introduction to the topic.

It is important to understand that the concept of using fuse and circuit breaker time–current curves to determine the opening time is applicable to any type of overcurrent. For overcurrent protection, fuses and circuit breakers sense and respond to actual current magnitude, whether the current of a given magnitude is a bolted or an arcing fault current.

Fuse Clearing Times

A circuit supplies a load protected by a 100-ampere fuse under different overcurrent scenarios, designated as A to E. **See Figure 7-8**. An ammeter reads the circuit current for each circuit condition and the 100-A fuse has a clearing time associated with each magnitude of overcurrent.

One characteristic of fuse technology is that the higher the percentage overcurrent, the faster the fuse opens. This relationship is referred to as an inverse time–current characteristic: The higher the current level during a fault, the faster the reaction time or interrupting time of the overcurrent device.

The opening time of the 100-A fuse is determined from the time–current characteristic curve for this specific fuse. **See Figure 7-9**. Note that on the time–current curve, the horizontal axis is the current in amperes and the vertical axis is the time in seconds.

For a given overcurrent value, the fuse time–current curve can be used to determine the fuse clearing (interrupting or opening) time. **See Figure 7-10**. Note that in the example with an 80-A continuous current, the fuse will carry this current indefinitely. The 100-A fuse protects the circuits for the scenarios designated as A to E; note the different overcurrent values and clearing times. **See Figures 7-11, 7-12, 7-13, and 7-14**.

The speed of response of an OCPD when interrupting an overcurrent greater than the OCPD's ampere rating can vary depending on the magnitude of overcurrent. To explore that concept, begin with the principle that OCPDs are intended to continuously carry the load current, which is typically below the ampere rating of the OCPD. If the overcurrent is a light overload, slightly higher than the OCPD's ampere rating, it may be permissible to allow the current to flow for many minutes. Some circuit components—such as motors, primary windings of transformers, and capacitors—have a harmless high starting or energizing inrush

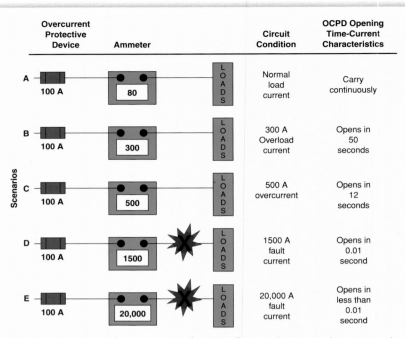

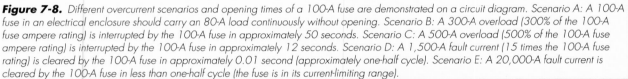

Figure 7-8. *Different overcurrent scenarios and opening times of a 100-A fuse are demonstrated on a circuit diagram. Scenario A: A 100-A fuse in an electrical enclosure should carry an 80-A load continuously without opening. Scenario B: A 300-A overload (300% of the 100-A fuse ampere rating) is interrupted by the 100-A fuse in approximately 50 seconds. Scenario C: A 500-A overload (500% of the 100-A fuse ampere rating) is interrupted by the 100-A fuse in approximately 12 seconds. Scenario D: A 1,500-A fault current (15 times the 100-A fuse rating) is cleared by the 100-A fuse in approximately 0.01 second (approximately one-half cycle). Scenario E: A 20,000-A fault current is cleared by the 100-A fuse in less than one-half cycle (the fuse is in its current-limiting range).*

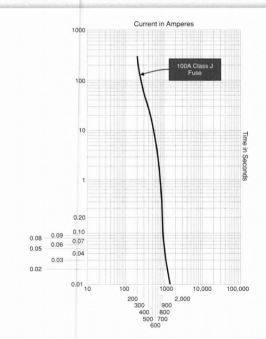

Figure 7-9. *The time–current curve characteristic for the 100-A fuse (100-A Class J fuse) has logarithmic scale* for the current on the horizontal axis and for the time on the vertical axis.*

*Both the current axis and the time axis are presented on a logarithmic scale, which is traditionally used to represent OCPD time–current characteristics. Logarithmic scales are not linear. For instance, on the current axis, compare the distance between 100 A and 200 A with the distance between 900 A and 1,000 A. Do the same on the time axis by comparing the distance between 0.01 second and 0.02 second to the distance between 0.09 second and 0.1 second. This logarithmic difference is repeated from 1,000 A to 10,000 A, from 0.1 second to 1.0 second, and so on.

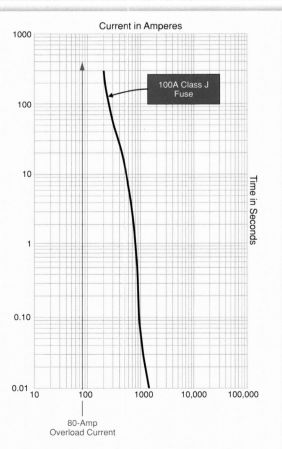

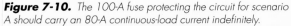

Figure 7-10. *The 100-A fuse protecting the circuit for scenario A should carry an 80-A continuous-load current indefinitely.*

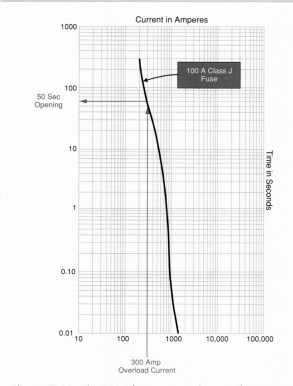

Figure 7-11. *The 100-A fuse protecting the circuit for scenario B clears the 300-A overload current in approximately 50 seconds.*

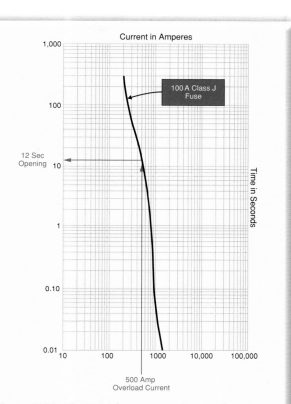

Figure 7-12. *The 100-A fuse protecting the circuit for scenario C clears the 500-A overload current in approximately 12 seconds.*

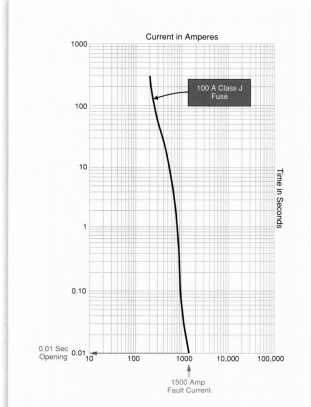

Figure 7-13. *The 100-A fuse protecting the circuit for scenario D interrupts the 1,500-A fault current in approximately 0.01 second (one-half cycle).*

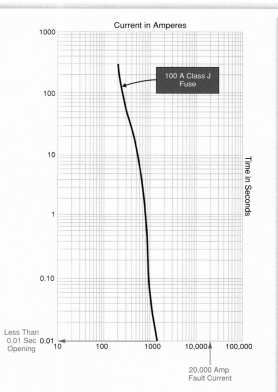

Figure 7-14. *The clearing time for a 100-A fuse protecting the circuit for scenario E with a 20,000-A fault current cannot be determined by reading the time–current curve because the fuse is in its current-limiting range and clears in less than 0.01 second (less than one-half cycle).*

current that can be many times greater than the normal full-load current. The OCPDs on these circuits must permit intentional short-duration overload currents for a period of time without opening. If the overcurrent is a faulted circuit, rapid OCPD response is desired to minimize circuit component or equipment damage.

Fuse time–current curves can vary based on the manufacturer and the type of fuses. For example, non-time-delay, time-delay, dual-element time-delay, and high-speed fuses are each associated with unique time–current curves. Nevertheless, all fuse curves demonstrate an inverse time relationship: The greater the overcurrent magnitude, the faster the clearing time.

Current-Limiting Fuse Clearing Times

The previous section on reading time–current curves provided basic information on how to determine fuse clearing times. Current-limiting fuses do not have the adjustability of overcurrent protection settings, which are found in some molded-case circuit breakers and low-voltage power circuit breakers. In most applications, there are choices in the fuse type that could be used. The choice of fuse type/characteristic may affect the arc-flash hazard results. For some types of loads, such as a motor branch circuit or transformer primary protection, different fuse types need to be sized differently.

Consider the time–current curves for four types of 400-A current-limiting fuses: Class RK5 dual-element time-delay, Class RK1 dual-element time-delay, Class J time-delay, and Class J non-time-delay. **See Figure 7-15.** All three time-delay fuse types have similar time-delay characteristics for times greater than 1 second. For times less than approximately 0.2 second, the 400-A Class RK5 fuse has a longer clearing time than the other fuses. For times less than 0.2 second, the Class J time-delay fuse has the most advantageous time–current characteristic of all the time-delay fuses. The Class J non-time-delay fuse provides significantly faster clearing times for low-level overcurrents compared to the other fuses. However, for circuits with normal inrush currents, such as motor branch circuits and transformer primary protection, Class J non-time-delay fuses typically must have a larger ampere rating than Class J time-delay fuses.

One significant advantage of current-limiting fuses compared to standard molded-case circuit breakers and low-voltage power circuit breakers is that these fuses can provide reduced clearing times when the arcing current is within the fuse's current-limiting range. The current-limiting range of a fuse is where the current exceeds the current value at which the clearing time

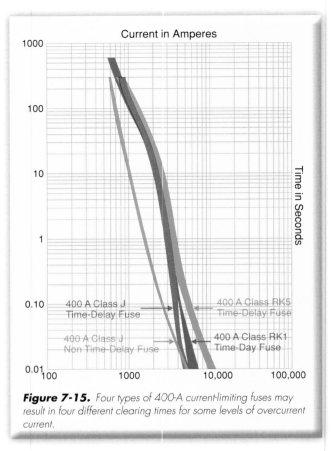

Figure 7-15. Four types of 400-A current-limiting fuses may result in four different clearing times for some levels of overcurrent current.

of the fuse is less than 0.01 second. This point is typically approximately 15 times the ampacity rating of the Class J, RK1, CF, and T current-limiting fuses, but can be verified by examining the fuse time–current curve.

Current-limiting fuses are not always current-limiting, however: Similar to circuit breakers, they have an overload region in which their clearing time can be several seconds or minutes. When the arcing current is less than the current-limiting range of a fuse, extended clearing times and increased incident energy can occur. If the fault is within the current-limiting range, however, clearing times of less than 0.01 second (assumed to be one-half cycle or 0.008 second) occur. **See Figure 7-16.** Based on the testing conducted for IEEE 1584, the clearing time for current-limiting fuses can be assumed to be 0.004 second if the arcing current value is more than twice the current at 0.01 second. Note that a specific manufacturer's Class J 400-A fuse enters its current-limiting range at approximately 5,200 A (where the total clearing curve crosses 0.01 second); at 10,400 A, it is assumed to clear in 0.004 second or less. In addition, when a fuse operates in its current-limiting range, the let-through current is reduced from what is available. The combination of

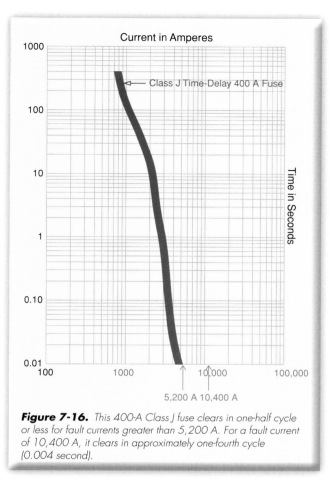

Figure 7-16. *This 400-A Class J fuse clears in one-half cycle or less for fault currents greater than 5,200 A. For a fault current of 10,400 A, it clears in approximately one-fourth cycle (0.004 second).*

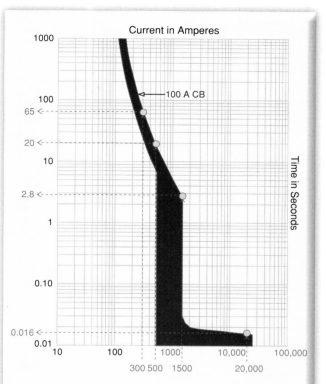

Figure 7-17. *Clearing times for various overcurrents for a 100-A thermal magnetic molded-case circuit breaker with a fixed instantaneous trip are shown. The dotted lines and yellow dots assist in determining the clearing times for various overcurrent values.*

Figure 7-18. *The 100-A circuit breaker results in these clearing times for the various overcurrent magnitudes.*

Overcurrent Magnitude	Clearing Time
300 A	65 seconds
500 A	20 seconds
1,500 A	2.8 seconds
20,000 A	0.016 second

decreased clearing times and reduced arcing current can greatly reduce the incident energy when the arcing current is within the current-limiting range of a fuse.

Circuit Breaker Clearing Times

Circuit breaker time–current characteristic curves have a different shape than fuse time–current characteristic curves. Nevertheless, in most situations, similar methods can be used for reading circuit breaker and fuse time–current curves. **See Figures 7-17** and **7-18**. For the purposes of clearing time for arc-flash analysis, the greatest opening time for a given current should always be used; that is, use the line to the right and top of the curve. This is the maximum clearing time curve for a circuit breaker.

Several types of circuit breakers are available, and each may have time–current curve characteristics that differ in the shape or total clearing times for certain overcurrent values. To illustrate this point, the clearing time for a 20,000-A fault is shown for three different types of 600-A circuit breakers. Note the differences in the opening time. Given these variations, it is important to know which type of circuit breaker is being analyzed as well as which settings are in use.

Interpreting the time–current curve for a typical 600-A thermal-magnetic (TM) molded-case circuit breaker, the 20,000-A fault intersects the circuit breaker maximum clearing time at 0.024 second. **See Figure 7-19**. This circuit breaker has an adjustable instantaneous trip (IT) setting and in this case is set at 10 times the circuit breaker 600-A current rating—that is, at 6,000 A. A ± tolerance is associated with this setting, and for this particular case, the threshold of where the instantaneous trip should start to operate when set at 10 times the circuit breaker ampere rating is between 5,400 A and 6,600 A.

For molded-case circuit breakers with an adjustable instantaneous trip, the IT setting can typically be set as low as 3 to 5 times the circuit breaker ampere rating, but generally no higher than 8 to 12 times this rating. In addition, in the past decade, newer-style molded-case circuit breakers have become available with IT settings as high as 20 or more times the circuit breaker ampere rating.

The electronic-sensing unit of the 600-A electronic-sensing molded-case circuit breaker can offer increased accuracy and more curve shaping (adjustments). **See Figure 7-20.** Interpreting the time–current curve of a typical 600-A electronic sensing molded-case circuit breaker, the 20,000-A fault intersects the circuit breaker maximum clearing time at 0.04 second.

Electronic-sensing molded-case circuit breakers may have three electronic sensing regions: long-time (LT), short-time delay (STD), and instantaneous trip (IT). The "long-time" (LT) setting determines the ampere rating of the circuit breaker, whereas the "long-time delay" (LTD) setting determines the overload region performance. The short-time delay is a sensing region between the long-time region and the instantaneous region; it provides a means to not open the circuit breaker instantaneously for some levels of fault current, yet to open it faster than with the long-time delay. The short-time delay for electronic-sensing molded-case circuit breakers is typically available at 0.1 second (6 cycles), 0.2 second (12 cycles), or 0.3 second (18 cycles). In this case, the STD is set at 0.1 second.

The instantaneous trip for this electronic-sensing molded-case circuit breaker is set at 10 times the circuit breaker's 600-A current rating—that is, at 6,000 A. Note that with an electronic-sensing molded-case circuit breaker, the maximum IT setting cannot typically be more than 8 to 10 times the ampere rating of the circuit breaker. The 20,000-A fault current is in the instantaneous region and the clearing time is 0.04 second—slower than the clearing time (0.024 second) for the thermal-magnetic molded-case circuit breaker example. In some instances, electronic-sensing circuit breakers might open faster than a thermal-magnetic circuit breaker, but this behavior depends on the specific circuit breaker type and characteristics.

Many of the low-voltage power circuit breakers (LVPCB) are similar to electronic trip molded-case circuit breakers with regard to the adjustability of their overcurrent protection settings. The difference is that these devices are larger in size than molded-case circuit breakers, offer draw-out capabilities (for increased maintenance capabilities), and can be provided with

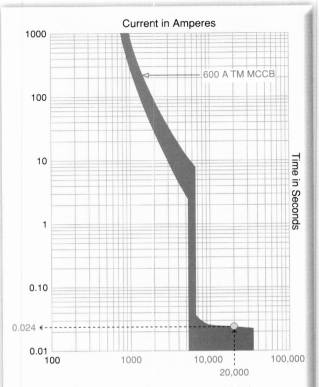

Figure 7-19. *This 600-A thermal-magnetic molded-case circuit breaker clears a 20,000-A fault in 0.024 second.*

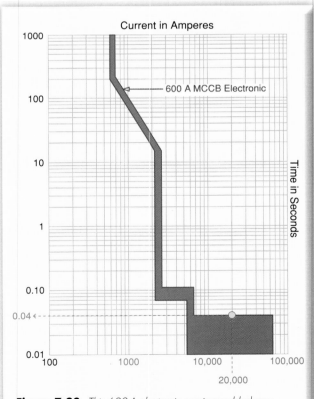

Figure 7-20. *This 600-A electronic-sensing molded-case circuit breaker can clear a 20,000-A fault in 0.04 second.*

or without an instantaneous trip. When the instantaneous trip is omitted (to increase their selective coordination capabilities), LVPCBs utilize a short-time delay. **See Figure 7-21**. In this case, the 20,000-A fault is cleared by the short-time delay in 0.5 second (30 cycles).

If equipped without an IT, the low-voltage power circuit breaker is set to "hold off opening" faults for a period of time equal to the short-time delay setting. The STD setting can range from 0.1 second (6 cycles) to 0.5 second (30 cycles). Its purpose is to allow the downstream OCPD to clear faults before the LVPCB trips, thereby avoiding unnecessary power loss to other loads. This increase of clearing time by using a short-time delay can result in an increase in arc-flash incident energy compared to the thermal-magnetic and electronic-sensing molded-case circuit breakers. As a consequence, for installations complying to the 2011 *NEC* or later edition, technology such as zone-selective interlocking, arc-reduction maintenance switches, and arc-flash relays is required when STD settings are used

(without instantaneous trip) on low-voltage power circuit breakers (reference *NEC* 240.87).

Circuit breaker time–current characteristics may be either adjustable or nonadjustable. For instance, many 150-A or less rated molded-case circuit breakers used in branch panels do not have settings that can be adjusted; the instantaneous trip settings are fixed. For circuit breakers with adjustable settings, the most common adjustments that can affect arc-flash hazard incident energy are changes made to the instantaneous trip settings and to the short-time delay settings. **See Figure 7-22**. A 400-A circuit breaker with an adjustable instantaneous trip is shown. Two different instantaneous trip settings are illustrated: 5 times (which is 2,000 A with a red tolerance band) and 10 times (which is 4,000 A with a blue tolerance band). An instantaneous trip setting can influence the clearing time for low-level faults. In this case, for a 3,000-A fault, if the IT is set at 5 times the circuit breaker ampere rating, the circuit breaker clears in 0.024 second; by comparison, if the IT is set at 10 times this rating, the circuit breaker clears in 35 seconds.

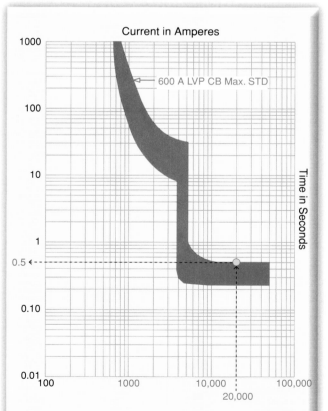

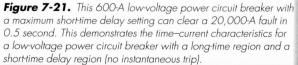

Figure 7-21. *This 600-A low-voltage power circuit breaker with a maximum short-time delay setting can clear a 20,000-A fault in 0.5 second. This demonstrates the time–current characteristics for a low-voltage power circuit breaker with a long-time region and a short-time delay region (no instantaneous trip).*

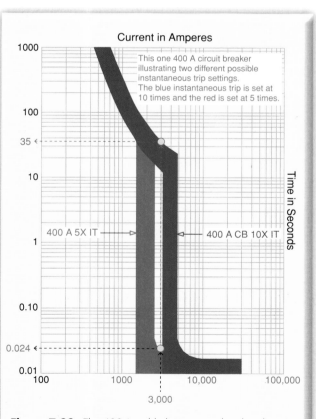

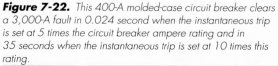

Figure 7-22. *This 400-A molded-case circuit breaker clears a 3,000-A fault in 0.024 second when the instantaneous trip is set at 5 times the circuit breaker ampere rating and in 35 seconds when the instantaneous trip is set at 10 times this rating.*

Generalized OCPD Clearing Times

As the previous sections have illustrated, fault clearing time can vary based on OCPD type, ampere rating, and settings as well as the magnitude of fault current. In determining the fault clearing time, it is recommended to use the time–current curve for the specific manufacturer, type, ampere rating, and setting of the OCPD. **See Figure 7-23**. This table provides general minimum OCPD clearing times. Typical short-time delay settings could be 6, 12, 18, 24, or 30 cycles.

OCPD Condition of Maintenance's Effects on Clearing Time

As noted in *NFPA 70E* 130.5, the arc-flash hazard analysis must take into consideration the design of the OCPD and its clearing time, including its condition of maintenance. 130.5 Informative Note No. 1 provides the rationale. The design and condition of maintenance of OCPDs may directly impact the clearing time of OCPDs, which in turn affects the incident energy. Poorly maintained OCPDs may take longer to clear, or may not clear at all, resulting in higher actual arc-flash incident energies than those expected from an arc-flash hazard analysis based on an OCPD operating

as originally specified. If the design of an OCPD requires periodic testing and maintenance as well as inspection and testing after fault interruption (*NFPA 70E* 225.3) to ensure proper operation, the inspection, maintenance, and testing must be performed and documented (*NFPA 70E* 205.4). If an OCPD's condition of maintenance cannot be assured to be acceptable, it may not be suitable to rely on the results of an arc-flash hazard analysis. For more information on this topic, see the chapter "OCPD Work Practices and Maintenance Considerations."

Ramifications of OCPD Clearing Times

The topics covered so far in this chapter include the available (bolted) short-circuit current, arcing current, and clearing time for OCPDs based on the OCPD type and characteristics and fault current. This section provides an example of how these parameters function with the incident energy calculation method. It demonstrates why, when using the incident energy calculation method, it is important to know the type, ampere rating, and settings for OCPDs protecting the equipment.

Figure 7-23. *Generalized minimum clearing times depend on the OCPD type.*

Device Type (600 V or less)	Clearing Time (seconds)[†]	
Current-limiting fuse	0.004 to 0.008*	
Molded-case circuit breaker: • Instantaneous trip • Short-time delay	0.025** Setting of short-time delay***	
Insulated-case circuit breaker: • Instantaneous trip • Short-time delay	0.05** Setting of short-time delay***	
Low-voltage power circuit breaker (integral trip) • Instantaneous trip • Short-time delay	0.05** Setting of short-time delay***	
Current-limiting molded-case circuit breaker	0.008 or less*	

[†]*These are approximate clearing times for short-circuit currents within the current-limiting range of a fuse or a current-limiting circuit breaker or within the instantaneous region or short-time delay region of a circuit breaker.*

These are approximate clearing times for current-limiting fuses and current-limiting circuit breakers when the fault current is in their current-limiting range.

**These circuit breaker instantaneous trip times are based on Table 1 in 2002 IEEE 1584.*

***Molded-case and insulated-case circuit breakers with a short-time delay setting typically also include an instantaneous trip override that operates above a certain fault current level.*
Note: Lower current values may cause the overcurrent device to operate more slowly. Arc-flash energy may actually be highest at lower levels of available short-circuit current. This requires that arc-flash energy calculations be completed for the range of sustainable arcing currents. This is also noted in NFPA 70E 130.5 IN No. 2.

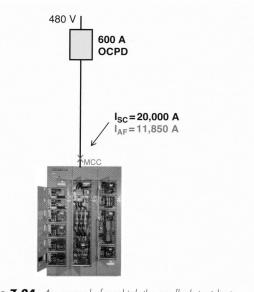

Figure 7-24. *An example for which the arc-flash incident energy is determined for protection by four different 600-A OCPD types; the motor control center has a 20,000-A available bolted short-circuit current, resulting in potentially an 11,850-A arcing fault current.*

Assume that work must be done in the main disconnect section of a motor control center (MCC) when it is not in an electrically safe work condition. Also assume that the bolted 3-phase short-circuit current at the line terminals of the MCC is calculated to be 20,000 amperes. Using the 2002 IEEE 1584 arcing fault equation for a 20,000-A available bolted short-circuit current, the arcing fault current is calculated as 11,850 A. **See Figure 7-24**.

Here, the arc-flash incident energy will be calculated for four different 600-A OCPD types. First, the clearing time for each of the four different OCPD types is determined for the 11,850-A arcing current using the OCPD time–current curves. **See Figures 7-25, 7-26, 7-27, and 7-28**. Then, by using the 2002 IEEE 1584 basic equations with these clearing times and the other parameters, the incident energy at an 18-inch working distance is determined. **See Figure 7-29**.

OCPD Incident Energy Profiles

If the arc-flash hazard is determined by *NFPA 70E* 130.5(B)(1) Incident Energy Analysis Method, the OCPD protective clearing time for the anticipated arcing current is a key factor in the end result. Each type of overcurrent protective device and associated ampere rating, specific settings, and time–current characteristics has an incident energy profile. This section provides generalized incident energy profiles for some widely used overcurrent protective devices.

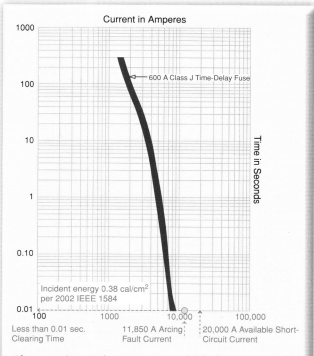

Figure 7-25. *With a 20,000-A available short-circuit current, the arcing current is 11,850 A and the 600-A Class J time-delay fuse clears the arcing fault current in less than one-half cycle, resulting in an incident energy of 0.38 cal/cm².*

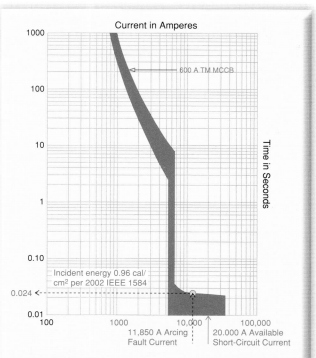

Figure 7-26. *With a 20,000-A available short-circuit current, the arcing current is 11,850 A and the 600-A thermal-magnetic molded-case circuit breaker clears the arcing fault current in 0.024 second, resulting in an incident energy of 0.96 cal/cm².*

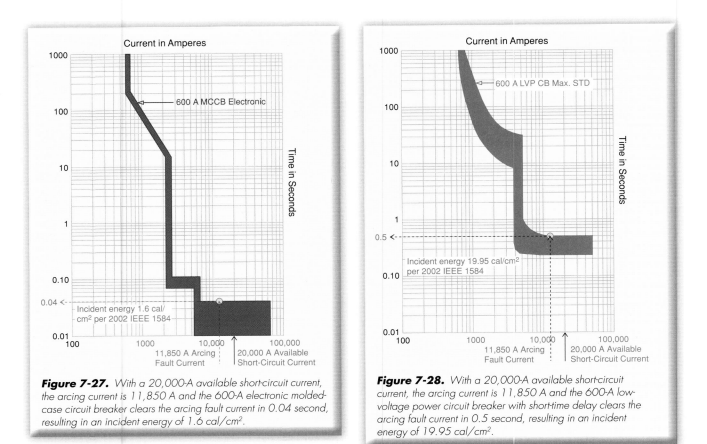

Figure 7-27. *With a 20,000-A available short-circuit current, the arcing current is 11,850 A and the 600-A electronic molded-case circuit breaker clears the arcing fault current in 0.04 second, resulting in an incident energy of 1.6 cal/cm².*

Figure 7-28. *With a 20,000-A available short-circuit current, the arcing current is 11,850 A and the 600-A low-voltage power circuit breaker with short-time delay clears the arcing fault current in 0.5 second, resulting in an incident energy of 19.95 cal/cm².*

Figure 7-29. *For the four different 600-A OCPD type scenarios, the incident energy at an 18-inch working distance can be determined using 2002 IEEE 1584 basic equations and the clearing times listed here.*

OCPD Type	OCPD Clearing Time for 11,850-A Arcing Fault Current	Incident Energy at 18" Working Distance
600-A Class J fuse	½ cycle or less	0.38 cal/cm²
600-A thermal-magnetic molded-case circuit breaker	0.024 second	0.96 cal/cm²
600-A electronic-sensing molded-case circuit breaker	0.04 second	1.60 cal/cm²
600-A low-voltage power circuit breaker with short-time delay set at 0.5 second	0.5 second	19.95 cal/cm²

The intent of this section is to provide a better understanding of how bolted fault current, arcing fault current, OCPD clearing times, and incident energy are related, and specifically how the OCPD characteristics can affect the incident energy. Recognize, however, that these profiles are a broad generalization. Even so, understanding the OCPD incident energy profiles may aid in better understanding of arc-flash hazards.

Circuit Breaker Incident Energy Profiles

A curve of arcing fault incident energy (at a working distance) versus arcing fault current varies for different types of circuit breakers. Such a profile curve can be generated by taking a specific OCPD and then sampling possible bolted short-circuit currents over the full range of overcurrent to determine (1) the arcing fault current for each bolted short-circuit current sample point (calculated using IEEE 1584); (2) for each sample point, the OCPD clearing time for the associated arcing fault current; (3) using the arcing current and associated OCPD clearing time, calculate the incident energy (using IEEE 1584); and (4) plot of the incident energy versus the arcing current (a graph).

To illustrate this concept, this process will be carried out for three separate points for a 600-A molded-case circuit breaker. A circuit breaker time–current curve and an adjacent arcing current versus incident energy profile curve will be used for each of these three scenarios. **See Figure 7-30**. The circuit time–current curve is shown on the right and the arcing current versus incident energy profile for this circuit breaker is shown on the left. The available bolted short-circuit current is 6,000 A; it is shown on the time–current current with an arrow. The arcing current that might flow if a 3-phase arcing fault incident occurred is depicted with the arcing icon (approximately 4,000 A) on the time–current curve and results in the circuit breaker clearing time at the yellow dot (approximately 8 seconds). With these parameters, the incident energy is determined to be more than 100 cal/cm^2 at 18 inches working distance (shown on the left image as a yellow dot). For the graph on the left, arcing current is shown on the horizontal axis; incident energy is shown on the vertical axis.

When arcing currents are less than the instantaneous trip setting of a circuit breaker, as in this scenario, the clearing time can take seconds, resulting in high incident energies. (*NFPA 70E* 130.5 Informational Note 2 mentions that in some cases low fault currents may result in higher incident energies than will occur with higher fault currents.) In this case, the arcing current is in the long-time opening range of the circuit breaker. For this outcome to occur, the arcing current would have to be this low and sustained for this full time duration.

For the same 600-A circuit breaker, but with the arcing current resulting in the instantaneous trip operating, the incident energy will be greatly reduced. **See Figure 7-31**. This point coincides with the point where the arcing current magnitude is at this circuit breaker's instantaneous trip threshold point. With the available bolted short-circuit current at 12,000 A, the resulting arcing current is 7,600 A and the circuit breaker clearing time is 0.024 second. The incident energy at 18 inches working distance

is approximately 0.60 cal/cm^2, which is the incident energy low point for this molded-case circuit breaker incident energy profile. It shows that the lowest incident energy occurs at the threshold of the circuit breaker instantaneous trip.

As the arcing current increases to levels greater than the point where the 600-A circuit breaker operates at more than its instantaneous trip threshold, the incident energy also increases. **See Figure 7-32**. With the available short-circuit current at 60,000 A, the resulting arcing current is 30,000 A and the circuit breaker clearing time is 0.024 second. The incident energy at 18 inches is then approximately 2.64 cal/cm^2 at an 18-inch working distance.

In summary, in the three scenarios for the 600-A circuit breaker generalized incident energy profile, the graph on the left is the incident energy profile. For this circuit breaker, the lowest incident energy occurs at the threshold of the circuit breaker's instantaneous trip. Higher fault currents that exceed the instantaneous trip threshold result in higher incident energy. Arcing fault currents less than the circuit breaker instantaneous trip threshold result in drastically higher incident energy. Molded-case circuit breakers with different characteristics and settings can have significantly different incident energies and profile.

When a current-limiting circuit breaker is used, incident energy will not increase as much with increases in the available short-circuit current; as available short-circuit current increases, incident energy will stay flat or increase to a lesser extent. **See Figure 7-33**.

Note that a low-voltage circuit breaker with a short-time delay and no instantaneous trip will have higher incident energies than another circuit breaker that has an instantaneous trip, because the clearing time for the arcing fault current in the short-time delay range can be much greater. **See Figure 7-34**. The graph on the left shows generalized incident energy profiles for three different types of circuit breakers on one profile: current-limiting, instantaneous trip, and short-time delay. In the long-time region (labeled A), all three of these circuit breakers will have similar incident energies (dependent on the actual time–current characteristics for each).

To summarize the incident energy profile for the three different types of circuit breakers, the current-limiting circuit breakers have the lowest incident energy, followed by standard non-current-limiting circuit breakers with an instantaneous trip, followed by circuit breakers with a short-time delay (no instantaneous trip). Note that for

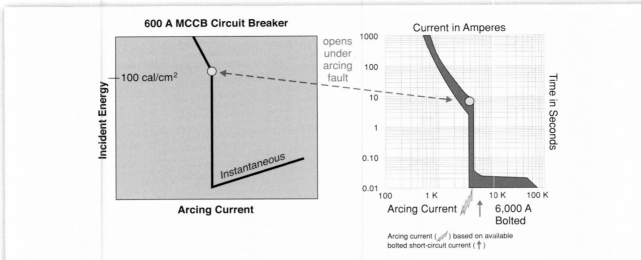

Figure 7-30. *The graph on the right shows this circuit breaker clears an arcing fault of approximately 4,000 A in approximately 8 seconds, which results in an incident energy of more than 100 cal/cm² at 18" working distance (shown in the graph on the left).*

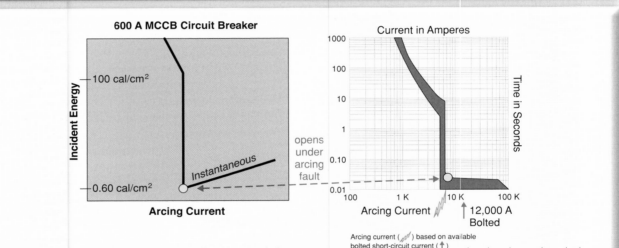

Figure 7-31. *This graph shows arcing current that just exceeds the instantaneous trip threshold of the circuit breaker; this results in the lowest incident energy at this point for this circuit breaker.*

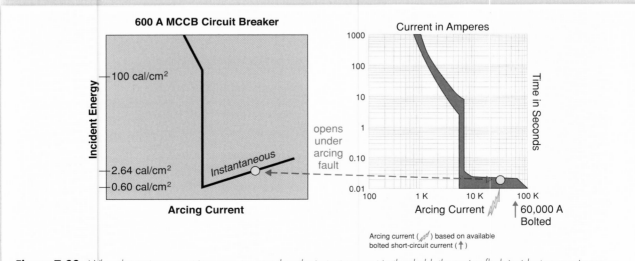

Figure 7-32. *When the arcing current increases to more than the instantaneous trip threshold, the arcing flash incident energy increases.*

low-level faults (the region labeled A in the graphs), the incident energy can be quite high. This situation may occur when the arcing current is less than the current-limiting threshold, instantaneous trip-setting threshold, or short-time delay setting threshold of a circuit breaker.

Fuse Incident Energy Profiles

Current-limiting fuses have a somewhat different generalized incident energy profile from circuit breakers.

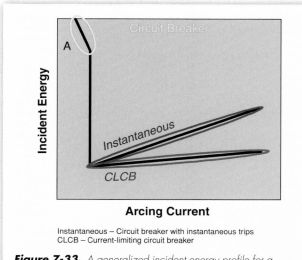

Instantaneous – Circuit breaker with instantaneous trips
CLCB – Current-limiting circuit breaker

Figure 7-33. *A generalized incident energy profile for a circuit breaker with an instantaneous trip and a current-limiting circuit breaker is shown. Both types of circuit breakers have the same profile in the long-time region but different profiles in the instantaneous trip region.*

Here, a fuse generalized incident energy profile will be reviewed similar to the process done for the 600-A circuit breaker. Three separate points for a 600-A Class RK1 dual-element, time-delay fuse will be investigated to understand a generalized fuse incident energy profile. **See Figure 7-35.** In this scenario, an available short-circuit current of 6,000 A results in an arcing fault current of approximately 4,000 A. The arcing current that might flow if a 3-phase arcing fault incident occurred is depicted with the arcing icon (approximately 4,000 A) on the time–current curve (the graph on the right) and results in the fuse total clearing time of approximately 8 seconds (yellow dot). With these parameters, the incident energy is determined to be more than 100 cal/cm² at 18 inches working distance (shown on the left graph as a yellow dot).

Note that there is a very high arc-flash energy for a low-level arcing current—one that takes seconds to open. In this case, the arcing current is in the long-time opening range of the fuse. For this outcome to occur, the arcing current would have to be this low and be self-sustaining for the full opening time. This scenario is the reason why *NFPA 70E* 130.5 Informational Note 2 mentions that low fault currents may result in higher incident energies than higher fault currents.

For the same 600-A fuse, but with a greater arcing current, which is slightly less than the fuse's current-limiting threshold, the incident energy will be

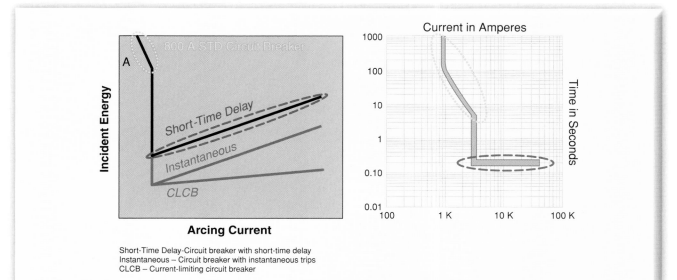

Short-Time Delay-Circuit breaker with short-time delay
Instantaneous – Circuit breaker with instantaneous trips
CLCB – Current-limiting circuit breaker

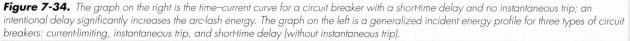

Figure 7-34. *The graph on the right is the time–current curve for a circuit breaker with a short-time delay and no instantaneous trip; an intentional delay significantly increases the arc-lash energy. The graph on the left is a generalized incident energy profile for three types of circuit breakers: current-limiting, instantaneous trip, and short-time delay (without instantaneous trip).*

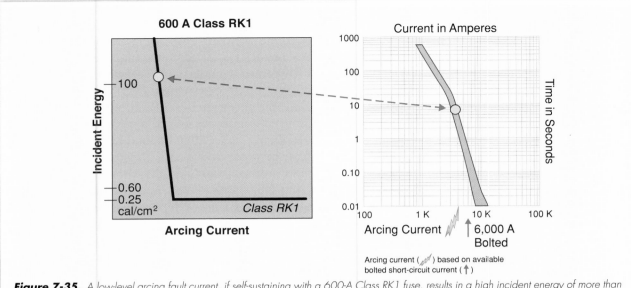

Figure 7-35. *A low-level arcing fault current, if self-sustaining with a 600-A Class RK1 fuse, results in a high incident energy of more than 100 cal/cm² at 18" working distance because the fault current is in the long-time region of the fuse.*

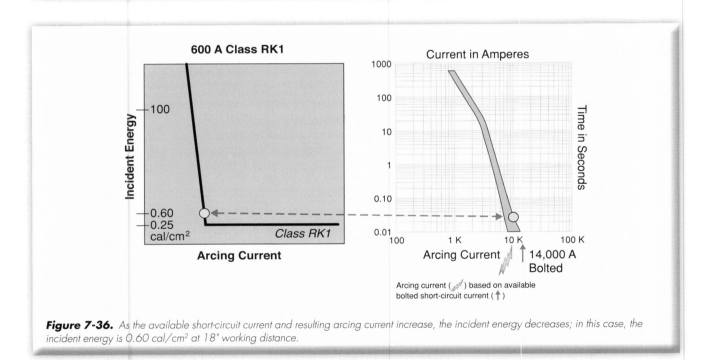

Figure 7-36. *As the available short-circuit current and resulting arcing current increase, the incident energy decreases; in this case, the incident energy is 0.60 cal/cm² at 18" working distance.*

greatly reduced. **See Figure 7-36.** When a higher arcing current occurs, the arc-flash incident energy decreases. The reason for this is that the higher the fault current, the faster a fuse operates.

When the arcing current is greater than the current-limiting threshold of the fuse, then the fuse is in its current-limiting range. **See Figure 7-37.** It will open in less than one-half cycle and reduce the magnitude of the arcing fault current, both of which help reduce the incident energy. In general, arcing currents beyond this value (at the current-limiting range) will not be associated with an increased arc-flash incident energy level, as arcing current increases even further. Once the arcing current is large enough to reach the current-limiting threshold of the fuse, the incident energy will fall to a low level and remain

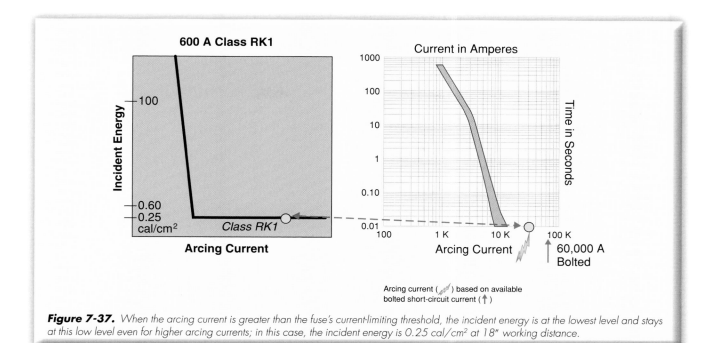

Figure 7-37. *When the arcing current is greater than the fuse's current-limiting threshold, the incident energy is at the lowest level and stays at this low level even for higher arcing currents; in this case, the incident energy is 0.25 cal/cm² at 18" working distance.*

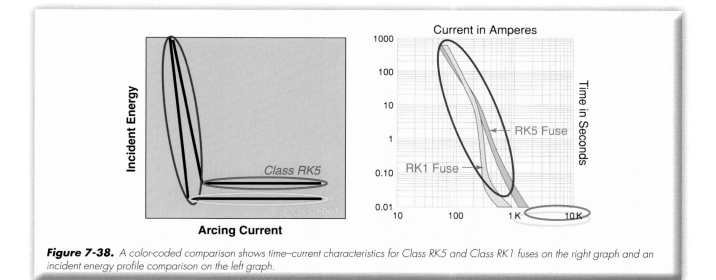

Figure 7-38. *A color-coded comparison shows time–current characteristics for Class RK5 and Class RK1 fuses on the right graph and an incident energy profile comparison on the left graph.*

at that level as the available arcing current continues to increase.

Different fuse types can have different incident energy profiles. **See Figure 7-38.** Two types of dual-element, time-delay fuses—Class RK5 fuses (red curve) and Class RK1 fuses (yellow curve)—can be compared in a generalized manner. Both are current-limiting devices, but the Class RK1 fuses have a higher degree of current limitation and the RK1 fuses enter their current-limiting range at a lower fault current than the Class RK5 fuses. For most arcing fault currents, the incident energy is less with Class RK1 fuses than with Class RK5 fuses. This is reflected in the incident energy profiles. Class J and Class CF fuses typically have time–current characteristics equal to or faster than the Class RK1 fuses, so the incident energies for Class J and Class CF time-delay dual-element fuses will be equal or lower at each arcing fault current level than for the Class RK1 fuses.

Summary

An understanding of available short-circuit (fault) current, arcing current, OCPD time–current curves, and OCPD clearing times provides the basis for properly applying 130.5 Arc-Flash Hazard Analysis. Different OCPD types can have differently shaped time–current characteristic curves, and some OCPDs have adjustable settings that can alter the shape of their time–current characteristic curve. The potential arc-flash incident energy determined from an arc-flash hazard analysis is directly related to the OCPD type and characteristics.

REVIEW QUESTIONS

1. For __?__ fault incident energy calculations (130.5(B)(1)), the OCPD clearing time depends on the OCPD characteristics and the arcing fault current magnitude.
   - a. arcing
   - b. bolted
   - c. full
   - d. short-circuit

2. A bolted fault condition represents a(n) __?__ connection of relatively low (or assumed "zero") impedance at the point of the fault.
   - a. arcing
   - b. dangling
   - c. solid
   - d. weak

3. For a given point in the electrical system, the greater the available short-circuit current, the __?__ the arcing fault current.
   - a. faster
   - b. greater
   - c. less
   - d. slower

4. Arc-flash incident energy varies by the fault current __?__ .
   - a. duration and magnitude
   - b. instantaneous peak
   - c. region
   - d. vector analysis

5. Fuses and circuit breakers sense and __?__ to actual current magnitude, whether the current of a given magnitude is a bolted or an arcing fault current.
   - a. close
   - b. feel
   - c. hesitate
   - d. respond

6. The speed of response of an OCPD when interrupting an overcurrent greater than the OCPD's __?__ rating can vary depending on the magnitude of overcurrent.
   - a. ampere
   - b. power
   - c. voltage
   - d. wattage

7. The clearing time of OCPDs is a(n) __?__ factor in the severity of an arc-flash incident.
   - a. important
   - b. incidental
   - c. insignificant
   - d. minor

8. The arc-flash hazard analysis takes into consideration the design of the OCPD and its clearing time, __?__ its condition of maintenance.
   - a. concerning
   - b. eliminating
   - c. excluding
   - d. including

9. Each type and ampere rating OCPD can be represented by a generalized incident energy profile curve.
   - a. True
   - b. False

10. *NFPA 70E* 130.5, Informational Note __?__ states that low fault currents may result in higher incident energies than higher fault currents.
    - a. 1
    - b. 2
    - c. 3
    - d. 4

8 Arc Flash Hazard Analysis Methods

Courtesy of James R. White, Shermco Industries.

OBJECTIVES

1. Understand the advantages and limitations of various methods available to determine the arc flash boundary (AFB) and selection of protective equipment.

2. Understand the conditions of use necessary to be permitted to use the hazard/risk category (HRC) method.

3. Recognize that *NFPA 70E* Annex D provides various calculation methods to determine the AFB and incident energy.

4. Understand other considerations for conducting an arc flash hazard analysis (AFHA).

REFERENCES

1. *NFPA 70E*®, 2012 Edition
2. 2002 IEEE 1584, *IEEE Guide for Performing Arc-Flash Hazard Calculations*

3. Cooper Bussmann Arc Flash Incident Energy Calculator: available at www.CooperBussmann.com and in the *Selecting Protective Devices* (SPD) publication

An electrician died as a result of burns he received while he was making a connection on a circuit breaker. The victim worked for an electrical contractor that employed 130 people. Safety functions were assigned to the firm's two project managers as an additional duty. The firm had no written safety policy or established safety program, and only on-the-job training was provided. The managers met with the foremen monthly to discuss safety issues.

On the day of the accident, the victim and his helper were to install three new circuit breakers into a panelboard that supplied power to various parts of an industrial complex. Three additional circuit breakers had previously been installed; they supplied power to the occupied portion of the building. The circuit breakers to be installed on the day of the accident were expected to control systems in the unoccupied portion of the building.

Before beginning the job, the victim was instructed by the foreman to install the top breaker first and the bottom breaker last. Although the victim had worked for the firm for only four days, he had approximately 16 years of experience as an electrical worker.

The victim did not follow the foreman's instructions. He secured the three breakers in the panelboard, and began to wire the bottom breaker first. The wires to be connected to the breakers were fed through a conduit in the bottom of the panelboard. The wires then had to be fed to the upper portion to make the connections. The line side of the bottom breaker became energized once the connections were completed. The victim then began to make the connections on the middle breaker without replacing the cover on the bottom breaker. He started to feed the wires for the middle breaker through the panelboard. He successfully fed four of the six wires to the upper portion. As the fifth wire was being fed upward, its uninsulated tip contacted one of the exposed energized connection points. The contact caused an arc, and the bottom breaker exploded. The victim was standing immediately in front of the breaker. His clothes caught fire, and he received massive burns to the upper portion of his body.

When the helper heard the explosion, he ran. As he was running away, a second explosion occurred. The helper received second- and third-degree burns to his left shoulder and back and to the back of his left arm.

The foreman was headed toward the room when he heard the explosions. He found the victim lying on the floor with his clothes on fire. He pulled the victim outside the room and extinguished the flames with his hands, receiving second- and third-degree burns in the process. The victim was conscious and talking at this time. The victim received burns to 80% of his body. Emergency rescue transported the victim to the hospital. He was later transferred to a burn center, where he died the following morning. The cause of death was attributed to massive burns to the victim's upper body.

Source: For details of this case, see FACE 8644. Accessed September 14, 2012.

For additional information, visit qr.njatcdb.org Item #1198.

Introduction

Prior to a worker being exposed to electrical hazards, an electrical hazard analysis consisting of a shock hazard analysis and an arc flash hazard analysis (AFHA) is required if energized electrical conductors or circuit parts at 50 volts or more are not placed in an electrically safe work condition. The arc flash hazard analysis, per *NFPA 70E* Section 130.5, is required to determine the arc flash boundary (AFB) and the personal protective equipment (PPE) necessary when people are within the AFB. *Arc flash hazard analysis* and *arc flash boundary* are defined by *NFPA 70E* as follows:

Arc Flash Hazard Analysis. A study investigating a worker's potential exposure to arc flash energy, conducted for the purpose of injury prevention and the determination of safe work practices, arc flash boundary, and the appropriate levels of personal protective equipment (PPE). **Boundary, Arc Flash.** When an arc flash hazard exists, an approach limit at a distance from a prospective arc source within which a person could receive a second degree burn if an electrical arc flash were to occur.

Reprinted with permission from NFPA-70E *2012,* Electrical Safety in the Workplace *Copyright 2011, National Fire Protection Association, Quincy, MA 02169. This reprinted material is not the complete and official position of the NFPA on the referenced subject, which is represented only by the standard in its entirety.*

The primary focus of this chapter is on common methods available for determining the arc flash boundary and incident energy (IE) or hazard/risk category (HRC) required for an arc flash hazard analysis as detailed in *NFPA 70E* 130.5. The limitations and advantages of each method are discussed here, as are examples for some of the methods.

Material in other chapters should be understood and used to properly comprehend and apply the material in this chapter. This includes calculating bolted short-circuit current and arcing current as well as determining overcurrent protective device (OCPD) clearing time.

Common AFHA Methods Summary Tables

Several methods can be used to determine the AFB and incident energy or HRC as required by the arc flash hazard analysis. **See Figures 8-1** and **8-2.** Each table is subdivided into methods for AC systems and DC systems, and identifies advantages, comments, and limitations for these methods. The HRC method is located in the main body of *NFPA 70E*. The other methods are found in *NFPA 70E* Annex D. Table D.1 in *NFPA 70E*

Annex D summarizes these other calculation methods available for determining the AFB and incident energy. The HRC and 2002 IEEE 1584 *Guide for Performing Arc-Flash Hazard Calculations* methods are the two most widely used methods. Each method is valid only for certain ranges of variables and certain other criteria.

Note that each method has specific parameters that determine its applicability. If a specific situation does not meet the parameters for a given method, then that method cannot be used. For instance, most of the methods are valid only for 3-phase alternating current (AC) systems. Some methods are valid only for systems of 600 VAC or lower or only for systems of more than 600 VAC, whereas other methods are valid for both. In addition, some methods address only arcs in open air, while others address arcs in open air and arcs-in-a-box.

HRC Method

The hazard/risk categories (HRC) method or "table method" is a method to determine the AFB, the required PPE, and the need for rubber insulating gloves and insulated tools. Its use is permitted by *NFPA 70E* 130.5 Exception and 130.5(B)(2). 130.7(C)(15) identifies conditions of use that must be met to use Table 130.7(C)(16) in conjunction with Table 130.7(C)(15)(a) or Table 130.7(C)(15)(b). A closer examination of the conditions of use appears in the next two sections, *AC Systems HRC Method* and *DC Systems HRC Method*.

The basis of the HRC method is stated in 130.7(C)(15) Informational Notes No. 1, No. 2, and No. 3, which include the phrases "based on the collective experience of the task group," "generally based on determination of estimated exposure levels," and "the premise used by the task group … considered to be reasonable, based on the consensus judgment of the full *NFPA 70E* Technical Committee."

The HRC method is considered by many to be the easiest method to use, but its application does require determining that the conditions of use have been met. If these conditions of use are not met, then the HRC method cannot be used and the AFHA must be utilized instead in compliance with 130.5(B)(1), which outlines the incident energy analysis method.

Consider the overcurrent protective device condition of maintenance per 130.5. If the OCPD providing the protection for arc flash hazard analysis has not been maintained properly or adequately, then other means should be investigated. *NFPA 70E* Chapter 2, including 205.4, 210.5, and Article 225 and the chapter, "OCPD Work Practices and Maintenance Considerations" in this publication provides insight into requirements and work practices related to equipment maintenance.

Figure 8-1. *Methods that can be used to determine AFB as required by NFPA 70E 130.5.*

Value	Method	Advantages	Comments/Limitations
AC system AFB	HRC (Tables Method) [*NFPA 70E* 130.7(C)(15) and Tables 130.7(C) (15)(a) and 130.7(C)(16)]	Easiest method	Conditions of use in 130.7(C)(15) and Table 130.7(C)(15)(a) must be met: • Equipment type listed in table • System voltage within voltage specified for equipment • Task in table • Comply with Parameters under equipment type/voltage ▪ Maximum available short-circuit current (bolted) does not exceed max. value in applicable table parameters ▪ Class of overcurrent protective device does not exceed the maximum fault clearing time value in applicable table parameters for the maximum short-circuit (bolted) current available value in the applicable table parameters ▪ Working distance is equal or greater than table min. value in the applicable table parameters
	IEEE 1584 *Guide for Performing Arc Flash Hazard Calculations* (2002) [exerpts in *NFPA 70E* Annex D.7.5]	Most used calculation method • Industry consensus guide for calculations • Based on most up-to-date testing available • Developed from extensive testing and statistical analysis	• Applicable for ▪ 3-phase systems ▪ 208 V through 15 kV (theoretical equation above 15 kV) ▪ Short-circuit currents between 700 and 106,000 amperes ▪ Conductor gaps 13 mm through 152 mm ▪ Arcs in open air or arcs-in-a-box • Most complicated equations (more variables)
	Ralph Lee IEEE Paper (1982) [*NFPA 70E* Annex D.2]	Least complicated calculation method	• Applicable for ▪ 3-phase ▪ Arcs in open air • Not as accurate as IEEE 1584 method
DC system AFB	HRC (Tables Method) [*NFPA 70E* 130.7(C)(15) and Tables 130.7(C) (15)(b) and 130.7(C)(16)]	Only industry consensus document method (2002 IEEE 1584 does not have method for dc)	Conditions of use in 130.7(C)(15) and Table 130.7(C)(15)(b) must be met: • Equipment type listed in table • Task listed in table • Comply with Parameters under equipment type/voltage ▪ System voltage within voltage value ▪ Arcing current within range ▪ Overcurrent protective device clearing time for arcing current less than 2 sec. ▪ Working distance 18 in. or greater

HRC Method for AC Systems

Two *NFPA 70E* tables are used for this method: Table 130.7(C)(15)(a) and Table 130.7(C)(16). **See Figures 8-3** and **8-4.** For a specific task on a specific type of equipment, Table 130.7(C)(15)(a) provides the associated HRC and AFB, and identifies whether rubber insulating gloves or insulated tools are required. Table 130.7(C)(16) lists the additionally required PPE, including the arc rating for that equipment, based on the HRC determined from Table 130.7(C)(15)(a).

NFPA 70E Table 130.7(C)(15)(a) is formatted as follows under the column *Tasks Performed on Energized Equipment*:

1. Several types of electrical equipment at a specific voltage, such as *Panelboards or other equipment rated > 240 V and up to 600 V.*

2. Specific tasks to be performed for each type of equipment, such as *Work on energized electrical conductors and circuit parts, including voltage testing.*

3. Parameters:
   a. Maximum available short-circuit current, such as *maximum of 25 kA short-circuit current available.*
   b. Maximum fault clearing time, such as *maximum of 0.03 sec (2 cycle) fault clearing time.*
   c. Minimum working distance, such as *minimum 18 in working distance.*

4. Potential arc flash boundary is provided: such as *30 in*

The user of Table 130.7(C)(15)(a) first selects the type of equipment/voltage. The user must then ensure that all of the parameters of the specific application being investigated comply with the parameters in the table directly under the equipment type/voltage. That is:

   a. The determined available short-circuit current at the actual equipment is equal to or less than the maximum short-circuit current available parameter value.

Figure 8-2. *Methods that can be used to determine incident energy or HRC as required by NFPA 70E 130.5.*

Value	Method	Advantages	Comments/Limitations

AC system HRC	HRC (Tables Method) and [*NFPA 70E* 130.7(C)(15), Tables 130.7(C)(15)(a) and 130.7(C)(16)]	Easiest method	Conditions of use in 130.7(C)(15) and Table 130.7(C)(15)(a) must be met:
			• Equipment type listed in table
			• System voltage within voltage specified for equipment
			• Task in table
			• Comply with Parameters under equipment type/voltage
			  ▪ Maximum available short-circuit current (bolted) does not exceed max. value in applicable table parameters
			  ▪ Class of overcurrent protective device does not exceed the maximum fault clearing time value in applicable table parameters for the maximum short-circuit (bolted) current available value in the applicable table parameters
			  ▪ Working distance is not less than table min. value in the applicable table parameters

AC system Incident Energy Analysis	Incident Energy Analysis IEEE 1584 *Guide for Performing Arc Flash Hazard Calculations* (2002) [*NFPA 70E* Annex D.7.5]	• Industry consensus guide for calculations
		• Based on significant amount and most up-to-date testing available: most accurate
			• Applicable for
			  ▪ 3-phase systems
			  ▪ 208 V through 15 kV (theoretical equation above 15 kV)
			  ▪ Short-circuit currents between 700 and 106,000 amperes
			  ▪ Conductor gaps 13 mm through 152 mm
			  ▪ Arcs in open air or arcs-in-a-box
			• Most complicated formulae (more variables)
			• PPE levels of 1.2, 8, 25, and 40 are adequate for 95% of the arc flash tests

	Incident Energy Analysis Doughty, Neal, Floyd IEEE Paper (1998) [*NFPA 70E* Annex D.5, D.5.1, D.5.2]	• Less complicated calculation method	• Applicable for
			  ▪ 3-phase
			  ▪ 600 V or less
			  ▪ Short-circuit currents from 16,000 through 50,000 A
			  ▪ Arcs in open air or arcs-in-a box
			  ▪ Working distances 18 in. or greater
			• Calculations at the maximum short-circuit current and at the minimum sustainable arcing current (38% of maximum short-circuit current at 480 V per paper)

	Incident Energy Analysis Ralph Lee IEEE Paper (1982) [*NFPA 70E* Annex D.6]	• Least complicated calculation method	• Applicable for
			  ▪ 3-phase
			  ▪ Above 600 V
			  ▪ Arcs in open air.
			• Conservative as voltage increases

	Incident Energy Analysis National Electrical Safety Code 2012 (NESC) [*NFPA 70E* Annex D Table D.1: *NFPA 70E* Errata deleted entire D.8 row in Table D.1]	• Utility oriented
		• Uses tables based on certain assumptions and parameters
			• Tables for 1.1 kV to 800 kV applicable for
			  ▪ Arcs in open air
			  ▪ Phase-to-ground arcing fault
			  ▪ Max. OCPD clearing time
			  ▪ 1.1kV to 800 kV
			• Table for 1kV or less applicable for
			  ▪ Three phase
			  ▪ Arcs-in-a-box
			  ▪ Specific conditions or assumptions

DC system HRC	HRC [*NFPA 70E* 130.7(C)(15), Tables 130.7(C)(15)(b) and 130.7(C)(16)]	Only industry consensus document with method (2002 IEEE 1584 does not have methods for dc)	Conditions of use in 130.7(C)(15) and Table 130.7(C)(15)(b) must be met:
			• Equipment type listed in table
			• Task listed in table
			• Comply with Parameters under equipment type/voltage
			  ▪ System voltage within voltage
			  ▪ Arcing current within range
			  ▪ Maximum arc duration (OCPD clearing time for arcing current)
			  ▪ Working distance 18 in. or greater

DC system Incident Energy Analysis	Incident Energy Analysis Dan Doan IEEE Paper (2010) [*NFPA 70E* Annex D.7.8]	• A method to calculate incident energy for dc systems	• Applicable for
			  ▪ 1000 VDC or less
			  ▪ Arcs in open air (for arcs in box, advise calculate for arc in open air and multiple results by 3)
			• Based on concept maximum power possible in dc arc occurs when arcing voltage is ½ of the system voltage
			• 2002 IEEE 1584 does not include dc incident energy calculation methods

Figure 8-3. NFPA 70E Table 130.7(C)(15)(a), Hazard/Risk Category Classifications and Use of Rubber Insulating Gloves and Insulated and Insulating Hand Tools—Alternating Current Equipment, is one of the tables used for the AC systems HRC method. (shown in part)

Tasks Performed on Energized Equipment	Hazard/Risk Category	Rubber Insulating Gloves	Insulated and Insulating Hand Tools
Panelboards or other equipment rated > 240 V and up to 600 V Parameters: Maximum of 25 kA short-circuit current available; maximum of 0.03 sec (2 cycle) fault clearing time; minimum 18 in. working distance Potential arc flash boundary with exposed energized conductors or circuit parts using above parameters: 30 in.			
Work on energized electrical conductors and circuit parts; including voltage testing	2	Y	Y

Y = Yes (required); N: No (not required).
Refer to notes following NFPA 70E Table 130.7(C)(15)(a).

Reprinted with permission from NFPA-70E 2012, Electrical Safety in the Workplace Copyright 2011, National Fire Protection Association, Quincy, MA 02169. This reprinted material is not the complete and official position of the NFPA on the referenced subject, which is represented only by the standard in its entirety.

Figure 8-4. NFPA 70E Table 130.7(C)(16), Protective Clothing and Personal Protective Equipment (PPE), is one of the tables used for the AC and DC systems HRC method. (shown in part)

Hazard/Risk Category	Protective Clothing and PPE
2	**Arc-Rated Clothing, Minimum Arc Rating of 8 cal/cm²** (See Note 3.) Arc-rated long-sleeve shirt and pants or arc-rated coverall Arc-rated flash suit hood or arc-rated face shield (See Note 2) and arc-rated balaclava Arc-rated jacket, parka, rainwear, or hard hat liner (AN)
	Protective Equipment Hard hat Safety glasses or safety goggles (SR) Hearing protection (ear canal inserts) Heavy-duty leather gloves (See Note 1.) Leather work shoes

AN: as needed (optional). AR: As required. SR: Selection required. Refer to notes following NFPA 70E Table 130.7(C)(16).

Reprinted with permission from NFPA-70E 2012, Electrical Safety in the Workplace Copyright 2011, National Fire Protection Association, Quincy, MA 02169. This reprinted material is not the complete and official position of the NFPA on the referenced subject, which is represented only by the standard in its entirety.

This means that the 3-phase available (bolted) short-circuit current at the equipment cannot exceed this value. If a parameter value in the HRC table is exceeded, the HRC method cannot be used.

For instance, assume a piece of equipment is on a 480-V system supplied by a transformer. A simple approach would be to determine the available short-circuit current on the 480-V secondary of the trans-former, assuming infinite available short-current on the primary. Next, compare this result to the table parameter value. The available short-circuit current at the 480-V equipment cannot exceed this value. If the result is equal to or less than the table parameter value, then this one condition of use is satisfied.

b. The clearing time for the class of OCPD protecting the equipment for the maximum short-circuit current available parameter value does not exceed the maximum fault clearing time parameter value.

This means that the clearing time for the class of OCPD providing the protection for the equipment does not exceed the maximum fault clearing time value shown under the parameters for the maximum short-circuit current available value shown under the parameters.

c. The working distance must be equal to or greater than the minimum working distance identified in the parameters provided in Table 130.7(C)(15)(a).

The arc flash boundary, for this equipment type/voltage and these specific parameters, is provided in the table immediately below the parameters. Notes (5) and (6) of Table 130.7(C)(15)(a) explain how the AFB for each equipment type/voltage in Table 130.7(C)(15)(a) was determined.

Select the row with the task that will be performed, and read across the row to determine the HRC and to see whether rubber insulating gloves and insulated and insulating hand tools are required. However, the HRC method cannot be used if the task is not listed in the table.

Once this information has been determined from *NFPA 70E* Table 130.7(C)(15)(a), Table 130.7(C)(16) is used to select the additional required PPE with the designated minimum arc rating for the task. Note that the protective clothing selected for the corresponding HRC must have an arc rating of at least the value listed in the descriptor beside the HRC numeral designation in Table 130.7(C)(16). For example, a task identified as a HRC 2 would require protective clothing with an arc rating of 8 cal/cm² or greater. A task identified as a HRC 4 would require protective clothing with an arc rating of 40 cal/cm² or greater. The HRC category for the equipment is required to be marked on the energized electrical work permit and on the label on the equipment.

Note (4) to Table 130.7(C)(15)(a) may apply in some cases, however; it allows the HRC required to be reduced by one HRC number if the equipment is protected by current-limiting fuses with arcing current in their current-limiting range (one-half cycle fault clearing time or less). To use this provision, it is necessary to determine the arcing current and verify the fuse is in its current-limiting range.

HRC Method for DC Systems

DC systems have significantly different arcing fault dynamics and outcomes than 3-phase systems; in turn, the AC system AFHA methods are not suitable for DC systems. The 2012 *NFPA 70E* introduced the DC systems HRC method. The steps in this method are similar to those just covered for the AC system HRC; however, some differences exist.

NFPA 70E 130.5 permits the requirements of 130.7(C)(15), Table 130.7(C)(15)(b), and Table 130.7(C)(16) to be used for the DC systems HRC method if the conditions of use are met. The use of 130.7(C)(15) and Table 130.7(C)(16) is the same as for the AC system. The difference is that Table 130.7(C)(15)(b) is used for the DC system HRC; this table has some formatting and conditions of use differences from the corresponding table for the AC system HRC. **See Figures 8-5.** This table, which is a partial excerpt from Table 130.7(C)(15)(b), lists the following items:

1. Types of electrical equipment are segmented into two specific DC voltage ranges:
   a. >100 V but <250 V
   b. ≥250 V but ≤600 V
2. Tasks: the same tasks are repeated, with each having a different arcing current range, such as "where arcing current is ≥1 kA and <4 kA"
3. Parameters:
   a. DC voltage not to exceed either 250 V or 600 V (depends on what part of table is being utilized)
   b. Maximum arc duration of 2 seconds
   c. Working distance at 18 in.

4. Under each listed task, the AFB is provided based on parameters for the respective equipment type at the highest value in the arcing current range for that task, such as *Potential arc flash boundary using above parameters at 4 kA: 36 in.*

Methods to determine the DC available short-circuit current and DC arcing current for specific applications are not presented in this publication. *NFPA 70E* Annex D, D.8, provides some information and reference papers that may assist in this area.

AFB and Incident Energy Calculation Methods Discussion

The *NFPA 70E* mandatory text does not provide the AFB or incident energy calculation methods. The most commonly used AFHA calculation methods are located in *NFPA 70E* Annex D, *Incident Energy and Arc Flash Boundary Calculation Methods*, which is provided for informational purposes only (nonmandatory). This annex contains useful information and references for these various AFHA calculation methods. The following sections will cover these calculation methods in Figure 8-1 for determining the AFB and Figure 8-2 for determining incident energy.

The 2002 IEEE 1584, *Guide for Performing Arc-Flash Hazard Calculations*, is the most up-to-date and widely used industry consensus standard on this topic for systems up to 15 kilovolts, especially for work involving enclosed electrical equipment that requires arc-in-the-box calculations, and is based on extensive arc fault testing and analysis.

IEEE 1584 AFHA Calculation Methods and Tools

The 2002 IEEE 1584 is based on extensive testing and analysis to develop equations to calculate the AFB and incident energy. Because it is the most up-to-date method, with the widest range of system characteristics and variables, it is considered by many to be the preferred calculation method for the AFHA. In addition, the arcing current equations in *NFPA 70E* Annex D.7.2 and IEEE 1584 improve the ability to determine the arcing current. Note that these equations utilize units of J/cm² for incident energy and millimeters for distances. To convert units, use the following equivalencies: 1 cal/cm² = 4.184 J/cm² and 1 in. = 25.4 mm.

The 2002 IEEE 1584 includes three broad methods that are referred to in this chapter as basic equations, simplified fuse equations, and simplified circuit breaker equations, respectively. **See Figure 8-6.** In addition, equations for these methods have been incorporated into spreadsheets, programs, and commercial power system analysis packages resulting in simpler and quicker analysis results

Figure 8-5. *NFPA 70E Table 130.7(C)(15)(b), Hazard/Risk Category Classifications and Use of Rubber Insulating Gloves and Insulated and Insulating Hand Tools—Direct Current Equipment. (shown in part)*

Tasks Performed on Energized Equipment	Hazard/Risk Category[a]	Rubber Insulating Gloves[b]	Insulated and Insulating Hand Tools
Storage batteries, direct-current switchboards, and other direct-current supply sources >100 V <250 V			
Parameters: Voltage: 250 V Maximum arc duration and working distance: 2 sec @ 18 in.			
Work on energized electrical conductors and circuit parts, including voltage testing where arcing current is ≥1 kA and <4 kA Potential arc flash boundary using above parameters at 4 kA: 36 in.	1	Y	Y

Y: Yes (required).
Refer to footnotes following NFPA 70E Table 130.7(C)(15)(b).

Reprinted with permission from NFPA-70E 2012, Electrical Safety in the Workplace Copyright 2011, National Fire Protection Association, Quincy, MA 02169. This reprinted material is not the complete and official position of the NFPA on the referenced subject, which is represented only by the standard in its entirety.

Figure 8-6. *IEEE 1584 arc fault boundary and arc flash incident energy method comparison. This gives a high-level overview of the three methods.*

2002 IEEE 1584 Method	Brief Description (Key Differentiating Points)	Comments

Basic Equations	• Based on testing
	• Ability to change many variables
	• For any overcurrent protective device (OPCD)
	• Process:
	  1. Determine the available (bolted) short-circuit current
	  2. Use the IEEE 1584 arcing current equations to determine the arcing current
	  3. Using OCPD time–current characteristics, determine the OCPD clearing time for the arcing current
	  4. Repeat step 3 for 85% of arcing current
	  5. Use IEEE 1584 AFB and I.E. basic equations to calculate AFB and incident energy at specified working distance for both steps 3 and 4; select the worst case of the two results
		If doing manual calculations, this is the more tedious method
		Method suggested for specific circuit breaker if time–current curve is available
		Method suggested for a fuse that is not one where the simplified fuse equations are applicable

Simplified Fuse Equations	• Equations based on testing
	• For specific types of current-limiting fuses or fuses that open in times equal or faster than the fuse types used for IEEE 1584 testing
	• Process:
	  1. Determine the available (bolted) short-circuit current and fuse type/amp rating
	  2. Use the IEEE 1584 simplified fuse equations for the specific fuse amp rating to calculate the AFB and incident energy at 18 in. working distance (Tables and online calculator available—see Comments)
		• Suggested if the fuse type is one that is applicable to this method
		• Easy method: Cooper Bussmann provides simple published tables and an online calculator
		• Fuse time–current curve is not needed
		• No need to determine the arcing current, because the equations are based on actual arcing fault tests

Simplified Circuit Breaker Equations	• Equations based on testing and analysis
	• For generalized circuit breakers based on type, amp rating, trip unit type, and maximum settings
	• Process:
	  1. Determine the available (bolted) short-circuit current
	  2. Verify that the available short-circuit current does not exceed the circuit breaker's interrupting rating
	  3. Verify using the IEEE 1584 simplified circuit breaker equations that the available short-circuit current is high enough to generate an arcing fault current sufficient to cause circuit breaker instantaneous tripping or short-time delay tripping (if the circuit breaker has no instantaneous trip)
	  4. From the IEEE 1584 simple circuit breaker method table, select and use the appropriate AFB and incident energy (at 18 in. working distance) simplified circuit breaker equations based on circuit breaker ampere rating, type, and trip unit type
		• Easier than the basic equation method
		• Generalized for all circuit breaker manufacturers—not manufacturer specific
		• No need to determine the arcing current
		• Requires qualifying calculations: available short-circuit current must be within the range of use for circuit breaker settings and interrupting rating
		• Cooper Bussmann provides simple published tables and an online calculator based on specific circuit breaker settings

NFPA 70E Annex D, Section D.7.1 provides the system limits for using IEEE 1584 calculation methods, and these are highlighted in Figures 8-1 and 8-2.

IEEE1584 Basic Equations Method

The basic equation method includes the following steps:

1. **Calculate the available (bolted) short-circuit current.**

   This calculation should be as accurate as possible. Also, calculations should be run for all switching and other system scenarios that may result in different available short-circuit currents. The following steps should be repeated for each scenario to determine the worst-case AFHA scenario.

2. **Calculate the arcing current.**

   Using the available short-circuit current, system voltage, conductor gap, and constant for open arc or arc-in-a-box, calculate the arcing current using the equations in Annex D, Section D.7.2. **See Figure 8-7**. *NFPA 70E* Annex D, Section D.7.2 provides the equations for determining the arcing current, which was covered in the chapter, "Incident Energy Varies by Fault Current Magnitude and Duration."

3. **Determine the OCPD clearing time at the arcing current value.**

   The OCPD time-current curves are used to determine the clearing time for the arcing current.

4. **Repeat step 3 for 85% of the arcing current.**

   The IEEE 1584 testing demonstrated variance in the actual arcing current based on a test setup with a fixed calibrated available short-circuit current. For situations where the arcing current happens to be in the steep (highly inverse) portion of an OCPD time-current curve, a small variance in arcing current can result in a large change in the incident energy and AFB. Therefore, as a precaution for the basic equation method, IEEE 1584 requires determining the AFB and incident energy at both the calculated arcing current and 85% of the calculated arcing current.

5. **Calculate the AFB using the equation in Annex D, Section D.7.5.**

   Do so both for the calculated arcing current and for 85% of the arcing current (steps 3 and 4).

6. **Calculate the incident energy using equation in Annex D, Section D.7.4.**

   Do so both for the calculated arcing current and for 85% of the arcing current (steps 3 and 4).

7. **Consider the overcurrent protective device condition of maintenance per 130.5.**

   If the OCPD providing the protection for arc flash hazard analysis has not been maintained properly or maintained inadequately, then the AFHA results of steps 5 and 6 cannot be relied on. Other means should be investigated. *NFPA 70E* Chapter 2, including Section 205.4,

Figure 8-7. NFPA 70E *Table D.7.2, Factors for Equipment and Voltage Classes. This shows the typical conductor gap for an equipment type/voltage rating and the distance exponent factor X needed for the IEEE 1584 arc current calculation equation. This subject was discussed in the chapter, "Incident Energy Varies by Fault Current Magnitude and Duration."*

System Voltage (kV)	Type of Equipment	Typical Conductor Gap (mm)	Distance Exponent Factor X
0.208–1	Open air	10–40	2.000
	Switchgear	32	1.473
	MCCs and panels	25	1.641
	Cables	13	2.000
>1–5	Open air	102	2.000
	Switchgear	13–102	0.973
	Cables	13	2.000
>5–15	Open air	13–153	2.000
	Switchgear	153	0.973
	Cables	13	2.000

210.5, and Article 225 and the chapter, "OCPD Work Practices and Maintenance Considerations" in this publication provides insight into requirements and work practices related to equipment maintenance.

IEEE 1584 Basic Equations: Arc Flash Boundary

The equation to calculate the AFB using the basic equations per IEEE 1584 is found in *NFPA 70E* Annex D.7.5 and shown here. Remember that the arcing current must be calculated prior to this step.

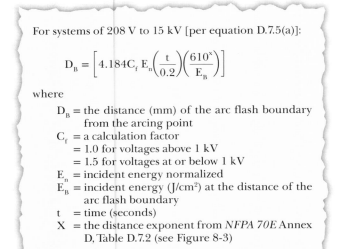

For systems of 208 V to 15 kV [per equation D.7.5(a)]:

$$D_B = \left[4.184 C_f\, E_n \left(\frac{t}{0.2} \right) \left(\frac{610^x}{E_B} \right) \right]$$

where

D_B = the distance (mm) of the arc flash boundary from the arcing point
C_f = a calculation factor
   = 1.0 for voltages above 1 kV
   = 1.5 for voltages at or below 1 kV
E_n = incident energy normalized
E_B = incident energy (J/cm²) at the distance of the arc flash boundary
t = time (seconds)
X = the distance exponent from *NFPA 70E* Annex D, Table D.7.2 (see Figure 8-3)

As indicated by the equations, to calculate the AFB, the values of "E_n" and "t" must be determined. To determine "E_n" and "t," the arcing current must first be found using the equations previously discussed in the chapter, "Incident Energy Varies by Fault Current Magnitude and Duration." The AFB is the distance from the arc source where a person can receive a second-degree burn. The threshold for a second-degree burn is where the incident energy is 1.2 cal/cm² or 5.0 J/cm² (1.2 × 4.184); therefore, E_B is 5.0 J/cm². The value for "t" can be found by determining the clearing time of the OCPD for the calculated arcing current. This step requires the OCPD time–current curve.

When determining the AFB, use both the I_a and 0.85 I_a arc currents to find the corresponding clearing times for the OCPD. The 0.85 I_a arc current accounts for variations in the arcing current that the IEEE 1584 testing program documented. This variation can affect the time for the OCPD to open. In some cases, the lower value of I_a may result in longer clearing times and a greater AFB.

To find E_n, the following equations can be used [per equation D.7.3(a)]:

$$\lg E_n = k_1 + k_2 + 1.081(\lg I_a) + 0.0011(G)$$

where

E_n = incident energy (J/cm²) normalized for time and distance
k_1 = −0.792 for open-air arcs; −0.555 for arcs-in-a-box
k_2 = 0 for ungrounded and high-resistance grounded systems; −0.113 for grounded systems
I_a = arcing current (kA; covered in the chapter, "Incident Energy Varies by Fault Current Magnitude and Duration")
G = the conductor gap (mm; see *NFPA 70E* Annex D, Table D.7.2)

Then, calculate E_n as follows per equation D.7.3(b):

$$E_n = 10^{\lg E_n}$$

For systems of more than 15 kilovolts, the IEEE 1584 equation cannot be used, as the system limits are valid only up to 15 kilovolts. However, the *NFPA 70E* Annex D.7.5(b) and IEEE 1584 standard reference the Ralph Lee Equation shown in *NFPA 70E* Annex D.6. For voltages of more than 15 kilovolts, the short-circuit current is considered equal to the arcing current. The equation shown here from *NFPA 70E* Annex D.7.5(b) has been modified from that shown in Annex D.6 to have distance in millimeters and incident energy in J/cm²:

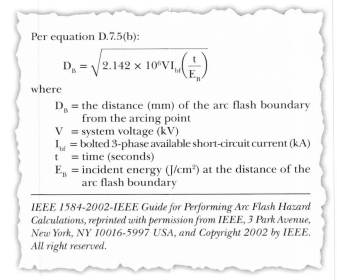

Per equation D.7.5(b):

$$D_B = \sqrt{2.142 \times 10^6 V I_{bf} \left(\frac{t}{E_B} \right)}$$

where

D_B = the distance (mm) of the arc flash boundary from the arcing point
V = system voltage (kV)
I_{bf} = bolted 3-phase available short-circuit current (kA)
t = time (seconds)
E_B = incident energy (J/cm²) at the distance of the arc flash boundary

Test 4 Example for IEEE 1584 Basic Equations: AFB Calculations

Per *NFPA 70E* Annex D Section D.7.5 and IEEE 1584, what would the calculated AFB be for Test 4 of staged tests that was covered in the chapter, "Electrical Hazard Awareness"? Following are the system characteristics. **See Figures 8-8 and 8-9.**

Bolted short-circuit current = 22.6 kilovolt-amperes at 480 volts

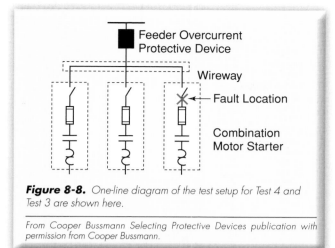

Figure 8-8. *One-line diagram of the test setup for Test 4 and Test 3 are shown here.*

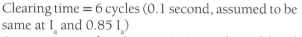

From Cooper Bussmann Selecting Protective Devices publication with permission from Cooper Bussmann.

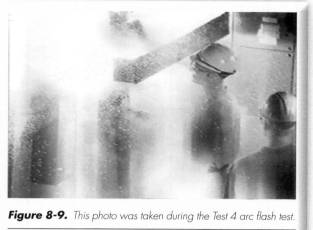

Figure 8-9. *This photo was taken during the Test 4 arc flash test.*

From the Cooper Bussmann Selecting Protective Devices publication with permission from Cooper Bussmann.

Clearing time = 6 cycles (0.1 second, assumed to be same at I_a and 0.85 I_a)

Assume arc-in-a-box (assume MCCs and panels) and ungrounded system.

Step 1. Find I_a (arcing current) (The chapter, "Incident Energy Varies by Fault Current Magnitude and Duration" covers calculation of the arcing current.)

Equation:

$$\lg I_a = K + 0.662 \lg I_{bf} + 0.0966\, V + 0.000526$$
$$+ 0.5588\, V\, (\lg I_{bf}) - 0.00304\, G\, (\lg I_{bf})$$

where
  $\lg = \log 10$
  I_a = arcing current (kA)
  K = −0.153 for open-air arcs; −0.097 for arcs-in-a-box
  I_{bf} = bolted 3-phase available short-circuit current (symmetrical rms) (kA)
  V = system voltage (kV)
  G = conductor gap (mm; see Table D.7.2)

Calculation:

$$\lg I_a = -0.097 + 0.662(\lg 22.6) + 0.0966\,(0.48)$$
$$+ 0.000526\,(25) + 0.5588\,(0.48)(\lg 22.6)$$
$$- 0.00304\,(25)\,(\lg 22.6)$$

$$\lg I_a = 1.119$$

Then:

$$I_a = 10^{\lg Ia}$$
$$I_a = 10^{1.119}$$
$$I_a = 13.15 \text{ kA}$$

Step 2. Find E_n (normalized incident energy).

Equation:

$$\lg E_n = k_1 + k_2 + 1.081(\lg I_a) + 0.0011(G)$$

where
  E_n = incident energy (J/cm²) normalized for time and distance
  k_1 = −0.792 for open-air arcs; −0.555 for arcs-in-a-box
  k_2 = 0 for ungrounded and high-resistance grounded systems; −0.113 for grounded systems
  I_a = arcing current (kA)
  G = the conductor gap (mm; see Table D.7.2)

Calculation:

$$\lg E_n = -0.555 + 0 + 1.081\,(\lg 13.15) + 0.0011\,(25)$$
$$\lg E_n = 0.682$$

Then:

$$E_n = 10^{\lg En}$$
$$E_n = 10^{0.682}$$
$$E_n = 4.81$$

Step 3. Calculate D_B (AFB).

Equation:

Per equation D.7.5(a):

$$D_B = \left[4.184 C_f E_n \left(\frac{t}{0.2} \right) \left(\frac{610^x}{E_B} \right) \right]^{\frac{1}{x}}$$

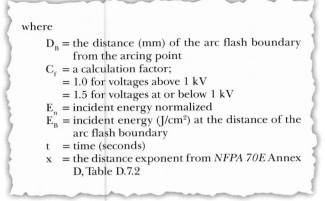

where

D_B = the distance (mm) of the arc flash boundary from the arcing point

C_f = a calculation factor;
= 1.0 for voltages above 1 kV
= 1.5 for voltages at or below 1 kV

E_n = incident energy normalized

E_B = incident energy (J/cm²) at the distance of the arc flash boundary

t = time (seconds)

x = the distance exponent from *NFPA 70E* Annex D, Table D.7.2

Note: $E_B = 1.2$ cal/cm² × $4.184 = 5.0$ J/cm²

Calculation:

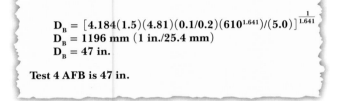

$$D_B = [4.184(1.5)(4.81)(0.1/0.2)(610^{1.641})/(5.0)]^{\frac{1}{1.641}}$$
$$D_B = 1196 \text{ mm } (1 \text{ in.}/25.4 \text{ mm})$$
$$D_B = 47 \text{ in.}$$

Test 4 AFB is 47 in.

IEEE 1584 Basic Equations: Incident Energy

Discussions of the IEEE basic equations for calculating the incident energy follow. The IEEE 1584 basic equations to calculate the incident energy are found in *NFPA 70E* Annex D.7.3 and Annex D.7.4 and shown here. Remember that the arcing current must be calculated prior to this step.

For system voltages ranging from 208 V to 15 kV [per equation D.7.3(c)]:

$$E = 4.184 C_f E_n \left(\frac{t}{0.2}\right)\left(\frac{610^x}{D^x}\right)$$

where

E = incident energy (J/cm²)

C_f = a calculation factor
= 1.0 for voltages above 1 kV
= 1.5 for voltages at or below 1 kV

E_n = incident energy normalized

t = arcing time (seconds)

D = distance (mm) from the arc to the person (working distance)

X = the distance exponent from *NFPA 70E* Annex D, Table D.7.2

To calculate E_n and t (and the value of I_a needed to determine E_n and t), use the equations in the previous section for calculating the AFB per the IEEE 1584 Basic Equation Method.

For systems of more than 15 kilovolts, the lower-voltage IEEE 1584 equation cannot be used, as the system limits are valid only up to 15 kilovolts. However, the *NFPA 70E* Annex D.7.4 and IEEE 1584 Standard reference the Ralph Lee Equation discussed later and shown in *NFPA 70E* Annex D.6. For voltages of more than 15 kilovolts, the short-circuit current is considered equal to the arcing current. The following equation has been modified from the equation shown in Annex D.6 to have distance in millimeters and incident energy in J/cm²:

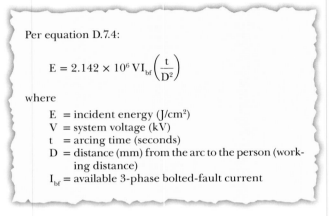

Per equation D.7.4:

$$E = 2.142 \times 10^6 \, VI_{bf} \left(\frac{t}{D^2}\right)$$

where

E = incident energy (J/cm²)

V = system voltage (kV)

t = arcing time (seconds)

D = distance (mm) from the arc to the person (working distance)

I_{bf} = available 3-phase bolted-fault current

Test 4 Example for IEEE 1584 Basic Equations: Incident Energy Calculation

Per *NFPA 70E* Annex D Section D.7.3 and IEEE 1584, what would the calculated incident energy be for Test 4?

Bolted short-circuit current = 22.6 kiloamperes at 480 volts

Clearing time = 6 cycles (0.1 second; assumed to be same at I_a and 0.85 I_a)

Assume arc-in-a-box (assume MCCs and panels) and an ungrounded system.

Step 1. Find I_a (arcing current).

$$I_a = 13.15 \text{ kA}$$

(See the prior AFB calculation example for Test 4, which provides details on this calculation.)

Step 2. Find E_n (normalized incident energy).

$$E_n = 4.81$$

(See the prior AFB calculation example for Test 4, which provides detail on this calculation.)

Step 3. Find E (incident energy).
Equation:

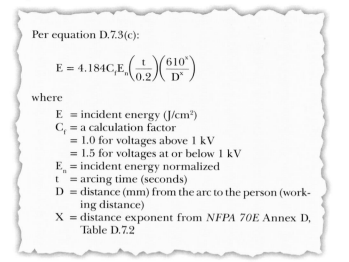

Per equation D.7.3(c):

$$E = 4.184 C_f E_n \left(\frac{t}{0.2}\right)\left(\frac{610^x}{D^x}\right)$$

where

E = incident energy (J/cm²)
C$_f$ = a calculation factor
    = 1.0 for voltages above 1 kV
    = 1.5 for voltages at or below 1 kV
E$_n$ = incident energy normalized
t = arcing time (seconds)
D = distance (mm) from the arc to the person (working distance)
X = distance exponent from *NFPA 70E* Annex D, Table D.7.2

Calculation:

$$E = 4.184(1.5)(4.81)(0.1/0.2)(610^{1.641})/(455^{1.641})$$
$$E = 24.4 \text{ J/cm}^2$$

Convert:

E = 24.4 J/cm² ÷ 4.184 J/cal
E = 5.84 cal/cm² at 18 in. (AFB = 47)

Test 4 incident energy is 5.84 cal/cm² at 18 in.

(The prior AFB calculation example for Test 4 provides details on the AFB calculation.)

Equipment must be marked with the results of the AFHA. One option is to mark the incident energy at the working distance and the AFB per *NFPA 70E* 130.5(C).

Select PPE using *NFPA 70E* 130.7(C)(1) through (C)(14) and nonmandatory Annex H, *Guidance on Selection of Protective Clothing and Other Personal Protective Equipment*, Tables H.3(a) and H.3(b).

IEEE 1584 Simplified Equations

To simplify the *NFPA 70E* Annex D.7 and IEEE 1584 calculation method, fuse and circuit "simplified equations" are available. These equations allow the incident energy to be calculated directly from the available 3-phase short-circuit current on a low-voltage system if the type and ampere rating of the OCPD are known. These simplified methods do not require the availability of time–current curves for the devices.

IEEE 1584 Simplified Equations: Fuses

The method for using IEEE 1584 simplified equations for fuses is revisited here. **See Figure 8-10**. The simplified fuse equations methods shown in *NFPA 70E* Annex D.7.6 are from IEEE 1584. These equations are not shown in this chapter. The simplified equations are applicable for a range of available bolted short-circuit currents. If the available short-circuit current is not within the applicable range for a specific fuse ampere rating, the simplified equations cannot be used. The IEEE 1584 basic equation method should be used.

IEEE 1584 Simplified Equations: Circuit Breakers

The method for using IEEE 1584 simplified equations for circuit breakers is revisited here. **See Figure 8-11**. The simplified circuit breaker equation method shown in *NFPA 70E* Annex D.7.7 is from IEEE 1584. If the short-circuit current falls outside the range of use for a circuit breaker setting or interrupting rating, the simplified equations cannot be used. The IEEE 1584 basic equations should be used. When equations are used for low-voltage power circuit breakers with short-time delay (without an instantaneous trip unit),

Figure 8-10. *This table gives a high-level overview of the IEEE 1584 simplified fuse equations method.*

2002 IEEE 1584 Method	Brief Description (Key Differentiating Points)	Comments

Simplified Fuse Equations	• Equations based on testing
	• For specific types of current-limiting fuses or fuses that open in times equal or faster than the fuse types used for IEEE 1584 testing
	• Process:
	  1. Determine the available (bolted) short-circuit current and fuse type/amp rating
	  2. Use the IEEE 1584 simplified fuse equations for the specific fuse amp rating to calculate the AFB and incident energy at 18 in. working distance (Tables and online calculator available—see Comments)
		• Suggested if the fuse type is one that is applicable to this method
		• Easy method: Cooper Bussmann provides simple published tables and an online calculator
		• Fuse time–current curve is not needed
		• No need to determine the arcing current, because the equations are based on actual arcing fault tests

Figure 8-11. *This table gives a high-level overview of the IEEE 1584 simplified circuit breaker equations method.*

2002 IEEE 1584 Method	Brief Description (Key Differentiating Points)	Comments
Simplified Circuit Breaker Equations	• Equations based on testing and analysis • For generalized circuit breakers based on type, amp rating, trip unit type, and maximum settings • Process: 　1. Determine the available (bolted) short-circuit current 　2. Verify that the available short-circuit current does not exceed the circuit breaker's interrupting rating 　3. Verify using the IEEE 1584 simplified circuit breaker equations that the available short-circuit current is high enough to generate an arcing fault current sufficient to cause circuit breaker instantaneous tripping or short-time delay tripping (if the circuit breaker has no instantaneous trip) 　4. From the IEEE 1584 simple circuit breaker method table, select and use the appropriate AFB and incident energy (at 18 in. working distance) simplified circuit breaker equations based on circuit breaker ampere rating, type, and trip unit type	• Easier than the basic equation method • Generalized for all circuit breaker manufacturers—not manufacturer specific • No need to determine the arcing current • Requires qualifying calculations: available short-circuit current must be within the range of use for circuit breaker settings and interrupting rating • Cooper Bussmann provides simple published tables and an online calculator based on specific circuit breaker settings

the maximum setting (30 cycles) is assumed. Because there are several equations for the various circuit breaker types and ampere ratings, as well as the range of short-circuit currents, the equations are not reprinted here.

IEEE 1584 Resources and Tools

A variety of resources and tools have been developed that utilize the IEEE 1584 information for arc flash hazard analysis, including the materials from IEEE and from vendors.

Resources and Tools with IEEE 1584 Purchase

When the IEEE 1584 guide is purchased, several useful Excel spreadsheets are provided with the actual document:

- Actual test data
- Bolted short-circuit current calculator
- Arc flash hazard calculator
- Current-limiting fuse test data
- Simplified circuit breaker method

IEEE 1584 Resources and Tools from Manufacturers

Other resources and tools are available from manufacturers, such as Cooper Bussmann. Cooper Bussmann has developed an online Arc Flash Incident Energy Calculator that makes it a simple matter to find the incident energy for low-voltage systems based on the available 3-phase short-circuit current and the type of OCPD. The calculated incident energy is valid for Cooper Bussmann Low-Peak Class RK-1 and Class L fuses and specific types of circuit breakers, based on the IEEE 1584 circuit breaker and fuse simplified equation methods. This information is available using

the interactive online calculator found at http://www.cooperbussmann.com. For convenience, it is also available in an easy-to-use tabular format in the *Cooper Bussmann Selecting Protective Devices (SPD)* publication.

The notes for the Arc Flash Incident Energy Calculator must be read before using either the online version or the tabular version of the calculator. **See Figures 8-12** and **8-13**.

Example: IEEE 1584 Resources and Tools from Manufacturers

The following examples demonstrate the use of IEEE 1584 fuse and circuit breaker simplified method resources and tools from manufacturers. Two examples are discussed here based on use of the *Cooper Bussmann Selecting Protective Devices* publication. **See Figures 8-14** and **8-15**.

Using the tables from *Cooper Bussmann Selecting Protective Devices* publication based on the IEEE 1584 simplified equation methods, the objective is to determine the incident energy for each of the following 480-volt circuits with an available short-circuit current of 24,000 amperes at the main lugs of a panelboard when protected by each of the following:

1. Cooper Bussmann Low-Peak KRP-C-800SP current-limiting Class L fuses.

2. 800-ampere low-voltage power circuit breaker with STD setting of 30 cycles.

To use Figure 8-15, find the ampere rating and type of OCPD in the header of the table. Select the available short-circuit current (bolted fault current) from the left column. Select the corresponding incident

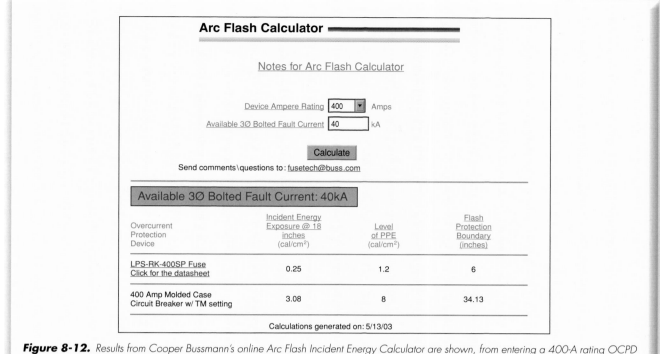

Figure 8-12. *Results from Cooper Bussmann's online Arc Flash Incident Energy Calculator are shown, from entering a 400-A rating OCPD and a 40,000-A available short-circuit current.*

Courtesy of Cooper Bussmann, Inc.

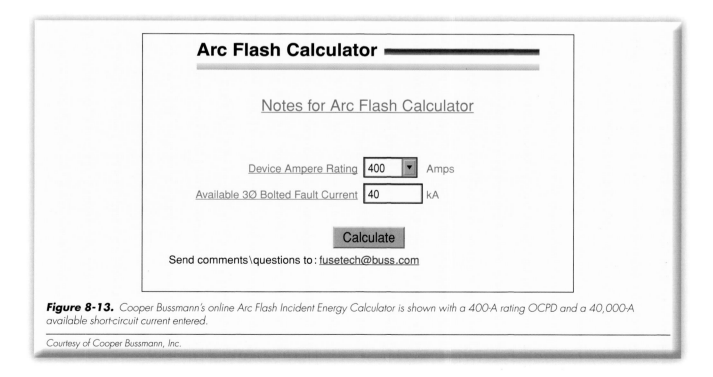

Figure 8-13. *Cooper Bussmann's online Arc Flash Incident Energy Calculator is shown with a 400-A rating OCPD and a 40,000-A available short-circuit current entered.*

Courtesy of Cooper Bussmann, Inc.

energy and AFB for the selected OCPD type and the amp rating for the available short-circuit current. For the KRP-C-800SP Fuse, the calculated incident energy is 1.46 cal/cm² with an AFB of 21 inches.

Note that if the IEEE 1584 basic equations were used, the calculated incident energy (assuming one-half cycle clearing time and a short-circuit current of 22.6 kiloamperes) would be 0.47 cal/cm² with an AFB of 10 inches for Test 3. Therefore, when the actual test data and equations for only fuses are used, the result can be higher (or in some cases lower) than the calculated value per the IEEE 1584 basic equation method. For this reason, it is recommended to use the IEEE 1584 simplified fuse equations (or Cooper Bussmann Arc Flash Calculator Program or Arc Flash Incident Energy Calculator Tables) when possible to ensure the most accurate result.

Select PPE using *NFPA 70E* Sections 130.7(C)(1) through (C)(14) and nonmandatory Annex H, *Guidance on Selection of Protective Clothing and Other Personal Protective Equipment*, Tables H.3(a) and H.3(b).

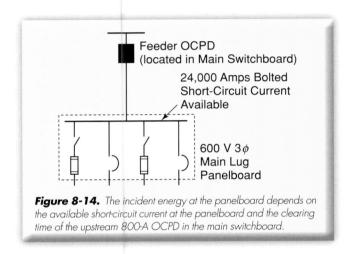

Figure 8-14. *The incident energy at the panelboard depends on the available short-circuit current at the panelboard and the clearing time of the upstream 800-A OCPD in the main switchboard.*

For the 800-ampere low-voltage power circuit breaker with STD (30 cycles),the calculated incident energy is 46.53 cal/cm² with an AFB of greater than 120 inches. The IEEE 1584 simplified circuit breaker equations were used for this calculation. These equations assume the maximum setting for a short-time delay of 30 cycles. While PPE manufacturers make arc flash suits with arc ratings greater than 40 cal/cm², *NFPA 70E* 130.7(A) Informational No. 3 recommends considering deenergizing the equipment before working on equipment within the limited approach boundary of the exposed electrical conductors or circuit parts when the incident energy exceeds 40 cal/cm² at the working distance.

If the IEEE 1584 basic equations were used with the same available short-circuit current and clearing time of 30 cycles, the incident energy would have been calculated at 30.86 cal/cm² with an AFB of 129.59 inches. In addition, note that if the IEEE 1584 basic equations were used, the calculated incident energy, assuming a reduced short-time delay setting of 6 cycles and an available short-circuit current of 22.6 kiloamperes, would be 5.8 cal/cm² with an AFB of 47 in (see the IEEE 1584 Basic Equation Example for Test 4). The decrease in the short-time delay from 30 to 6 cycles dramatically decreases the incident energy and the AFB.

Figure 8-15. *The Cooper Bussmann Arc Flash Incident Energy Calculator is shown.**

Fuses: Bussmann® Low-Peak® KRP-C_SP (601–1200 Amp), Circuit Breakers: Low-Voltage Power Circuit Breakers (with STD)
Incident energy (I.E.) values are expressed in cal/cm². Flash protection boundary (FPB) values are expressed in inches.

Bolted Fault Current (kA)	601–800 Amp				801–1200 Amp			
	Fuse		LVPCB		Fuse		LVPCB	
	I.E.	FPB	I.E.	FPB	I.E.	FPB	I.E.	FPB
1	>100	>120	>100	>120	>100	>120	>100	>120
2	>100	>120	>100	>120	>100	>120	>100	>120
3	>100	>120	>100	>120	>100	>120	>100	>120
—	—	—	—	—	—	—	—	—
—	—	—	—	—	—	—	—	—
—	—	—	—	—	—	—	—	—
20	1.70	23	38.87	>120	10.37	78	>100	>120
22	1.58	22	42.70	>120	9.98	76	>100	>120
24	1.46	21	46.53	>120	8.88	70	46.53	>120
26	1.34	19	50.35	>120	7.52	63	50.35	>120
28	1.22	18	54.18	>120	6.28	55	54.18	>120
30	1.10	17	58.01	>120	5.16	48	58.01	>120

*Full table is found in *Cooper Selecting Protective Devices*, Electrical Safety section.

Select PPE using *NFPA 70E* Sections 130.7(C)(1) through (C)(14). Nonmandatory Annex H, *Guidance on Selection of Protective Clothing and Other Protective Equipment*, Tables H.3(a) and H.3(b), provides additional insight.

IEEE 1584: Future Edition

At the time of this text's publication, the most recent edition of IEEE 1584 is dated 2002. It represents a major contribution to furthering electrical safety in that it provided the first industry consensus standard on calculating the incident energy and AFB based on extensive arc fault testing under specific test conditions. Although the 2002 IEEE 1584 is a valuable tool, it does not provide calculation methods for all possible situations confronted in the workplace. To enhance and expand the calculation methods for arcing faults, NFPA and IEEE created a joint effort, *IEEE/NFPA Collaboration on Arc Flash Research*. IEEE and NFPA are working together on a research and testing initiative to enhance understanding and further mitigate the effects of hazardous electrical arc phenomena on workers, with the goal of protecting people and saving lives. Several manufacturers have donated large sums of money or equivalent services to assist this effort. The test results and information derived from these tests will be used to revise future editions of IEEE 1584 and *NFPA 70E*.

Other AFHA Calculation Methods

A recent addition in *NFPA 70E* Annex D.7.8 focuses on DC incident energy calculations. In addition, other AC AFHA methods predate the release of the 2002 IEEE 1584; these other methods helped in the progression of quantifying arc flash hazards.

DC Incident Energy Analysis Method

NFPA 70E Annex D.8 provides information on a direct-current incident energy analysis method. This method is provided in a paper authored by Dan Doan and was presented at the 2010 IEEE Electrical Safety Workshop. Annex D.8.1.2 provides three reference sources on DC arcing current and energy.

Ralph Lee Method: AFB and Incident Energy
R. Lee AFB Method

The Ralph Lee AFB calculation method, detailed in *NFPA 70E* Annex D, Section D.2, is based on his seminal paper, "The Other Electrical Hazard: Electrical Arc Blast Burns," which appeared in *IEEE Transactions on Industrial Applications*, vol. 1A-18, no. 3, p. 246, May/June 1982. This important paper represented

one of the first efforts focusing on the quantification of arcing fault hazards. The method outlined in Lee's paper is not applicable for most commercial and industrial electrical work, as they typically involve enclosed equipment. The incident energy from open-air arcs is less than the energy from arcs-in-a-box. Owing to this difference, it would not be suitable to determine the AFB for an arc-in-a-box (such as arcs in electrical equipment as seen in Test 4, Test 3, and Test 1), because this equation would yield results lower than the expected values. The R. Lee AFB method predates IEEE 1584.

R. Lee Incident Energy Method

Ralph Lee's incident energy calculation method, detailed in *NFPA 70E* Annex D, Section D.6, is based on a modification of the AFB calculation found in Annex D, Section D.2. This equation has been modified to include incident energy as a variable in the equation in lieu of an assumed value of 1.2 cal/cm^2. This method can be used to determine the incident energy, but applies only to open-air arcs. The energy from open-air arcs is less than the energy from arcs-in-a-box. Owing to this difference, it would not be suitable to determine the incident energy for an arc-in-a-box (such as arcs in electrical equipment as seen in Test 4, Test 3, and Test 1), because this equation would yield results lower than the expected values.

Doughty/Neal/Floyd Method: Incident Energy

The Doughty/Neal/Floyd method, detailed in *NFPA 70E* Annex D, Section D.5, including D.5.1 and D.5.2, is based on the paper noted in Annex D, Section D.5.3, by R. L. Doughty, T. E. Neal, and H. L. Floyd II. This paper, which is entitled "Predicting Incident Energy to Better Manage the Electric Arc Hazard on 600V Power Distribution Systems," is part of the *Record of the Conference Papers* for the IEEE IAS 45th Annual Petroleum and Chemical Industry Conference held on September 28–30, 1998. The range of short-circuit current values, voltage, and working distances is limited compared to the IEEE 1584 method. This paper and method predate IEEE 1584.

Other Considerations

This section covers some other considerations for arc flash hazard analysis.

Updating AFHA

NFPA 70E 130.5, second paragraph, requires the arc flash hazard analysis to be updated when a major modification or renovation occurs. In addition, the arc flash hazard analysis must be reviewed periodically, not to exceed five years, because changes in the electrical distribution systems could affect the results of the arc flash hazard analysis.

Arcing Currents in the Long Time Characteristic of Overcurrent Protective Devices

In many short-circuit current studies, the focus is only on the highest short-circuit currents to evaluate OCPD interrupting ratings and equipment short-circuit current ratings. However, when performing an arc flash hazard analysis, it is necessary to consider both higher and lower available short-circuit currents that might occur due to various electrical distribution system operating scenarios. It is important to determine whether the highest or lowest available short-circuit current scenario will result in the worst-case calculated arc flash energy. Review *NFPA 70E* 130.5, Informational Note No. 2. The lower available short-circuit currents could result in an arcing current value that is less than a circuit breaker's instantaneous trip setting or the current-limiting range of a fuse, resulting in a clearing time of several seconds. This could result in higher incident energy than anticipated in the high fault current scenario.

On lower-ampere-rated circuits, this is not typically a problem. On higher-ampere-rated circuits (more than 1,200 amperes), however, it can become more of an issue. For some higher-ampere-rated OCPDs, the calculated incident energy and AFB can be extremely large for some lower available short-circuit currents. The chapter, "Calculation of Short-Circuit Currents" discusses different switching and system scenarios to consider that may impact the available short-circuit currents and, therefore, the AFHA.

Although some lower level arcing faults may not be self-sustaining, there is insufficient research or published guidance as to which arcing fault conditions will self-sustain and which will not.

Overcurrent Protective Device Condition of Maintenance

The reliability of overcurrent protection devices can directly impact arc-flash hazards. Poorly maintained OCPDs may take longer to clear, or may not clear at all, resulting in higher arc flash incident energies. The third paragraph of 130.5 requires the arc flash hazard analysis to "take into consideration the design of the overcurrent protective device and its opening time, including the condition of maintenance." 130.5 Informational Notes Nos. 1 and 4 provide insight into

this requirement. If the condition of maintenance of the OCPD intended to be used for an arc flash hazard analysis is not known to be acceptable, it is advisable to not use that OCPD for an analysis. Instead, other means should be used to determine the AFHA, such as reliance on the characteristics of an OCPD farther upstream toward the source that is known to be in an acceptable maintenance condition. *NFPA 70E* Chapter 2 provides requirements for OCPD maintenance. For example, 205.4 requires OCPDs to be maintained per the manufacturers' instructions or industry consensus standards. Other *NFPA 70E* requirements are found in 210.5, 225.1, 225.2, and 225.3; OSHA requirements include 1910.334(b)(2). See the chapter, "OCPD Work Practices and Maintenance Considerations" for more information.

Documentation

Two documentation requirements pertaining to an arc flash hazard analysis are specified. First, the last sentence of 130.5 requires that "The method of calculating and data to support the information for the label shall be documented." Second, 205.4 requires that OCPD "maintenance, tests, and inspections shall be documented." This documentation helps verify compliance with the 130.5 OCPD condition of maintenance requirement.

Labeling

This section discusses several *NEC* and *NFPA 70E* labeling requirements.

NEC 110.24 Marking Maximum Available Fault Current on Service Equipment

Marking the maximum available fault (bolted) current (synonymous with the maximum available short-circuit current) at the service is intended to ensure proper application of OCPD interrupting ratings and short-circuit current ratings (SCCR) of equipment. **See Figure 8-16.** When determining the maximum available short-circuit current to comply with *NEC* 110.24, methods are permitted that may result in values significantly higher than the actual values. This outcome is acceptable as long as the equipment short-circuit current rating and interrupting ratings for the OCPDs are equal to or greater than the determined maximum available fault current.

However, *NEC* 110.24 is not intended for incident energy analysis calculations. Arc flash hazard incident energy calculations should use available short-circuit current calculations that are as accurate as possible.

Figure 8-16. *The label shown here is in compliance with NEC 110.24, Marking Maximum Available Fault Current on Service Equipment.*

Figure 8-17. *The arc flash warning label shown is in compliance with NEC 110.16.*

The 130.5 Informational Note No. 2 provides the rationale for this approach. If the available short-circuit current is a rough estimate that is significantly higher than actual, it may result in an incident energy value that is lower or higher than what actually could occur.

In contrast, the maximum available short-circuit current marked on a *NEC* 110.24 label could be used for determining whether an actual situation is within the HRC method's conditions of use parameter.

NEC 110.16 Arc Flash Hazard Warning Label

The *NEC* requires an arc flash hazard warning label. **See Figure 8-17.** It is not required to provide the arc flash hazard detailed information. Any electrical equipment as described in *NEC* Section 110.16 that is likely to be serviced or worked on while not in an electrically safe work condition must be field labeled if this equipment has been installed according to the 2002 *NEC* or more recent editions. The intent of this requirement is for the arc flash hazard to be field marked, although many electrical equipment manufacturers add this basic arc flash hazard warning label on equipment prior to shipping.

130.5(C) Labeling Equipment with AFHA Information

If workers are involved in an environment where they are exposed to an arc flash hazard, an arc flash hazard analysis is required to be conducted in accordance with *NFPA 70E* 130.5; in addition, 130.5 requires marking the results of the arc flash hazard analysis on the equipment. This process correlates with OSHA regulations to document hazards. 29 Code of Federal Regulations (CFR) 1910.132(d), *Hazard Assessment and Equipment Selection*, requires that if a hazard is present, the employer must document it through written certification. This OSHA regulation requires

employers to assess for hazards that might require PPE, to select and have affected employees use the appropriate PPE for the hazards, and to communicate the PPE requirement to each affected employee.

The *NFPA 70E* Section 130.5(C) labeling requirement, along with the elements required in the *Energized Electrical Work Permit* in *NFPA 70E* Section 130.2(B), are a means to comply with a major portion of the OSHA written certification requirement for the hazard assessment. A benefit of the AFHA labels per 130.5(C) is that a qualified worker can readily know the hazard, which helps foster safe electrical work practices. **See Figures 8-18** and **8-19.**

130.5(C) requires marking three items on the AFHA label:

1. At least one of the following:
   a. Incident energy at the working distance
   b. Minimum PPE arc rating
   c. Required level of PPE
   d. Highest HRC for the equipment
2. Nominal system voltage.
3. Arc flash boundary.

There are different philosophies on how much information to include on a label. One perspective is to include only what is required on the label. An alternative perspective is to include more information than is required, such as the date the analysis was performed.

Premise-Wide AFHA Labeling

Many owners conduct arc flash hazard analysis and shock hazard analysis for their entire premises and label the electrical equipment. The labels used in such a case are similar to those shown. Premise-wide implementation of labeling can facilitate a more efficient operation. The typical process is to update or create a complete and accurate

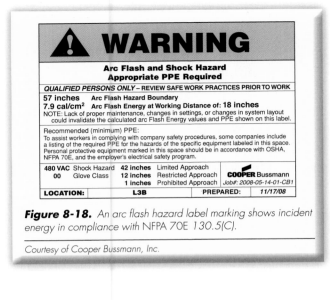

Figure 8-18. An arc flash hazard label marking shows incident energy in compliance with NFPA 70E 130.5(C).

Courtesy of Cooper Bussmann, Inc.

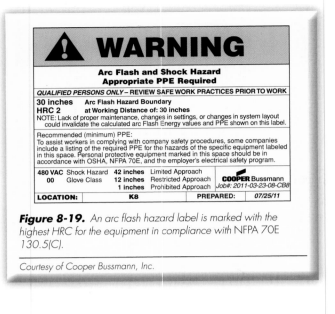

Figure 8-19. An arc flash hazard label is marked with the highest HRC for the equipment in compliance with NFPA 70E 130.5(C).

Courtesy of Cooper Bussmann, Inc.

single-line diagram of the electrical facilities, calculate the short-circuit currents throughout the premise, calculate the arc flash boundaries and incident energies throughout the premises, and affix labels on the electrical equipment.

Overcurrent Protective Device to Use for AFHA

Generally, if an AFHA is to be performed for a piece of electrical equipment, the upstream supply-side OCPD located in a separate enclosure is used as the protecting device for the AFHA. Even though electrical equipment might have a main OCPD and disconnecting means, an arcing fault can occur on the line side of the main within the electrical assembly.

An industrial control panel might have a main fusible disconnect switch or circuit breaker fed by an overhead busway plug-in OCPD/disconnect. When working on the industrial control panel, an arcing fault on the line side terminals of the main disconnect poses an arc flash hazard. Therefore, the OCPD in the busway plug-in disconnect must be considered in the arc flash hazard analysis.

Arc Blast Considerations

As noted in *NFPA 70E* 130.7(A) Informational Note No. 1, physical trauma injuries may occur due to the explosive effect of some arcing fault events. The PPE

requirements of 130.7 do not address protection against physical trauma other than arc flash. Little testing has been done to date on the effects and predictability of arc blast hazards. Generally, arcing fault events involving higher incident energy in short time durations result in higher arc blast energy.

Summary

This chapter provides information on the methods for determining the AFB and required PPE as part of an arc flash hazard analysis. Methods are available for both AC and DC systems, and for both arcs-in-the-open and arcs-in-a-box. The HRC methods use tables, where conditions of use must be satisfied before these methods can be applied. The 2002 IEEE 1584 guide provides calculation methods based on extensive testing, but the application must be within the various parameter ranges specified in IEEE 1584. For practical application of these methods, other arc flash hazard analysis requirements and other considerations must be taken into account, such as the OCPD condition of maintenance, documentation, labeling, and updating of the analysis if system changes warrant changes.

REVIEW QUESTIONS

1. An arc flash hazard analysis is defined as a study investigating a worker's potential exposure to arc flash energy, conducted for the purpose of __?__ prevention and determination of __?__ work practices.
   a. arcing/*National Electrical Code*
   b. blister/safe
   c. excessive work/project
   d. injury/safe

2. To use the HRC method for determining an arc flash hazard requires that the __?__ have been met.
   a. clearance distances
   b. conditions of use
   c. foreman's tailgate safety training tips
   d. human resource policies

3. For AC systems, once the HRC has been determined from *NFPA 70E* Table 130.7(C)(15)(a), Table 130.7(C)(16) is used to select the required __?__ with the designated minimum arc rating for the task.
   a. hearing protection
   b. PPE
   c. rain gear
   d. screwdrivers

4. DC systems have significantly __?__ arcing fault dynamics and outcomes than 3-phase systems; therefore, AC system arc flash hazard analysis methods are not suitable for DC systems.
   a. different
   b. larger
   c. more robust
   d. smaller

5. The __?__ is an industry consensus document based on testing and analysis to develop equations for calculating the AFB and incident energy.
   a. Hazard/Risk categories method in 130.5(B)(2)
   b. IEEE 1584 *Guide for Performing Arc-Flash Hazard Calculations*
   c. *NFPA 70E*
   d. NJATC *Electrical Safety-Related Work Practices* publication

6. The AFB is the distance from the arc source where a person can receive a __?__ burn.
   a. first-degree
   b. fourth-degree
   c. second-degree
   d. third-degree

7. The calculated arc flash boundary for Test 4 is equal to __?__.
   a. 45"
   b. 47"
   c. 49"
   d. 51"

8. The calculated incident energy for Test 4 is equal to __?__ at 18 inches.
   a. 1.73 cal/cm^2
   b. 4.58 cal/cm^2
   c. 5.84 cal/cm^2
   d. 8.45 cal/cm^2

9. Once the arc flash hazard analysis has been determined, equipment must be __?__ according to 130.5(C).
   a. maintained
   b. marked
   c. painted
   d. repaired

10. For doing an arc flash hazard analysis, 130.5(B) has two methods for the selection of protective clothing and other PPE: __?__ and __?__.
    a. incident energy analysis/arcing fault current calculator
    b. incident energy analysis/HRC
    c. OCPD clearing time/HRC
    d. short-circuit current calculations/HRC

9 Fundamentals of 3-Phase Bolted Fault Currents

Courtesy of James R. White, Shermco Industries

OBJECTIVES

1. Understand why short-circuit studies are required and when they are needed.
2. Understand the effect of short-circuit current on arc-flash hazards.
3. Understand the procedures and calculation methods used to perform a 3-phase short-circuit current study.

REFERENCES

1. *National Electrical Code® (NEC®)*, 2011 Edition
2. *NFPA 70E®*, 2012 Edition
3. Cooper Bussmann® SPD (Selecting Protective Devices) section on Calculating Short-Circuit Currents
4. Cooper Bussmann downloadable smart phone app FC^2 Available Fault Current Calculator

A customer needs to have a 200-ampere, 480-volt, 3-phase panel installed to feed a new addition. The new panel will be fed from an existing 600-A panel that has a 10,000-A short-circuit current rating (SCCR) and circuit breakers with 10,000-A interrupting ratings (IR). A few years ago, the existing panel was installed, presumably compliant with the *NEC*. The customer is asking contractors to quote a price for the required electrical equipment and installation.

Is it as simple as buying and installing a 3-phase, 480-V, 200-A, 10,000-A SCCR panel (with 10,000-A interrupting rated breakers) and installing it with the feeder from the existing 600-A panel (including installing a new 200-A circuit breaker in the existing panel)? Or is there more to this task?

What are some of the things that need to be considered? Considerations include the need to perform a short-circuit and arc flash study to verify compliance of the existing installation, select the proper new electrical equipment, and select shock and arc flash protection to install the new system. The new electrical equipment must be selected and installed in compliance with *NEC* requirements, including 110.9 and 110.10. Existing electrical equipment will require an examination for *NEC* compliance and must be replaced to comply with OSHA 1910.302(b)(1), 1910.303(b)(4), and 1910.303(b)(5) should analysis results indicate that the available fault current is now higher than the interrupting and short-circuit current rating of the existing equipment. Additional considerations may include electrical hazard analysis and work practices, as well as selection of personal protective equipment (PPE).

An analysis was performed that determined the existing installation has an available fault current greater than the existing panel SCCR and IRs for the circuit breakers. This situation may occur when either a lower impedance or larger size utility transformer is installed to replace an existing transformer.

Uncovering safety hazards of inadequate SCCR- and IR-rated existing equipment is unwelcome news to the facility owner but important as part of the effort to protect persons and property. Now the scope of work includes more than scheduling a shutdown so that the new 3-pole 200-A breaker can be installed in the existing panel. Not only is a panel with a more robust SCCR and IRs required than originally anticipated, but a decision now needs to be made by the customer on what to do with the existing equipment.

In this scenario, the situation was discovered before inadequately rated equipment became a factor. All too often, however, it is only after an incident occurs that the extent of the hazard is realized. Recognition, acceptance, and application of *NEC*, *NFPA 70E*, and OSHA requirements are helping to improve the safety levels of today's new and existing electrical systems.

All parties play an important role in ensuring a safe and reliable electrical installation, both today and in the future. Data collection is an essential part of analysis and must not be overlooked. The calculation and software options available today make it easier than ever to specify, install, and maintain *Code*- and OSHA-compliant installations.

Source: Steve Abbott, Stark Safety Consultants

For additional information, visit qr.njatcdb.org Item #1199.

Introduction

The available short-circuit current is identified through a short-circuit study. The purpose of a short-circuit current study is to determine the available short-circuit current (bolted) at one or multiple points in an electrical system so as to ensure safety compliance with OSHA, NEC, and NFPA 70E requirements, as well as for other electrical system analysis purposes. For arc flash hazard analysis, whether using the incident energy analysis method (130.5(B)(1)) or the hazard/risk categories method (130.5(B)(2)), typically the available short-circuit current is a necessary value that must be determined.

This chapter provides some insight into the variables that affect the available short-circuit current magnitude at various points in an AC electrical system. In addition, it provides the procedures, including equations and tables, for performing simple short-circuit current calculations for 3-phase systems referred to as the point-to-point method. Two examples of using this calculation method are provided.

Short-Circuit Calculation Requirements

Knowledge of and the ability to calculate the short-circuit current at various points in an electrical system is required to comply with various sections of OSHA regulations, the NEC, and NFPA 70E. The following list details a number of those requirements and explains why the short-circuit current is needed for compliance:

- NFPA 70E 130.5: Typically, the available short-circuit current is needed to perform an arc flash incident energy calculation or to verify the available short-circuit current is within the conditions of use parameter for the hazard/risk categories (HRC) table method.
- NEC 110.24: The maximum available fault current must be marked on service entrance equipment (with some exceptions) to help ensure the proper application of overcurrent protective device (OCPD) interrupting ratings and electrical equipment short-circuit current ratings.
- OSHA §1910.303(b)(4), NEC 110.9, and NFPA 70E 210.5: Overcurrent devices must have adequate interrupting ratings for the short-circuit current available.
- OSHA §1910.303(b)(5) and NEC 110.10: System components must have adequate short-circuit current ratings for the short-circuit current available.
- NEC 240.12, 517.26, 620.62, 695.3(C), 700.27, 701.27, and 708.54: OCPDs must be selectively coordinated for the full range of overcurrents, including the available short-circuit current.

Short-Circuit Current Calculation Basics

The short-circuit current study for a system starts from the service point, extends to the feeder equipment, and then continues to the branch-circuit devices. This study determines the short-circuit current at all critical points in the system. Various types of equipment, such as service equipment, switchboards, panelboards, transfer switches, motor control centers, and motor starters, are analyzed in this study. The level of short-circuit current available depends on many factors, such as the utility system, generators, and motor contribution. In general, the short-circuit current decreases as distance from the source increases due to the increased impedance that is added by such circuit components that have impedance, such as transformers, conductors, and busway.

Normally, short-circuit studies involve calculating a bolted 3-phase short-circuit condition. This scenario can be characterized as all three phases "bolted" together to create a zero-impedance connection. It establishes a circuit faulted condition that results in maximum thermal and mechanical stress in the system for evaluation of the OCPD interrupting rating and the equipment short-circuit current ratings. In addition, as covered in the chapter "Incident Energy Varies by Fault Current Magnitude and Duration," the arcing fault current magnitude depends on the available short-circuit current. The greater the available short-circuit current at a point in the system, the greater the arcing fault current at that point.

Caution

Short-circuit currents can change over time due to system changes. To ensure protection of both people and equipment, short-circuit current studies and arc flash hazard analysis studies must be updated as necessary.

The bolted short-circuit current determines the highest amount of current that the electrical system can deliver during a short-circuit condition.

A short-circuit current study can be carried out to determine the available short-circuit current at just one point in a system or it can be done on a system-wide basis. Short-circuit current calculations should be completed for all critical points in the system. **See Figure 9-1.** These would include, but are not limited to, the following:

- Switchboards
- Panelboards
- Motor control centers (MCC)
- Motor starters
- Disconnect switches
- Transfer switches

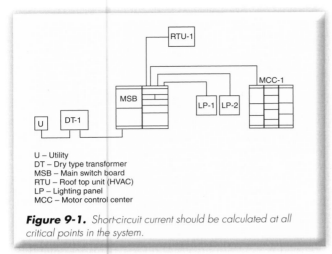

U – Utility
DT – Dry type transformer
MSB – Main switch board
RTU – Roof top unit (HVAC)
LP – Lighting panel
MCC – Motor control center

Figure 9-1. *Short-circuit current should be calculated at all critical points in the system.*

Sources of Short-Circuit Current

Sources of short-circuit current that are normally taken under consideration include utility generation, local generation, synchronous motors, and induction motors. In addition, alternative energy sources, such as batteries, photovoltaic systems, fuel cells, wind turbines, and flywheels, may also feed into an AC fault in some cases through a DC-to-AC converter. When electrical systems are supplied from a utility- or customer-owned transformer, the amount of short-circuit current depends on the size (kVA) and impedance (%Z) of the transformer. The larger the size and/or the lower the impedance, the higher the available short-circuit current.

The available short-circuit current cannot be determined or estimated just by considering the ampere rating of an electrical panel or the type of facility. For instance, one 800-ampere distribution panelboard in a facility may have an available short-circuit current of 15,000 amperes at its line terminals, whereas another 800-ampere distribution panel in the same facility or in another facility may have an available short-circuit current of 80,000 amperes. Calculations should be done to determine the available short-circuit current.

With the qualifier of the previous paragraph kept firmly in mind, this paragraph provides some broad generalizations about available short-circuit currents. Small residential building systems (100-A to 200-A service) typically have short-circuit currents of 10,000 A to 15,000 A or less. Small commercial building systems (400-A to 800-A service) typically have short-circuit currents of 15,000 A to 30,000 A. Larger commercial and manufacturing building systems (2,000-A to 3,000-A service) typically have short-circuit currents in the range of 50,000 A to 65,000 A. These short-circuit current values can be much higher where low-impedance transformers are

used to increase efficiency. When commercial buildings are directly connected to a utility "low-voltage grid system," such as in major metropolitan cities (for example, New York, Chicago, Dallas), the short-circuit currents can exceed 200,000 A. The busway plug-in drops in auto production plants may have 150,000 A available short-circuit currents. Some hospitals with large alternate supply generators that can operate in parallel may have short-circuit current available exceeding 100,000 A.

Caution

Short-circuit currents can be very high (65,000 amperes to more than 200,000 amperes). For this reason, it is important to verify short-circuit current levels when applying overcurrent devices and equipment. It is also necessary to determine the available short-circuit current when doing an arc flash hazard analysis.

Short-Circuit Current Factors

The amount of short-circuit current depends on many factors, including the short-circuit current contribution of the utility, local generation, and motors as well as the kVA rating and impedance of the transformer, the size and length of the wire, and more. **See Figure 9-2.**

The short-circuit current is typically the highest at the service point. Further in an electrical system, the circuit conductors and additional transformers will decrease the available short-circuit current at downstream equipment. **See Figure 9-3.**

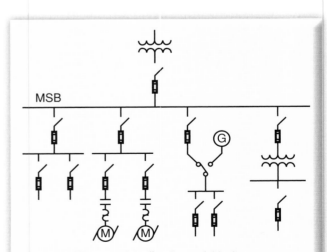

Figure 9-2. *Factors that affect the available short-circuit current include the utility, transformer (kVA and %Z), generators, motors, voltage, and conductor size and length.*

Courtesy of Cooper Bussmann, Inc.

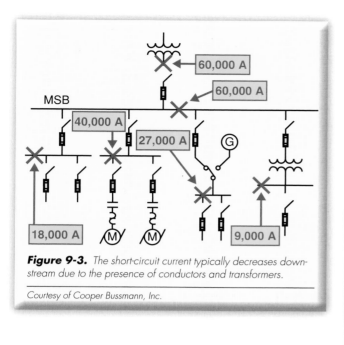

Figure 9-3. *The short-circuit current typically decreases downstream due to the presence of conductors and transformers.*

Courtesy of Cooper Bussmann, Inc.

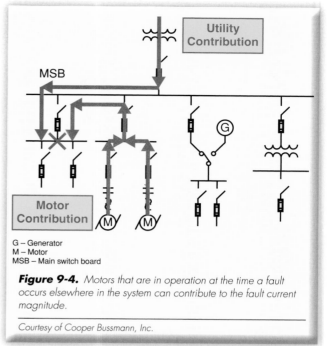

G – Generator
M – Motor
MSB – Main switch board

Figure 9-4. *Motors that are in operation at the time a fault occurs elsewhere in the system can contribute to the fault current magnitude.*

Courtesy of Cooper Bussmann, Inc.

Effects of Generators and Motors on Short-Circuit Current

Generators and motors in a premise system that are operating at the time of a fault condition can contribute short-circuit current in addition to the short-circuit current the utility delivers. Typically, the range of a short-circuit current from a generator is 7 to 14 times the full-load ampacity of the generator, with a general rule of thumb being 10 times. If the generator is operated in parallel with the utility during transfer (closed-transition transfer), the short-circuit current of the generator is added to the short-circuit current of the utility. If not, the generator short-circuit current is typically less than the normal source (utility) and is not used unless the Electrical Worker is examining arc-flash hazards when the system is under generator power.

When a fault occurs, a motor in operation acts like a generator and will contribute to the short-circuit current. **See Figure 9-4.** The motor contribution is usually four to six times the full-load ampacity of the motor. The additional motor contribution normally must be added to the utility contribution to determine the required ratings of overcurrent devices and equipment as well as to be considered in the various arc flash hazard analysis scenarios.

Effect of Transformers on Short-Circuit Current

Generally, the transformer is the single most important component affecting the available short-circuit currents throughout an electrical system. This relationship is especially notable for service equipment. It is important to realize that the available short-circuit current at the service may change, and often increase, due to changes in utility contribution or the transformer's kVA rating and percentage impedance. If the available short-circuit current changes at the service equipment, it will also change further downstream in the system.

The transformer size (kVA), impedance (%Z), and secondary voltage affect short-circuit current. **See Figure 9-5.** Assuming an "infinite" primary current, the short-circuit current at point A for a 5% impedance transformer with a secondary voltage of 480/277 volts is equal to the full-load ampacity of the transformer (601 A) × 100/%Z = (601 A) × (100/5), which equals 12,020 A (601 A × 20). This value may be 10% higher due to the tolerance of the transformer impedance (± 10). For instance, the available fault current at point A would be calculated as 601 A × 100/(5 × 0.9) = 13,356 A. Continuing with the same example:

- If the transformer is replaced with a lower impedance transformer of 2% (to increase transformer efficiency), as shown at point B, the short-circuit current increases by 2.5 times (100/%Z increases from 20 to 50) to 30,050 A (601 A × 100/2) = 30,050 A).
- If the kVA is increased to 1,500 kVA (1,804 A full-load ampacity), as shown at points C and D, the short-circuit current triples to 36,080 A and 90,200 A, respectively.
- If the transformer secondary voltage is 208/120 V, as shown in points E and F, the available short-circuit currents are 83,280 A and 208,200 A, respectively.

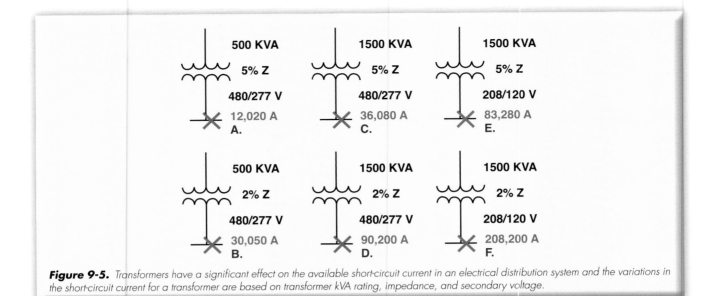

Figure 9-5. *Transformers have a significant effect on the available short-circuit current in an electrical distribution system and the variations in the short-circuit current for a transformer are based on transformer kVA rating, impedance, and secondary voltage.*

It is evident from the different available short-circuit currents shown that transformer characteristics have a major impact on the available short-circuit current. The transformer kVA rating and percentage impedance are very important for determining the available short-circuit current for the initial installation. However, it is also important to reassess the available short-circuit current if a transformer is replaced. If a utility transformer fails and the utility replaces the transformer, the percentage impedance can be higher or lower than the original by a significant amount. Most often, the percentage impedance of the newer transformers will be lower, resulting in a higher short-circuit current. Many utilities have a policy that if a transformer needs to be replaced due to failure, the utility can use a transformer with the next larger kVA rating. This practice is necessary in the event that the company's inventory of transformers with the same kVA rating is depleted at the time of replacement. Thus, the available short-circuit current could increase due to both a kVA rating increase and a percentage impedance decrease.

Effect of Conductors on Short-Circuit Current

The conductor (or bus duct) size (ampacity), length, and number per phase may also affect the short-circuit current. **See Figure 9-6.** Assuming a short-circuit current at point A of 40,000 A [at the beginning of a 25-foot run of a 1 American Wire Gauge (AWG) conductor] at 480/277 V, the short-circuit current at the end of the conductor at point B would be 27,000 A. If the length of a 1 AWG conductor was increased to 50 feet, as shown

from point C to D, the short-circuit current at point D would decrease to 20,380 A due to the increased impedance of the extra 25 feet of conductor. If the size of the conductor was then increased to 500 thousand circular mils (kcmil), as shown from point E to F, the short-circuit current at point F would be 31,490 A, which is higher than the resulting fault current for the 50 feet of 1 AWG at point D due to the decrease in the impedance per foot of the larger conductor.

Caution

Short-circuit currents can greatly increase when transformer impedance is decreased or when transformer kVA increases. When service transformers are changed, short-circuit studies and arc-flash studies must be updated as well. In addition, the interrupting ratings of OCPDs and short-circuit current ratings for equipment must be verified as sufficient: if these ratings are found to not be adequate, a safety hazard is present. Immediate action to remedy this problem should be taken.

Effect of Short-Circuit Current on Arc-Flash Hazards

The available short-circuit current is a key factor that determines the extent of the arc-flash hazard. An integral part of protecting workers from an arc-flash hazard includes, but is not limited to, conducting an arc flash hazard analysis. An arc flash hazard analysis will determine, among other things, the AFB and incident energy or the required level of PPE using the HRC method.

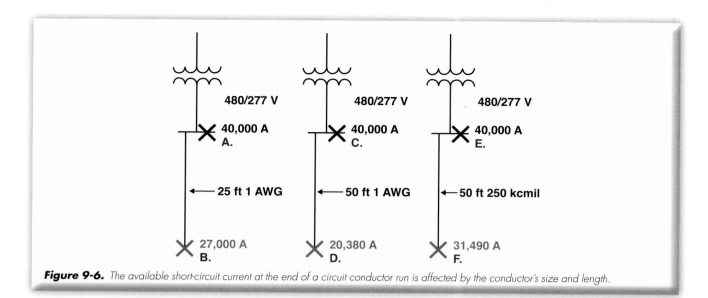

Figure 9-6. *The available short-circuit current at the end of a circuit conductor run is affected by the conductor's size and length.*

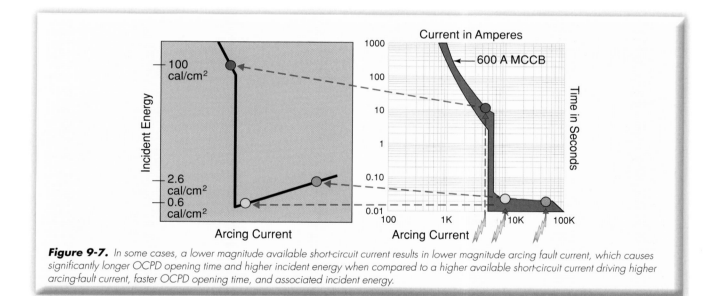

Figure 9-7. *In some cases, a lower magnitude available short-circuit current results in lower magnitude arcing fault current, which causes significantly longer OCPD opening time and higher incident energy when compared to a higher available short-circuit current driving higher arcing-fault current, faster OCPD opening time, and associated incident energy.*

When using the 130.5(B)(1) method of arc flash hazard analysis (incident energy analysis), it is important to determine both the low and high values of short-circuit current that may occur at each point in the system. In some cases, a lower value of available short-circuit current means a lower arcing current may result in a longer clearing time of the overcurrent device. This longer clearing time may, in turn, cause a high incident energy, greater AFB, and higher level of PPE required. See *NFPA 70E* 130.5 Informational Note No. 2. Lower arcing fault currents can result in increased opening times of OCPDs and higher incident energy compared to higher arcing fault currents, which can result in decreased OCPD opening time. **See Figure 9-7.** With both circuit breakers and fuses, a low arcing-fault current might potentially be in the OCPD's time-current operating region requiring many seconds to interrupt, which may result in the incident energy exceeding that for a high available short-circuit current. Thus, the analysis should consider various electrical distribution system scenarios and variables to determine the lowest and highest available short-circuit current, with an incident energy analysis then being performed for each. The chapter "Incident Energy Varies by Fault Current Magnitude and Duration" provides a more in-depth discussion of this topic.

Switching Scenarios

Often a facility's electrical distribution system can be configured in various ways by opening or closing disconnects. For instance, a double-ended unit substation may include a low-voltage tie disconnect between the switchgear. One scenario might include both transformers energized and the low-voltage tie disconnect open between the switchgear. An alternative scenario might be the same as the first but with the tie disconnect closed. This second scenario might possibly have twice the magnitude of available short-circuit current at the switchgear as compared to the first scenario. This variation can have a dramatic effect on the incident energy analysis; however, the incident energy calculation depends on the OCPD characteristics and the available short-circuit current.

Power Source Scenarios

In addition, the available short-circuit current at a specific point in a system may depend on the power sources that can feed into the fault. Is there only a normal source or can there be an alternate generator operating either by itself or in combination with the normal source? Are there parallel generators, and are different scenarios associated with the specific number of generators that may be in operation? These potential scenarios can result in different magnitudes of available short-circuit current, such as one finds with switching scenarios.

Motor Contribution Scenarios

Motor contribution was discussed earlier in this chapter as a possible factor determining the magnitude of short-circuit current at a fault, which can affect the incident energy analysis. However, different scenarios may apply to different levels of motor contribution. The two extremes occur when all the motors are operating at the time of a fault and when none of the motors is operating at the time of a fault.

Effect of System Changes on Arc-Flash Hazard

Short-circuit currents can change over time due to changes in the electrical distribution system. If the electrical system is upgraded, expanded, or reduced, such a change may affect the available short-circuit current throughout the system. Transformer changes were discussed earlier in this chapter. Other changes could include, but are not limited to, the utility system, generators, motors, or other building distribution system modifications. In recognition of these pos-

sibilities, the second paragraph of *NFPA 70E* 130.5 requires the arc flash hazard analysis to be reviewed whenever a major modification or renovation takes place. This information is also required to be reviewed periodically—no longer than every five years—to account for changes that may affect the results of the arc flash hazard analysis.

Procedures and Methods

To determine the short-circuit current at any point in the system, first draw a one-line diagram showing all of the sources of short-circuit current. Then include the system components, such as conductors and busway (sizes, lengths, material, number per phase, and conduit type or no conduit), transformers (sizes, voltages, and percentage impedances), and OCPDs (ampere, voltage, and interrupting ratings).

Short-circuit calculations are performed without OCPDs in the system. These calculations assume that these devices are replaced with copper bars, so as to determine the maximum "available" short-circuit current. After the calculation is completed throughout the system, current-limiting devices can be used to show reduction of the available short-circuit current at a single location only. Current-limiting devices do not operate in series to produce a compounding current-limiting effect.

Various methods have been developed to calculate the available short-circuit current, but all are based on Ohm's Law. These methods include the ohmic method, per-unit method, and point-to-point method, along with computer-based versions of all three methods.

The application of the point-to-point method permits the quick determination of available short-circuit currents with a reasonable degree of accuracy at various points for either 3-phase or single-phase electrical distribution systems. The procedures and examples in this chapter focus only on 3-phase systems. The point-to-point calculation methods for 3-phase and single-phase systems are available from Cooper Bussmann in its *Selecting Protective Devices* (SPD) publication, which is available online at www.CooperBussmann.com. The point-to-point method is provided courtesy of Cooper Bussmann.

Point-to-Point Method

The point-to-point method uses equations to calculate the short-circuit current at the secondary of a transformer assuming infinite primary short-circuit current available, fault at the end of a run of conductor or busway, and fault at the secondary of a transformer with primary short-circuit current known. Following the equations and notes shown here are two examples of how the point-to-point method is used.

At the end of this chapter are reference tables that are resources for using the point-to-point method. Review these tables along with their associated notes to become familiar with how this information is applicable in performing short-circuit current calculations. The tables appear as Figures 9-11 to 9-17.

3-Phase Short-Circuit Current Calculation Procedure

Equations to Calculate Short-Circuit Current on Transformer Secondary with Infinite Primary Available: Steps 1 to 3

Use the following procedure to determine the short-circuit current (I_{SCA}) at 3-phase transformer secondary terminals assuming infinite primary short-circuit current available:

Step 1: Determine the transformer full-load amperes (I_{FLA}) from either the nameplate, following equation, or Figure 9-11.

$$3\varnothing \text{ transformer: } I_{FLA} = \frac{kVA \times 1000}{E_{L-L} \times 1.732}$$

Where:
E_{L-L} = line-to-line voltage
kVA = transformer kVA

Step 2: Determine the transformer multiplier

$$\text{Multiplier} = \frac{100}{\%Z}$$

Note 1: The marked (nameplate) transformer impedance (%Z) value may vary ±10% from the actual values determined by the American National Standards Institute/Institute of Electrical and Electronics Engineers (ANSI/IEEE) test. See UL Standard 1561. For the high-fault condition scenario, multiply the transformer %Z by 0.9 and for the low-fault condition scenario, multiple the %Z by 1.1.

Step 3: Determine the available short-circuit current (I_{SCA}) at 3-phase transformer secondary (Figure 9-16 provides a generalized estimate based on lowest %Z transformer survey.)

$3\varnothing$ transformer (see Notes 2 for motor contribution):
$I_{SCA (L-L-L)} = I_{FLA} \times \text{Multiplier}$

Note 2: The motor short-circuit contribution, if significant, may be determined and added to all fault locations throughout the system. A practical estimate of the motor short-circuit contribution is to multiply the total motor load current by 4 or 6. A good general rule is to calculate the total motor load current, multiply by 4, and add the result to step 3.

Note 3: The utility voltage may vary ±10% for power, and ±5.8% for 120 V lighting services. For worst-case conditions, multiply the values calculated in step 3 by 1.1 and/or 1.058, respectively.

Note 4: For 3-phase systems, line-to-line-to-line (L-L-L), line-to-line (L-L), line-to-ground (L-G), and line-to-neutral (L-N) bolted faults can be found in Figure 9-17 based on the calculated 3-phase line-to-line-to-line (L-L-L) bolted fault. Use IEEE 1584, *Guide for Performing Arc-Flash Hazard Calculations*, equations to calculate the 3-phase arcing current when performing an arc flash hazard analysis.

Equations to Calculate Short-Circuit Current at End of Conductor Run: Steps 4 to 6

Use the following procedure to determine the short-circuit current (I_{SCA}) at the end of a run of conductor or busway:

Step 4: Calculate the "f" factor
$3\varnothing$ line-to-line-to-line (L-L-L) fault

$$f = \frac{1.732 \times L \times I_{SCA\,(L-L-L)}}{C \times n \times E_{L-L}}$$

Where:
L = length (feet) of conduit to the fault
C = constant for conductors or busway (from Figure 9-12, 9-13, or 9-14)
n = number of conductors per phase (adjusts the "C" value for parallel runs)
I_{SCA} = available short-circuit current in amperes (A) at the beginning of the circuit
E_{L-L} = line-to-line voltage

Step 5: Determine the "M" (multiplier) from Figure 9-15 or the following equation:

$$\text{Equation: } M = \frac{1}{1 + f}$$

Step 6: Calculate the available short-circuit current (I_{SCA}) at the end of the circuit
$3\varnothing$ line-to-line-to-line (L-L-L) fault (see Notes 2 for motor contribution):

$$I_{SCA\,(L-L-L)} = I_{SCA\,(L-L-L)} \times M$$
$$\text{\small AT END OF CIRCUIT} \quad \text{\small AT BEGINNING OF CIRCUIT}$$

Note 5: See Note 4.

Determining Short-Circuit Current on 3-Phase Transformer Secondary with Primary Short-Circuit Current Known: Steps A to C

When the primary short-circuit current at a transformer is known, either through a previous calculation (if it is the second transformer in the system) or if the short-circuit current of the utility connection is provided, then a more accurate calculation can be performed. Use the following procedure Steps A to C to calculate the short-circuit current at the secondary when the short-circuit current at the transformer primary is known. **See Figure 9-8.**

Step A: Calculate the "f" factor for the transformer

3ϕ transformer: $f = \dfrac{I_{P(SCA)\,L-L-L} \times V_P \times 1.732 \times \%Z^*}{100,000 \times kVA}$

*Transformer Z is multiplied by 0.9 to establish the high-fault condition scenario. See Note 1.

Where:
$I_{P(SCA)}$ = primary short-circuit current
V_P = primary voltage (L-L)
kVA = transformer kVA

Step B: Determine the multiplier "M" for the transformer from **Figure 9-15** or the following equation:

$M = \dfrac{1}{1 + f}$

Step C: Determine the short-circuit current at the transformer secondary (see Notes 2 for motor contribution):

$I_{S(SCA)} = \dfrac{V_P}{V_S} \times M \times I_{P(SCA)}$

Where:
$I_{S(SCA)}$ = secondary short-circuit current
V_S = secondary voltage (L-L)

To determine the available short-circuit current at the end of a conductor or busway run connected to the secondary of this transformer, follow Steps 4, 5, and 6.

The point-to-point method is provided by Cooper Bussman.

Point-to-Point Method Examples

The following examples show how to use the point-to-point method equations and notes for 3-phase systems. The tables in Figures 9-11 through 9-17 at the end of the chapter are needed for some data, such as the conductor C values. For simplicity, the motor contribution and voltage variance are not included. See Notes 2 and 3 in the section 3-Phase Short-Circuit Current Calculation Procedure.

Example 1

This example has three points in the system to calculate the available short-circuit current. **See Figure 9-9.** This example assumes the utility available on the transformer primary is infinite.

Calculation Example for Fault #1
Step 1: Determine the transformer full-load amperes (I_{FLA})

$I_{FLA} = \dfrac{kVA \times 1000}{E_{L-L} \times 1.732} = \dfrac{300 \times 1000}{208 \times 1.732} = 833\ A$

Step 2: Find the transformer multiplier

$\text{Multiplier} = \dfrac{100}{*0.9 \times \text{Transf. }\%Z} = \dfrac{100}{1.8} = 55.55$

*Transformer Z is multiplied by 0.9 to establish a high fault condition scenario. See Note 1 in the section 3-Phase Short-Circuit Current Calculation Procedure.

Step 3: Determine the short-circuit current (I_{SCA}) at the transformer secondary

$**I_{SCA\,(L-L-L)} = I_{FLA} \times M = 833 \times 55.55 = 46,273\ A$

**For simplicity, the motor contribution and voltage variance were not included. See Notes 2 and 3 in the section 3-Phase Short-Circuit Current Calculation Procedure.

Calculation Example for Fault #2
(Use $I_{SCA\,(L-L-L)}$ at Fault #1 in Steps 4 and 6.)

Step 4: Calculate the "f" factor

$f = \dfrac{1.732 \times L \times I_{SCA(L-L-L)}}{C \times E_{L-L}} = \dfrac{1.732 \times 20 \times 46,273}{22,185 \times 208} = 0.35$

Step 5: Determine the "M" (multiplier)

$M = \dfrac{1}{1 + f} = \dfrac{1}{1 + 0.35} = 0.74$

Step 6: Calculate the available short-circuit current (I_{SCA}) at Fault #2

$I_{SCA\,(L-L-L)} = I_{SCA\,(L-L-L)} \times M$

AT END OF CIRCUIT   AT BEGINNING OF CIRCUIT

$= 46,273 \times 0.74 = 34,242A$

Calculation Example for Fault #3
(Use $I_{SCA\,(L-L-L)}$ at Fault #2 in steps 4 and 6.)

Step 4: Calculate the "f" factor

$f = \dfrac{1.732 \times 20 \times 34,242}{5,907 \times 208} = 0.97$

Step 5: Determine the "M" (multiplier)

$M = \dfrac{1}{1 + 0.97} = 0.51$

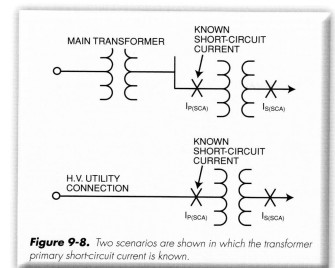

Figure 9-8. *Two scenarios are shown in which the transformer primary short-circuit current is known.*

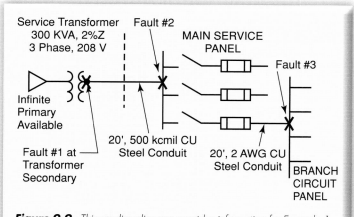

Figure 9-9. *This one-line diagram provides information for Example 1, which has three short-circuit currents to be calculated.*

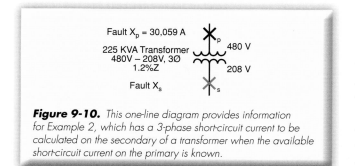

Figure 9-10. *This one-line diagram provides information for Example 2, which has a 3-phase short-circuit current to be calculated on the secondary of a transformer when the available short-circuit current on the primary is known.*

Step 6: Calculate the available short-circuit current (I_{SCA}) at Fault #3

$$I_{SCA\,(L-L-L)\,AT\,FAULT\,\#3} = 34{,}242 \times 0.51 = 17{,}463\ A$$

Example 2

The following is an example of the procedure for the calculation of short-circuit current on a 3-phase transformer secondary with the primary short-circuit current known. **See Figure 9-10.**

Step A: Calculate the "f" factor for the transformer

$$f = \frac{30{,}059 \times 480 \times 1.732 \times 0.9* \times 1.2}{100{,}000 \times 225} = 1.2$$

*Transformer Z is multiplied by 0.9 to establish a high-fault condition scenario. See Note 1 in the section 3-Phase Short-Circuit Current Calculation Procedure.

Step B: Determine the multiplier "M" for the transformer

$$M = \frac{1}{1 + 1.2} = 0.45$$

Step C: Determine the short-circuit current at the transformer secondary

$$I_{S(SCA)} = \frac{480}{208} \times 0.45 \times 30{,}059 = 31{,}215\ A$$

Short-Circuit Calculation Software Programs

Several short-circuit calculation software programs are available. Some of these programs not only calculate the available short-circuit current, but also draw one-line diagrams, perform coordination studies, and determine the arc flash hazard. These programs are available for purchase, and the price of the software program can vary greatly based on its capabilities.

Cooper Bussmann provides a smartphone app of the point-to-point method referred to as FC² Available Fault Current Calculator, which can be accessed from the QR Code shown. This application runs on Android and Apple devices. For web-based version, see www.cooperbussmann.com/FC2.

For additional information, visit qr.njatcdb.org Item #1215.

Summary

Conducting a short-circuit current study is crucial in increasing electrical safety and is necessary to comply with the requirements of the *NEC* and *NFPA 70E*. This analysis helps ensure that the interrupting rating of overcurrent devices and the short-circuit current rating of equipment are adequate. In addition, a short-circuit study must be completed before an arc flash hazard analysis can be performed, which improves workplace safety by determining the AFB and the necessary PPE.

Short-circuit current calculations help the Electrical Worker to determine the short-circuit current, which can vary depending on the utility, transformers, generators, motors, voltage, conductors, and busway. Becoming familiar with the equations, tables, and notes associated with the point-to-point short-circuit calculation method described in this chapter will enable the Electrical Worker to make the proper calculations and use them to ensure proper protection of equipment and people.

Figure 9-11. *3-Phase Transformer: Full Load Current Rating (in amperes) values can be used in lieu of the equation shown in Step 1 of the 3-phase short-circuit current calculation procedure.*

3-Phase Voltage (Line-to-Line)	3-Phase Transformer kVA Rating								
	150	**167**	**225**	**300**	**500**	**750**	**1000**	**1500**	**2000**
208	417	464	625	833	1388	2080	2776	4164	5552
220	394	439	592	788	1315	1970	2630	3940	5260
240	362	402	542	722	1203	1804	2406	3609	4812
440	197	219	296	394	657	985	1315	1970	2630
460	189	209	284	378	630	945	1260	1890	2520
480	181	201	271	361	601	902	1203	1804	2406
600	144	161	216	289	481	722	962	1444	1924

Figure 9-12. *"C" values for copper conductors are used in the Step 4 equation of the 3-phase short-circuit current calculation procedure.*

AWG or kcmil	Copper Conductors											
	Three Single Conductors Conduit						Three-Conductor Cable Conduit					
	Steel			Nonmagnetic			Steel			Nonmagnetic		
	600 V	5 kV	15 kV	600 V	5 kV	15 kV	600 V	5 kV	15 kV	600 V	5 kV	15 kV
14	389	—	—	389	—	—	389	—	—	389	—	—
12	617	—	—	617	—	—	617	—	—	617	—	—
10	981	—	—	982	—	—	982	—	—	982	—	—
8	1557	1551	—	1559	1555	—	1559	1557	—	1560	1558	—
6	2425	2406	2389	2430	2418	2407	2431	2425	2415	2433	2428	2421
4	3806	3751	3696	3826	3789	3753	3830	3812	3779	3838	3823	3798
3	4774	4674	4577	4811	4745	4679	4820	4785	4726	4833	4803	4762
2	5907	5736	5574	6044	5926	5809	5989	5930	5828	6087	6023	5958
1	7293	7029	6759	7493	7307	7109	7454	7365	7189	7579	7507	7364
1/0	8925	8544	7973	9317	9034	8590	9210	9086	8708	9473	9373	9053
2/0	10,755	10,062	9390	11,424	10,878	10,319	11.245	11,045	10,500	11,703	11,529	11,053
3/0	12,844	11,804	11,022	13,923	13,048	12,360	13,656	13,333	12,613	14,410	14,119	13,462
4/0	15,082	13,606	12,543	16,673	15,351	14,347	16,392	15,890	14,813	17,483	17,020	16,013
250	16,483	14,925	13,644	18,594	17,121	15,866	18,311	17,851	16,466	19,779	19,352	18,001
300	18,177	16,293	14,769	20,868	18,975	17,409	20,617	20,052	18,319	22,525	21,938	20,163
350	19,704	17,385	15,678	22,737	20,526	18,672	22,646	21,914	19,821	24,904	24,126	21,982
400	20,566	18,235	16,366	24,297	21,786	19,731	24,253	23,372	21,042	26,916	26,044	23,518
500	22,185	19,172	17,492	26,706	23,277	21,330	26,980	25,449	23,126	30,096	28,712	25,916
600	22,965	20,567	17,962	28,033	25,204	22,097	28,752	27,975	24,897	32,154	31,258	27,766
750	24,137	21,387	18,889	29,735	26,453	23,408	31,051	30,024	26,933	34,605	33,315	29,735
1000	25,278	22,539	19,923	31,491	28,083	24,887	33,864	32,689	29,320	37,917	35,749	31,959

See Note at bottom of Figure 9-13.

Figure 9-13. *"C" values for aluminum conductors are used in the Step 4 equation of the 3-phase short-circuit current calculation procedure.*

AWG or kcmil	Aluminum Conductors											
	Three Single Conductors Conduit						Three-Conductor Cable Conduit					
	Steel			Nonmagnetic			Steel			Nonmagnetic		
	600 V	5 kV	15 kV	600 V	5 kV	15 kV	600 V	5 kV	15 kV	600 V	5 kV	15 kV
14	237	—	—	237	—	—	237	—	—	237		
12	376	—	—	376	—	—	376	—	—	376		
10	599	—	—	599	—	—	599	—	—	599		
8	951	950	—	952	951	—	952	951	—	952	952	
6	1481	1476	1472	1482	1479	1476	1482	1480	1478	1482	1481	1479
4	2346	2333	2319	2350	2342	2333	2351	2347	2339	2353	2350	2344
3	2952	2928	2904	2961	2945	2929	2963	2955	2941	2966	2959	2949
2	3713	3670	3626	3730	3702	3673	3734	3719	3693	3740	3725	3709
1	4645	4575	4498	4678	4632	4580	4686	4664	4618	4699	4682	4646
1/0	5777	5670	5493	5838	5766	5646	5852	5820	5717	5876	5852	5771
2/0	7187	6968	6733	7301	7153	6986	7327	7271	7109	7373	7329	7202
3/0	8826	8467	8163	9110	8851	8627	9077	8981	8751	9243	9164	8977
4/0	10,741	10,167	9700	11,174	10,749	10,387	11,185	11,022	10,642	11,409	11,277	10,969
250	12,122	11,460	10,849	12,862	12,343	11,847	12,797	12,636	12,115	13,236	13,106	12,661
300	13,910	13,009	12,193	14,923	14,183	13,492	14,917	14,698	13,973	15,495	15,300	14,659
350	15,484	14,280	13,288	16,813	15,858	14,955	16,795	16,490	15,541	17,635	17,352	16,501
400	16,671	15,355	14,188	18,506	17,321	16,234	18,462	18,064	16,921	19,588	19,244	18,154
500	18,756	16,828	15,657	21,391	19,503	18,315	21,395	20,607	19,314	23,018	22,381	20,978
600	20,093	18,428	16,484	23,451	21,718	19,635	23,633	23,196	21,349	25,708	25,244	23,295
750	21,766	19,685	17,686	25,976	23,702	21,437	26,432	25,790	23,750	29,036	28,262	25,976
1000	23,478	21,235	19,006	28,779	26,109	23,482	29,865	29,049	26,608	32,938	31,920	29,135

Note: The values for Figure 9-12 and 9-13 are equal to 1 over the impedance per 1000 feet and based upon resistance and reactance values found in IEEE Std 241-1990 (Gray Book), IEEE Recommended Practice for Electric Power Systems in Commercial Buildings, and IEEE Std 242-1986 (Buff Book), IEEE Recommended Practice for Protection and Coordination of Industrial and Commercial Power Systems. Where resistance and reactance values differ or are not available, the Buff Book values have been used. The values for reactance in determining the "C" value at 5 and 15 kV are from the Gray Book only (values for 14-10 AWG at 5 kV and 14-8 AWG at 15 kV are not available, and values for 3 AWG have been approximated).

Figure 9-14. *"C" values for busway are used in the Step 4 equation of the 3-phase short-circuit current calculation procedure.*

Ampacity	Plug-In		Feeder		High Impedance	
	Copper	Aluminum	Copper	Aluminum	Copper	
225	28,700	23,000	18,700	12,000	—	
400	38,900	34,700	23,900	21,300	—	
600	41,000	38,300	36,500	31,300	—	
800	46,100	57,500	49,300	44,100	—	
1000	69,400	89,300	62,900	56,200	15,600	
1200	94,300	97,100	76,900	69,900	16,100	
1350	119,000	104,200	90,100	84,000	17,500	
1600	129,900	120,500	101,000	90,900	19,200	
2000	142,900	135,100	134,200	125,000	20,400	
2500	143,800	156,300	180,500	166,700	21,700	
3000	144,900	175,400	204,100	188,700	23,800	
4000	—	—	277,800	256,400	—	

Note: These values are equal to 1 over the impedance per foot for impedance in a survey of industry.

Figure 9-15. *This table can be used to determine "M" (multiplier)* rather than doing the calculation in Step 5 or Step B when "f" is known from Step 4 or Step A.*

f	M	f	M
0.01	0.99	1.5	0.4
0.02	0.98	1.75	0.36
0.03	0.97	2	0.33
0.04	0.96	2.5	0.29
0.05	0.95	3	0.25
0.06	0.94	3.5	0.22
0.07	0.93	4	0.2
0.08	0.93	5	0.17
0.09	0.92	6	0.14
0.1	0.91	7	0.13
0.15	0.87	8	0.11
0.2	0.83	9	0.1
0.25	0.8	10	0.09
0.3	0.77	15	0.06
0.35	0.74	20	0.05
0.4	0.71	30	0.03
0.5	0.67	40	0.02
0.6	0.63	50	0.02
0.7	0.59	60	0.02
0.8	0.55	70	0.01
0.9	0.53	80	0.01
1.0	0.5	90	0.01
1.2	0.45	100	0.01

* $M = 1/(1 + f)$

Figure 9-16. *These generalized worst-case or highest available short-circuit current for 3-phase transformers of given kVA and secondary voltage provide some guidance. The lowest impedance of transformers is assumed based on a survey, infinite primary short-circuit current available, as well as worst-case adjustment values for transformer impedance tolerance and system voltage fluctuations.*

Voltage and Phase	kVA	Full Load Amps	% Impedance [tt] (Nameplate)	Short-Circuit Amps [t]	Voltage and Phase	kVA	Full Load Amps	% Impedance [tt] (Nameplate)	Short-Circuit Amps [t]
277/480 3-phase**	112.5	135	1	15,000	120/208 3-phase**	25	69	1.6	4,791
	150	181	1.2	16,759		50	139	1.6	9652
	225	271	1.2	25,082		75	208	1.11	20,821
	300	361	1.2	33,426		112.5	278	1.11	27,828
	500	601	1.3	51,362		150	416	1.07	43,198
	750	902	3.5	28,410		225	625	1.12	62,004
	1000	1203	3.5	38,180		300	833	1.11	83,383
	1500	1804	3.5	57,261		500	1388	1.24	124,373
	2000	2406	5	53,461		750	2082	3.5	66,095
	2500	3007	5	66,822		1000	2776	3.5	88,167
	–	–	–	–		1500	4164	3.5	132,190
	–	–	–	–		2000	5552	5	123,377
	–	–	–	–		2500	6950	5	154,444

**3-phase short-circuit currents based on "infinite" primary.

[tt]UL-listed transformers 25 kVA or greater have a ±10% impedance tolerance. Short-circuit amps reflect a "worst-case" condition.

[t]Fluctuations in system voltage will affect the available short-circuit current. For example, a 10% increase in system voltage will result in a 10% increase in the available short-circuit currents shown in the table.

Figure 9-17. *To provide estimates, various types of short-circuit currents as a percentage of 3-phase bolted fault current are shown.*

3-phase (L-L-L) bolted fault	100%
Line-to-line (L-L) bolted fault	87%
Line-to-ground (L-G) bolted fault	25%–125%* (use 100% at transformer, 50% elsewhere)
Line-to-neutral (L-N) bolted fault	25%–125%* (use 100% at transformer, 50% elsewhere)
Three-phase (L-L-L) arcing fault (480 V)	89%** (maximum)
Line-to-line (L-L) arcing fault (480 V)	74%** (maximum)
Line-to-ground (L-G) arcing fault (480 V)	38%** (minimum)

*Typically line-ground bolted fault current much lower than the bolted 3-phase fault current, but the bolted line-ground fault current can exceed the 3-phase bolted fault current if it is near the transformer terminals. Bolted line-ground fault current will normally be between 25% to 125% of the 3-phase bolted fault value.

**These arcing fault values as a percentage of the 3-phase bolted short-circuit currents are rough estimates. See IEEE 1584, Guide for Performing Arc-Flash Hazard Calculations, for equations to calculate the arcing current when performing an arc flash hazard analysis.

Data source: IEEE Standard 241-1990, Recommended Practice for Electric Power Systems in Commercial Buildings, Table 63.

REVIEW QUESTIONS

1. The purpose of a short-circuit current study is to determine the available short-circuit current at various points in an electrical __?__ so as to ensure safety and compliance with OSHA, *NEC*, and *NFPA 70E* requirements.
   - a. battery
   - b. field
   - c. generator
   - d. system

2. The maximum available fault current __?__ be marked on new service entrance equipment (with some exceptions) to help ensure the proper application of OCPD interrupting ratings and electrical equipment short-circuit current ratings.
   - a. cannot
   - b. might
   - c. must
   - d. should

3. When doing an incident energy arc flash hazard assessment, the analysis should consider various electrical distribution system scenarios and variables to determine the __?__ available short-circuit current, with an incident energy analysis then being performed for that/those value(s).
   - a. average
   - b. highest
   - c. lowest
   - d. lowest and highest

4. The bolted short-circuit current determines the __?__ amount of current that the electrical system can deliver during a short-circuit condition.
   - a. fastest
   - b. highest
   - c. minimum
   - d. slowest

5. The actual transformer impedance (%Z) value may vary ±10% from the marked (nameplate) values and therefore, for the high-fault condition scenario, multiply the transformer %Z by __?__.
   - a. 0.1
   - b. 0.9
   - c. 1.0
   - d. 1.1

6. The short-circuit current is typically the highest at the __?__.
   - a. branch circuit panel
   - b. feeder panel
   - c. receptacle outlets
   - d. service point

7. If a utility transformer fails and the utility replaces the transformer, most often the percentage __?__ of the newer transformers will be lower, resulting in a higher short-circuit current.
   - a. impedance
   - b. volt-amperes
   - c. voltage
   - d. wattage

8. Generally, the __?__ is the single most important component affecting the available short-circuit currents throughout an electrical system.
   - a. distance from transformer to the service panel
   - b. facility square footage
   - c. largest motor
   - d. transformer

9. Various methods have been developed to calculate the available short-circuit current, but all are based on __?__.
   - a. *NFPA 70*
   - b. *NFPA 70E*
   - c. Ohm's Law
   - d. OSHA regulations and standards

10. For arc flash hazard analysis, whether using the incident energy analysis method (130.5(B)(1)) or the hazard/risk categories method (130.5(B)(2)), typically the __?__ is a necessary value that must be determined.
    - a. ambient temperature
    - b. available short-circuit current
    - c. distance to the exit door
    - d. square footage of electrical room

10 OCPD Work Practices and Maintenance Considerations

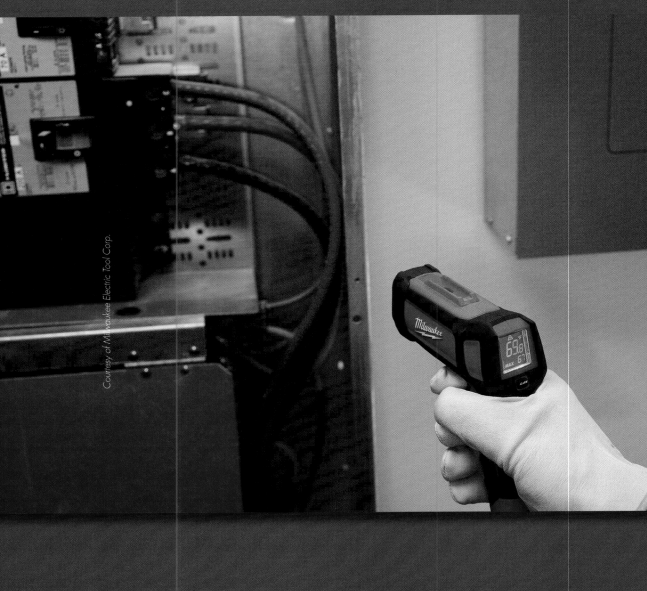

Courtesy of Milwaukee Electric Tool Corp.

- Overcurrent Protective Device Work Practices
- Maintenance: How It Relates to Electrical Safety

OBJECTIVES

1. Understand work practices with regard to overcurrent protective devices (OCPDs) that can enhance worker safety.
2. Understand that the arc-flash hazard incident energy is related to OCPD design and condition of maintenance.
3. Understand that electrical safety is directly tied to the condition of maintenance of other electrical equipment and the *NFPA 70E* Chapter 2 maintenance requirements.

REFERENCES

1. *NFPA 70E*, 2012 Edition
2. *NFPA 70B*, 2010 Edition, *Recommended Practice for Electrical Equipment Maintenance*
3. American National Standards Institute/InterNational Electrical Testing Association (ANSI/NETA) MTS-11, *Standard for Maintenance Testing Specifications for Electrical Power Equipment and Systems*

CASE STUDY

On October 22, 2006, at 7:44 a.m., a utility experienced an arc flash in a 4.16-kilovolt air/retrofit/vacuum circuit breaker. The circuit involved fed Forced Draft Fan 3; operations personnel tried to rack the breaker in, albeit unsuccessfully. Operations called an Electrical Worker, who removed the breaker and connected it to a Test Station, where it was determined that the circuit breaker had a mechanical problem and needed to be replaced. The spare breaker was connected to the Test Station and cycled several times to ensure it was working properly.

The spare breaker was then inserted into the cubicle and racked in. As the racking was taking place, the Electrical Worker heard a loud "clunk." The Electrical Worker assumed this noise was the floor trippers

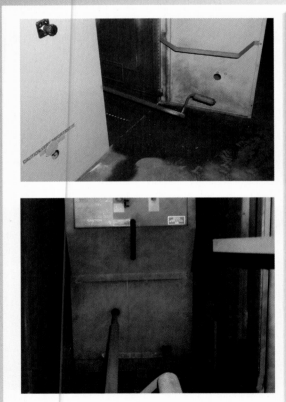

Figure C10-1. The photo at top shows burn marks on the floor of the enclosure and outside of the enclosure, indicating that the arc-flash incident occurred with the doors open. The photo at bottom shows arc-flash damage to the circuit breaker and racking handle after the incident.

Courtesy of the Lower Colorado River Authority

discharging the springs, which would be a normal event. The Electrical Worker did not verify the status of the circuit breaker. The breaker had actually closed instead of discharging the springs. Later discussions with the manufacturer revealed this situation could occur once in every 10,000 racking operations. As the Electrical Worker racked the now-closed breaker onto the switchgear stationary connectors, the circuit breaker tried to carry load current. However, because it was only partially racked in, the breaker did not have adequate contact to carry the full current. The resulting arc flash blew an arc plasma ball and molten copper out of the cubicle.

This utility had received a new 40 calories per square centimeter (cal/cm²) arc-rated suit the week before. The suit it replaced was an obsolete, unrated arc-flash suit. Fortunately, the Electrical Worker had elected to wear the new arc-rated suit to perform this task. **See Figure C10-1.** Note the condition of the breaker and cubicle as found after the incident. The door is open, even though it could have been closed during the racking operation. This step could have provided some shielding for the Electrical Worker, even though the door probably would have been forced open by the pressure wave (blast) of the arc. The arc flash was estimated at 33 cal/cm². **See Figure C10-2.** Note the inside of the cubicle where the breaker's stab connectors touched the switchgear stationary connectors. There are vaporized metal deposits. **See Figure C10-3.** Note the breaker's stab connectors after making contact (not fully engaged) with the energized switchgear stationary connectors; much of the stab connector metal vaporized during the arcing fault incident.

The Electrical Worker received minor burns across the back of his neck, due to the hood being worn incorrectly. He sustained no other injuries.

Source: This case study was contributed by Jim White of Shermco Industries.

For additional information, visit qr.njatcdb.org Item #1200.

Figure C10-2. *This view inside the switchgear cubicle shows damage from the incident to the stationary bus connectors, which accept the circuit breaker's stab connectors.*

Courtesy of the Lower Colorado River Authority

Figure C10-3. *This photo shows the damage to the stab connectors on the back of the circuit breaker; these connectors are intended to fully engage with the stationary connectors prior to being energized under load.*

Courtesy of the Lower Colorado River Authority

Introduction

Proper OCPD device work practices and maintenance of electrical equipment are important to worker safety. Workers should be trained in and use safe work practices regarding overcurrent protective devices (OCPDs). If a fault has resulted in opening an OCPD, OSHA and *NFPA 70E* do not permit resetting the circuit breaker or replacing fuses until it is safe to do so. After a circuit breaker interrupts a fault, testing the circuit breaker may be required. There are means to reduce the arc-flash hazard when racking circuit breakers. Fuses should be tested properly. Maintenance of electrical equipment is required in *NFPA 70E* and is important for worker safety. The arc-flash hazard can be negatively affected by improper or lack of OCPD maintenance. *NFPA 70E* requires that OCPDs be maintained and the maintenance, tests, and inspections documented.

Overcurrent Protective Device Work Practices

Electrical Workers deal with circuit breakers and fuses on a regular basis from installing these devices on new installations, to performing diagnostics, to replacing them. There are several important safe work practices that should be implemented when working with circuit breakers and fuses, including maintenance and diagnostic tips.

Resetting Circuit Breakers or Replacing Fuses

A circuit breaker should not be reset, nor should fuses be replaced, until the cause of the problem is known and rectified as well as it has been verified that it is safe to reenergize the circuit. *NFPA 70E* 2012 Edition Section 130.6(L) states the following:

> **Reclosing Circuits After Protective Device Operation.**
> After a circuit is de-energized by the automatic operation of a circuit protective device, the circuit shall not be manually reenergized until it has been determined that the equipment and circuit can be safely energized. The repetitive manual reclosing of circuit breakers or reenergizing circuits through replaced fuses shall be prohibited. When it is determined that the automatic operation of a device was caused by an overload rather than a fault condition, examination of the circuit or connected equipment shall not be required before the circuit is reenergized.
>
> *Reprinted with permission from NFPA-70E 2012, Electrical Safety in the Workplace Copyright 2011, National Fire Protection Association, Quincy, MA 02169. This reprinted material is not the complete and official position of the NFPA on the referenced subject, which is represented only by the standard in its entirety.*

Occupational Safety and Health Administration (OSHA) 1910.334(b)(2) is essentially the same requirement with slightly different wording.

This is an important safety practice. If an OCPD opens as a result of fault conditions, damage at the point of the fault could result. If the fault is not located and rectified, reclosing it into the fault situation might result in an even more severe fault than the first fault. If the protective device is a circuit breaker, it could have been damaged or degraded during the initial interruption. Reclosing a degraded circuit breaker onto a repaired circuit may energize the circuit, but the protection against future overcurrents may be diminished. Therefore, following proper procedures after an OCPD has interrupted a fault is important. If the cause of a circuit breaker or fuse opening is an overload, then it is permissible to merely reset the circuit breaker or replace the fuses.

Circuit Breakers

Important work practice considerations should be implemented when an electrical installation includes circuit breakers. These include what to do after a circuit breaker interrupts a fault, racking circuit breakers, or replacing or adding circuit breakers to a panel.

Circuit Breaker Evaluation After Fault Interruption

After fault interruption, it may be necessary to evaluate a circuit breaker for suitability of use before being placed back into service. This evaluation requires visual inspection and may require electrical testing to specifications according to the manufacturer's procedures. If the circuit breaker manufacturer's procedures are not available it is advisable to use industry standards.

70E Highlights

NFPA 70E requires the following with regard to circuit breakers after interrupting a fault:

225.3 Circuit Breaker Testing After Electrical Faults. Circuit breakers that interrupt faults approaching their interrupting ratings shall be inspected and tested in accordance with the manufacturer's instructions.

Reprinted with permission from NFPA-70E 2012, Electrical Safety in the Workplace Copyright 2011, National Fire Protection Association, Quincy, MA 02169. This reprinted material is not the complete and official position of the NFPA on the referenced subject, which is represented only by the standard in its entirety

After a circuit breaker interrupts a fault, it might not be suitable for further service. A fault might potentially erode a circuit breaker's contacts, damage the arc chutes, or weaken the circuit breaker's case. If the short-circuit (fault) current is high (near its interrupting rating), circuit breaker manufacturers recommend that a circuit breaker receive a thorough inspection, with replacement if necessary. Some difficulties in the evaluation process include not knowing a circuit breaker's service history, the level of short-circuit current that a circuit breaker interrupted, or the type of degradation that occurred on the inside of the circuit breaker. That is one reason why periodic maintenance and testing

by a qualified maintenance person are necessary as well as inspection and testing after interrupting a fault approaching a circuit breaker's interrupting rating.

The following is an insightful passage written by Vince A. Baclawski, Technical Director, Power Distribution Products, National Electrical Manufacturers Association (NEMA) in Electrical Construction & Maintenance Magazine, January 1995, p. 10. Copyright EC&M (January 1995). Reprinted by permission of Penton media.

> After a high level fault has occurred in equipment that is properly rated and installed, it is not always clear to investigating electricians what damage has occurred inside encased equipment. The circuit breaker may well appear virtually clean while its internal condition is unknown. For such situations, the NEMA AB4 "Guidelines for Inspection and Preventive Maintenance of MCCBs Used in Commercial and Industriaal Applications" may be of help. Circuit breakers unsuitable for continued service may be identified by simple inspection under these guidelines. Testing outlined in the document is another and more definite step that will help to identify circuit breakers that are not suitable for continued service.
>
> After the occurrence of a short circuit, it is important that the cause be investigated and repaired and that the condition of the installed equipment be investigated. A circuit breaker may require replacement just as any other switching device, wiring or electrical equipment in the circuit that has been exposed to a short circuit. Questionable circuit breakers must be replaced for continued, dependable circuit protection.

For more information, see the discussion on circuit breaker maintenance later in this chapter.

Racking Circuit Breakers

Numerous injuries and deaths have occurred among workers who rack in/out low-voltage power circuit breakers or medium-voltage vacuum circuit breakers. One safe work practice is to increase the working distance in such a setting. If the working distance from the potential arc flash is increased to 36 inches rather than 18 inches, the potential incident energy at 36 inches is typically about one-fourth what it would be at 18 inches. The following methods can be used to move the worker farther from the potential arc source or outside the flash boundary for hazardous operations:

1. Remote-controlled motorized devices that rack in and out low- and medium-voltage circuit breakers. **See Figure 10-1.**

2. Extended length, hand-operated racking tools. **See Figure 10-2.**

Figure 10-1. *A remote racking device permits a worker to rack circuit breakers in or out at a safe distance from the equipment.*

Courtesy of Eaton Corporation

Figure 10-2. *Racking tools of varying lengths are available to move the worker farther from the potential arc source.*

Courtesy of DuPont

Replacing Circuit Breakers

When adding a circuit breaker to a panel or replacing a circuit breaker, it is imperative to ensure that the circuit breaker is suitable for that panel and the circuit breaker has the proper ampere rating, voltage rating, interrupting rating, and equivalent short-circuit performance.

With molded case circuit breakers, there typically is a variety of circuit breakers from the same manufacturer which are physically interchangeable but have different ratings. **See Figure 10-3.** These three circuit breakers are the same frame size and are physically interchangeable. The 240-V circuit breaker could physically be installed in place of the 600-V circuit breaker, the 100-A circuit breaker could physically be installed in place of the 20-A circuit breaker, and the 10-kA interrupting rated circuit breaker could physically be installed in place of the 65-kA interrupting circuit breaker. In addition, a listed current-limiting molded case circuit breaker must be tested and marked as current limiting. Most molded case circuit breakers are not current limiting, yet a current-limiting circuit breaker and a standard (non-current-limiting) circuit breaker can be physically interchangeable.

Fuses

Several work practices should be evaluated when an electrical installation includes fuses. These include testing and replacing fuses.

Testing Fuses

When a fuse is suspected of having opened, safe work practices designated by the employer should be followed. One option is to deenergize the fuse from the source of power, including implementing lockout/tagout

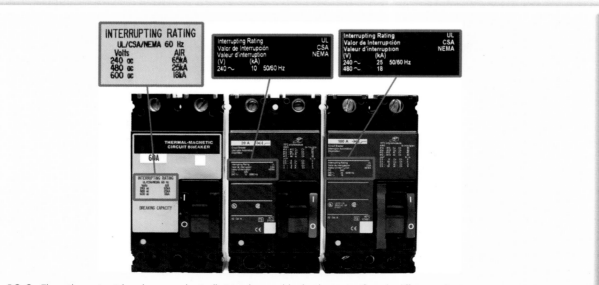

Figure 10-3. *These three circuit breakers are physically interchangeable, but have significantly different voltage ratings, interrupting ratings, and ampere ratings.*

Courtesy of Cooper Bussmann

procedures, followed by removing both indicating and nonindicating fuses from the circuit, and checking each fuse for continuity by resistance measurement. If a fuse is being replaced, the replacement fuse should be of a proper type and ampere rating.

When testing a fuse for resistance:

1. For ferrule fuses, place the test probes on the metal ferrules. **See Figure 10-4**.

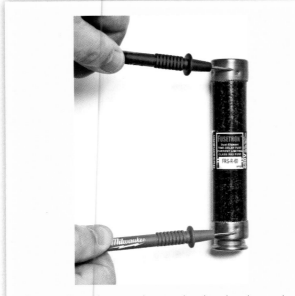

Figure 10-4. *The test probes must be placed on the metal ferrules for a ferrule fuse.*

Courtesy of Cooper Bussmann

2. For knife blade fuses, place the test probes on the metal blades, not on the fuse end caps. **See Figure 10-5**.

It is important to properly test knife-blade fuses. Fuse manufacturers do not generally design these types of fuses to ensure electrically energized fuse caps during normal fuse operation. For knife-blade fuses, electrical inclusion of the end caps into the circuit occurs as a result of the coincidental mechanical contact between the fuse cap and terminal extending through it. In most brands of knife-blade fuses, this mechanical contact is not guaranteed; therefore, electrical contact is not guaranteed. One fuse manufacturer has designed some knife-blade fuse versions so that the end caps are insulated to reduce the possibility of accidental contact with a live part. Thus a resistance reading to check for continuity (i.e., that the fuse is open or still usable) taken across the fuse caps is not indicative of whether the fuse is open. For knife-blade fuses, you should always have the test probes touch the metal knife blades. **See Figure 10-6**.

As part of diagnostic testing/troubleshooting, fuses can be checked via voltage testing while the equipment is energized. Safe work practices and proper personal protective equipment (PPE) must be utilized during this process. The test probes should contact the metal end caps of ferrule fuses and the blades on knife-blade fuses, much in the same way discussed earlier in this section.

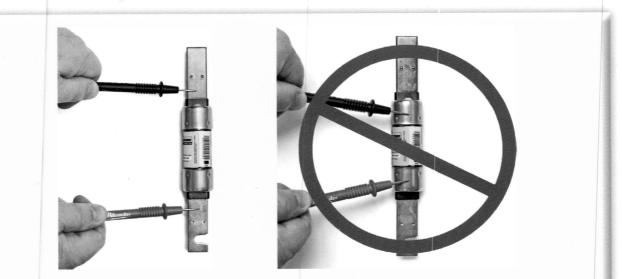

Figure 10-5. *For knife blade fuses always place test probes on the metal blades as shown on the left. Do not place test probes on fuse end caps as shown on the right.*

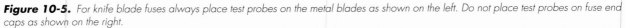

Courtesy of Cooper Bussmann

Insulated Caps   Noninsulated Caps

Figure 10-6. *The fuse on the left is designed to ensure that the end caps are electrically insulated. The fuse on the right does not have the designed-in end-cap insulation; however, an electrical connection between the fuse blades and end caps is not assured.*

Courtesy of Cooper Bussmann

Replacing Fuses

Fuses that interrupt a circuit should be replaced with the proper fuse in terms of both type and ampere rating. Modern current-limiting fuses are always recommended. When using modern current-limiting fuses, new factory-calibrated fuses are installed in the circuit, and the original level of overcurrent protection is maintained for the life of the circuit.

In most newer building systems and utilization equipment, the fuse mountings accept only current-limiting fuses of a specific Underwriters Laboratories Inc. (UL) Class fuse. For example, Class J fuse mountings will accept only UL Class J fuses; no other UL Class fuses can be installed in a Class J mounting. This standardization ensures a unique safety system in regard to voltage rating, interrupting rating, and short-circuit protection capabilities. The reason is that per the UL standard, all Class J fuses are rated as 600 V AC, have at least a 200,000-A interrupting rating, and provide a specific degree of current limitation (at a minimum) under short-circuit conditions. Thus, when a Class J fuse is replaced, this system ensures that only a Class J fuse can be inserted in its place—that is, it ensures that the fuse has a 600 VAC rating, at least a 200,000-kA rating, and a very high degree of current limitation. In essence, this standardization creates an electrical safety system for overcurrent protection

that is unique only for current-limiting fuse classes. Each UL fuse class has its own unique dimensions and specific voltage rating, interrupting rating, and current-limiting performance per the UL standard for fuses.

For older systems, where the fuse clips can accept older-style fuses (Class H), it is recommended to store and use only modern current-limiting fuses (Class RK1) that also can be used in Class H fuse clips. See the section in the chapter covering design and upgrades for improved arc-flash mitigation by upgrading existing fuses.

Maintenance: How It Relates to Electrical Safety

This section was contributed by Jim White, Training Director at Shermco Industries, who is a certified Level IV NETA Technician and a technical committee member on *NFPA 70E* and *NFPA 70B*, in addition to serving on several maintenance-related industry standard committees. Shermco Industries provides training, testing, repair, maintenance, and analysis of power distribution systems and related equipment. NETA (InterNational Electrical Testing Association) is an independent, third-party electrical testing association that establishes world standards in electrical maintenance and acceptance testing.

How does maintaining electrical power system equipment and devices become a worker safety issue? Many mid- and upper-level managers (especially those from a management or accounting background) would consider maintenance to be an overhead cost unrelated to worker safety. This is far from accurate and is why the 2012 edition of *NFPA 70E* Chapter 2, "Safety-Related Maintenance Requirements," specifically requires electrical protective devices to be maintained properly. Section 205.4 states,

Overcurrent Protective Devices. Overcurrent protective devices shall be maintained in accordance with the manufacturers' instructions or industry consensus standards. Maintenance, tests, and inspections shall be documented.

Reprinted with permission from NFPA-70E *2012, Electrical Safety in the Workplace Copyright 2011, National Fire Protection Association, Quincy, MA 02169. This reprinted material is not the complete and official position of the NFPA on the referenced subject, which is represented only by the standard in its entirety.*

In addition, *NFPA 70E* Section 200.1(3) states,

(3) For the purpose of Chapter 2, maintenance shall be defined as preserving or restoring the condition of electrical equipment and installations, or parts of either, for the safety of employees who work where exposed to electrical hazards. Repair or replacement of individual portions or parts of equipment shall be permitted without requiring modification or replacement of other portions or parts that are in a safe condition.
Informational Note: Refer to NFPA 70B, *Recommended Practice for Electrical Equipment Maintenance*, and ANSI/NETA MTS-2011, *Standard for Maintenance Testing Specification for Electrical Power Distribution Equipment and Systems*, for guidance on maintenance frequency, methods, and tests.

Reprinted with permission from NFPA-70E *2012,* Electrical Safety in the Workplace *Copyright 2011, National Fire Protection Association, Quincy, MA 02169. This reprinted material is not the complete and official position of the NFPA on the referenced subject, which is represented only by the standard in its entirety.*

Note that the preceding passage specifically calls out that maintenance is preserving or restoring the condition of electrical power systems equipment or devices *for the safety of employees* exposed to electrical hazards. Electrical safety is directly tied to the condition of electrical power systems equipment, especially protective devices, such as circuit breakers and protective relays. One topic of discussion that repeatedly came up during the *NFPA 70E* committee meetings was that enclosed energized equipment is safe if:

1. Properly designed and engineered.
2. Properly installed (using all applicable national and local codes and standards).
3. Properly maintained.

The consensus of the *NFPA 70E* committee was that if all three of these elements were not in place, no procedure or method to determine how to work safely on that system exists. If equipment and protective devices are not properly maintained, engineering analysis, such as short-circuit current studies and coordination studies, no longer have any relevance. This also includes the calculations performed as part of an arc flash hazard analysis. IEEE-1584, *NFPA 70E*, and other similar methods of determining the arc-flash hazard and choosing PPE presume the existence of properly maintained and operating electrical systems.

130.7(A)
Informational Note No. 2: It is the collective experience of the Technical Committee on Electrical Safety in the Workplace that normal operation of enclosed electrical equipment, operating at 600 volts or less,

that has been properly installed and maintained by qualified persons is not likely to expose the employee to an electrical hazard.

Reprinted with permission from NFPA-70E *2012,* Electrical Safety in the Workplace *Copyright 2011, National Fire Protection Association, Quincy, MA 02169. This reprinted material is not the complete and official position of the NFPA on the referenced subject, which is represented only by the standard in its entirety.*

How a Lack of Maintenance Can Increase Hazards

When workers operate electrical power equipment, it is assumed that when the handle is operated everything will work as advertised. The National Institute of Occupational Safety and Health (NIOSH) presented a paper at the 11th Annual IEEE/IAS Electrical Safety Workshop: "Non-contact Electric Arc-Induced Injuries in the Mining Industry: A Multi-disciplinary Approach," by Kathleen Kowalski-Trakofler, PhD. **See Figures 10-7** and **10-8**. Thirty-four percent of electrically related accidents are caused by component failure, and 19% of those accidents are caused by components that failed during normal operation.

Among Electrical Workers, 24% of accidents are the result of troubleshooting-type activities. This makes sense when the compactness of newer equipment is

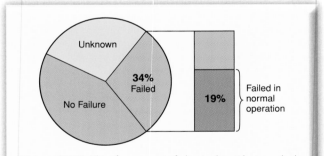

Figure 10-7. *Thirty-four percent of electrical accidents resulted from component failure, and 19% involved an electrical component that failed in routine operation.*

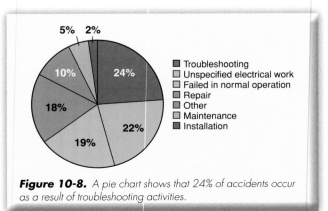

Figure 10-8. *A pie chart shows that 24% of accidents occur as a result of troubleshooting activities.*

considered, as well as the fact that the conductors often have to be pushed out of the way to get test probes or tools into position. Eighteen percent of accidents were the result of failure during repair or repair-related activities. In this study, five types of electrical devices were implicated in two-thirds of the cases studied:

Circuit breakers	17%
Conductors	16%
Nonpowered hand tools	13%
Electrical test meters and leads	12%
Connectors and plugs	11%

We can be fairly certain that most of the devices that failed in this study did so due to lack of maintenance.

Lack of Lubrication

Electrical power system protective devices require regular maintenance, and when they do not receive that maintenance, dangerous situations are created for Electrical Workers. Lack of maintenance can take on many forms. For example, the circuit breaker repair shop of Shermco Industries sees an astonishing number of circuit breakers and switches of all voltage ratings that suffer from a lack of lubrication. **See Figure 10-9**. Note that this operating mechanism has severely rusted components and spider webs hanging off it. Obviously, it has been neglected for years. Many equipment owners do not realize that the conductive (current-carrying) paths of circuit breakers and many switches are lubricated when they are built and that the lubrication dries out over time, even if the breaker (or switch) has not been operated. They think that the device should be like new, because it rarely operates.

Some components of a circuit breaker moving contact assembly are in the current path. **See Figure 10-10**.

These parts are normally lubricated at the factory and require periodic cleaning and relubrication to function properly. As the lubricant on these parts ages, it dries and becomes gummy, finally drying to the point where it flakes off the components. The rollers show deep gouging caused by lack of lubrication, and the arcing contact assembly also shows significant wear at the point where it contacts the rollers. This breaker operated very slowly and would occasionally seize in mid-operation.

Consider the effect on operating time if this occurs. According to *NFPA 70E* Annex D, incident energy is proportional to operating time. If a circuit breaker's clearing time doubles due to lack of maintenance, the resultant arc-flash incident energy also doubles. As an example, suppose a worker is about to service a device in a panel protected by a molded-case circuit breaker (approximately two-cycle clearing time) that has an incident energy of 8 cal/cm^2 [at 18 in.]. If the Electrical Worker wears arc-rated PPE for a maximum 10 cal/cm^2, the worker should be protected. If that device delays opening, the protection might not be adequate. The operating time of OCPDs is critical to worker safety.

Maintenance Can Affect Safety

NFPA 70E Section 130.5, Arc Flash Hazard Analysis, requires that the arc flash boundary (AFB) and PPE be determined for people who will be within the AFB. Section 130.5 also contains requirements for considering the condition of maintenance for OCPDs when conducting arc flash hazard analysis.

Figure 10-9. *This circuit breaker has rusted parts and spider webs; would you bet your life on this equipment working properly?*

Courtesy of James R. White, Shermco Industries

Figure 10-10. *Lack of lubrication on contact components caused damage, which affected the circuit breaker operation.*

Courtesy of James R. White, Shermco Industries

The reliability of OCPDs can directly impact arc-flash hazards. One portion of Section 130.5 reads as follows:

> The arc flash hazard analysis shall take into consideration the design of the overcurrent protective device and its opening time, including its condition of maintenance.
>
> *Reprinted with permission from NFPA-70E 2012, Electrical Safety in the Workplace Copyright 2011, National Fire Protection Association, Quincy, MA 02169. This reprinted material is not the complete and official position of the NFPA on the referenced subject, which is represented only by the standard in its entirety.*

In addition, Section 130.5 refers to two Informational Notes concerning the importance of overcurrent protective device maintenance:

> Informational Note No. 1: Improper or inadequate maintenance can result in increased opening time of the overcurrent protective device, thus increasing the incident energy. Informational Note No. 4: For additional direction for performing maintenance on overcurrent protective devices see Chapter 2, Safety-Related Maintenance Requirements.
>
> *Reprinted with permission from NFPA-70E 2012, Electrical Safety in the Workplace Copyright 2011, National Fire Protection Association, Quincy, MA 02169. This reprinted material is not the complete and official position of the NFPA on the referenced subject, which is represented only by the standard in its entirety.*

Poorly Maintained Circuit Breakers May Affect Incident Energy

A poorly maintained circuit breaker may result in an actual arc-flash event having higher incident energy than that determined by an arc-flash hazard analysis which assumed the circuit breaker was in properly maintained operating condition. **See Figure 10-11**. The panel is protected by an upstream 800-ampere circuit breaker that has a short-time delay setting of six cycles. This is an intentional delay feature of the circuit breaker so as to have this circuit breaker selectively coordinate with the circuit breakers in the panel it is supplying. The calculated arc-flash hazard uses the six-cycle opening time for the circuit breaker short-time delay. However, if the circuit breaker has not been maintained, it may take longer to interrupt an arcing fault. The arc-flash hazard is much higher if the circuit breaker took 30 cycles to open due to poor maintenance. Obviously, if the circuit breaker took even longer to open, the actual arc-flash hazard would be even worse. This scenario illustrates why maintenance of OCPDs is so important.

What if the ability of an OCPD to function properly is questioned? Often, as part of the hazard identification and a risk assessment, assuming that the OCPD will not function properly is safer. In these cases, it is best to deenergize. The time spent in a planned outage is far less than what would be required to repair and replace damaged equipment, not to mention the possible effects on the worker.

If the decision is made that the circuit or equipment cannot be placed in an electrically safe work condition, the next device upstream that is deemed reliable must be considered as the protective device that will operate. In the example in the preceding paragraph, note that because the 800-ampere OCPD is not maintained, the next device upstream that is maintained properly would be the one used in determining the arc-flash hazard. It is probable that, due to the increase in operating time of a larger ampacity OCPD, the incident energy will be substantially higher. If the next device upstream is not reliable, the device upstream from it that has been properly maintained would have to be used to assess the arc-flash hazard.

Figure 10-11. *The arc-flash hazard versus actual hazard that could occur due to poor or no maintenance is illustrated: Assuming the circuit breaker has been maintained and operates as specified by manufacturer's performance data, the incident energy would be 5.8 cal/cm² at 18 inches and the AFB is 47 inches (calculation per IEEE 1584) (left). If lack of maintenance causes the circuit breaker to clear in 30 cycles rather than in six cycles, the actual arc-flash hazard would be an incident energy of 29 cal/cm² at 18 inches, and the AFB would be 125 inches (calculation per IEEE 1584) (right).*

The effects on arc flash hazard analysis if the first-level upstream circuit breakers do not operate as intended has been examined for an industrial facility in "Prioritize Circuit Breaker and Protective Relay Maintenance Using an Arc Flash Hazard Assessment" (IEEE Paper No. ESW-2012-11), by Dan Doan. The results of the analysis included the following conclusion: "Based on this analysis, approximately 2/3 (91 out of 136) of the circuit breakers and relays identified in the arc-flash study can be designated as 'Critical.' This means that if they fail, the PPE labeled would be inadequate, by one or more classes."

Misapplied Fuse Application

Overcurrent protective devices must not be modified beyond approved means or rendered inoperative. **See Figure 10-12**. Note that a disconnect had renewable type (replaceable element) fuses installed. The fuses clearly state they have a maximum of a 200-ampere full-load current rating, but the equivalent of a 400-ampere element (double-linked) was used. The loading on the switch exceeded the switch rating for an extended time and resulted in severe overheating of the switch. To remove the switch from service, the technician had to pry the mechanism open with his screwdriver.

OSHA regulation 29 CFR 1910.334(b)(3) states:

"Overcurrent protection modification." Overcurrent protection of circuits and conductors may not be modified, even on a temporary basis, beyond that allowed by 1910.304(e).

Renewable fuses are an older Class H fuse type in which the body of the fuse can be disassembled and renewable fuse links inserted. These types of fuses are not recommended for use and should be replaced with modern Class RK1 current-limiting fuses, which will physically fit in Class H fuse mountings.

The 2011 edition of the *NEC* also weighs in on replaceable element fuses in 240.60(D):

Renewable Fuses. Class H cartridge fuses of the renewable type shall be permitted to be used only for replacement in existing installations where there is no evidence of overfusing or tampering.

Reprinted with permission from NFPA-70 2011, National Electrical Code® Copyright 2010, National Fire Protection Association, Quincy, MA 02169. *This reprinted material is not the complete and official position of the NFPA on the referenced subject, which is represented only by the Code in its entirety.*

General Maintenance Requirements

OCPD interrupting times are one of the key factors affecting the arc-flash incident energy. Circuit breakers are mechanical devices and require periodic maintenance to ensure their proper operation. Preventive maintenance for circuit breakers should include exercising the mechanism by periodically opening and closing the circuit breakers, periodic visual and mechanical inspections, and periodic calibration tests. The trip latch mechanism is also important to exercise, as it can seize due to lack of use. The trip latch mechanism can be exercised by primary injection testing or, if a circuit breaker is so equipped, by using a push-to-trip or similar button (usually red in color) that directly operates the trip latch. This is preferable to just opening and closing the breaker. In addition, conductor terminations should be periodically checked for signs of overheating, poor connections, and insufficient

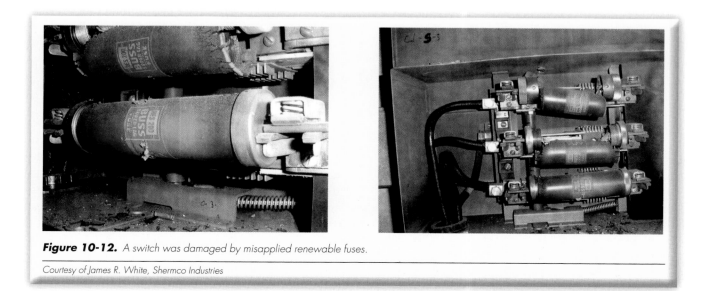

Figure 10-12. *A switch was damaged by misapplied renewable fuses.*

Courtesy of James R. White, Shermco Industries

conductor ampacity. Infrared thermographic scans are one method that can be used to monitor these conditions. Records of the maintenance inspections, tests, and conditions should be retained to establish a trend.

The internal parts of current-limiting fuses do not require maintenance. However, periodically checking fuse bodies, fuse mountings, and adjacent conductor terminations for signs of overheating, poor connections, or insufficient conductor ampacity is important. Fuses are typically used in conjunction with disconnects—and disconnects also require periodic inspection and maintenance. If a disconnect has lubricated mechanisms, then it is necessary to maintain the lubrication properly.

In applications where disconnects are equipped with a ground-fault protection relay, the disconnects should be periodically inspected and maintained, and the ground-fault relay calibrated. After a disconnect interrupts a ground fault, the disconnect should be inspected and maintained if necessary.

Often, engineering studies are performed analyzing protective device coordination and arc-flash hazard potential for an electrical system. As part of the engineering analysis, the type of OCPDs and the circuit breaker or relay settings and the fuse types (characteristics) are determined. OCPDs must be maintained and their settings not altered from those determined from these studies. If a circuit breaker or a fuse operates, it is a clear indication that there is a problem in the electrical system—not that there is a problem with the circuit breaker or fuse. Unfortunately, many production people do not understand that replacing correctly rated fuses with larger-ampacity fuses, automatically reclosing circuit breakers after they open, or changing a breaker's settings is dangerous to the equipment being protected and to the people working on or near it.

Maintenance Requirements for OCPDs in *NFPA 70E*

Important requirements for OCPD maintenance are contained in Chapter 2 of *NFPA 70E*:

205.4 Overcurrent Protective Devices. Overcurrent protective devices shall be maintained in accordance with the manufacturers' instructions or industry consensus standards. Maintenance, tests, and inspections shall be documented.

210.5 Protective Devices. Protective devices shall be maintained to adequately withstand or interrupt available fault current.

Informational Note: Failure to properly maintain protective devices can have an adverse effect on the arc flash hazard analysis incident energy values.

225.1 Fuses. Fuses shall be maintained free of breaks or cracks in fuse cases, ferrules, and insulators. Fuse clips

shall be maintained to provide adequate contact with fuses. Fuseholders for current-limiting fuses shall not be modified to allow the insertion of fuses that are not current-limiting.

225.2 Molded-Case Circuit Breakers. Molded-case circuit breakers shall be maintained free of cracks in cases and cracked or broken operating handles.

225.3 Circuit Breaker Testing After Electrical Faults. Circuit breakers that interrupt faults approaching their interrupting ratings shall be inspected and tested in accordance with the manufacturer's instructions.

Reprinted with permission from NFPA-70E 2012, Electrical Safety in the Workplace *Copyright 2011, National Fire Protection Association, Quincy, MA 02169. This reprinted material is not the complete and official position of the NFPA on the referenced subject, which is represented only by the standard in its entirety.*

Maintenance of Other Equipment

NFPA 70E Chapter 2 is dedicated to the maintenance requirements of electrical equipment for worker safety. In addition to maintenance requirements for OCPDs, *NFPA 70E* contains requirements for other parts of the electrical power system. The following are Articles of *NFPA 70E* Chapter 2:

200 Introduction

205 General Maintenance Requirements

210 Substations, Switchgear Assemblies, Switchboards, Panelboards, Motor Control Centers (MCC), and Disconnect Switches

215 Premises Wiring

220 Controller Equipment

225 Fuses and Circuit Breakers

230 Rotating Equipment

235 Hazardous (Classified) Locations

240 Batteries and Battery Rooms

245 Portable Electric Tools and Equipment

250 Personal Safety and Protective Equipment

Maintenance Programs and Frequency of Maintenance

The scope of *NFPA 70E* Chapter 2 clarifies that these maintenance requirements are only those associated with employee safety; the requirements are not prescriptive. The employer must choose the maintenance and methods that satisfy the requirements. Sources for guidance are provided for setting up maintenance programs, determining the frequency of maintenance, and providing prescriptive procedures. Equipment manufacturer's maintenance manuals should be one of the first sources to use.

NFPA 70B, Recommended Practice for Electrical Equipment Maintenance, provides some frequency of maintenance guidelines as well as guidelines for setting up an electrical preventive maintenance (EPM) program, including sample forms and requirements for electrical system maintenance. The appendix, *Maintenance Tests for Molded-Case and Insulated-Case Circuit Breakers*, of this book is an example of the maintenance and tests recommended for molded case circuit breakers from MTS-2011. ANSI/NETA MTS-2011, *Standard for Maintenance Testing Specifications for Electrical Power Equipment and Systems*, is another standard that is prescriptive regarding which maintenance and testing are required for electrical power systems devices and equipment. Mechanical, visual, and electrical maintenance and tests are specified by equipment type as well as in terms of which results are acceptable. ANSI/NETA MTS-2011 includes guidelines for the frequency of maintenance required for electrical system power equipment in its Appendix B, *Frequency of Maintenance Tests*. This is a very useful tool for determining how often testing and maintenance are required, based on the condition of the equipment and its criticality. This Maintenance Frequency Matrix is reproduced as the appendix, *ANSI/NETA Frequency of Maintenance*, in this book, courtesy of NETA.

Maintenance and Hazard Identification/ Risk Assessment

NFPA 70E Section 130.3(B)(1) directs a worker to perform a shock hazard analysis and arc flash hazard analysis prior to work beginning when the electrical equipment is energized. The hazard would be the voltage level (shock) and the expected incident energy exposure (arc flash). When blast hazard calculations become available, they should be included as well. This is where many people or organizations stop their assessment; however, the risk must also be evaluated.

In approaching a lineup of switchgear, several factors must be evaluated:

1. The environment in which the equipment is operating: Is it exposed to the elements? Is it in a positive-pressure, air-conditioned area that is generally clean? Equipment in clean, dry environments is much less likely to experience deterioration than equipment that is exposed to the elements or contaminants.
2. How heavily loaded is the equipment? Lightly loaded equipment is under much less thermal stress than equipment that is carrying its maximum rated load (or more).
3. What is the overall condition of the equipment? Clean, well-maintained equipment is more likely to operate in accordance to the manufacturer's specifications than equipment in distress.
4. What is the operating history of the equipment? If certain types or brands of equipment are known to be troublesome, take that history into account.
5. When was the equipment last maintained and tested? If the equipment is out of calibration or if there is no calibration sticker present, the chances of a problem increase greatly.
6. What is the configuration of the equipment and enclosure? Equipment with vents poses an arc-flash hazard to operators, even though the equipment's door might be closed. These vents cannot be closed, as the breakers and bus current ratings depend on the airflow through such vents. Closing them would cause rapid deterioration of the insulation or OCPD nuisance tripping in this type of switchgear. **See Figure 10-13**. Note that the expanded metal air vents expose arcing components directly to the operator. In either case, an arc flash would expose the operator to the arc-plasma ball, potentially causing severe burns if the worker was not properly protected. For the purposes of a hazard identification and a risk assessment, these doors are closed for shock hazards, but opened for the arc-flash hazard.

Typically, workers will find that more arc-flash PPE—not less—is needed after a hazard identification and risk assessment is performed. *NFPA 70E* is intended to make injuries survivable, not to prevent injury. If a worker is exposed to the maximum incident energy for which his or her PPE is rated, there is a 50% probability of a second-degree burn on bare skin under the arc-flash personal protective clothing. If the incident energy is greater than the arc rating of the PPE, the chances of injury increase.

The best course of action is to deenergize the equipment if there is serious doubt about the condition and functionality of the equipment to be worked on. This is not always possible, especially when troubleshooting electrical equipment and devices. In such a case, the following points should be considered:

1. Distance is your friend. The incident energy generally decreases by the inverse square of the distance. Simply put, if the distance from the arc source is doubled, the incident energy exposure to the worker is decreased by approximately one-fourth. If the distance is tripled, the incident energy exposure is decreased by approximately

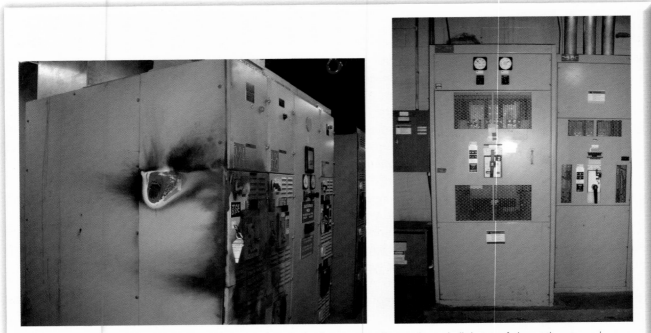

Figure 10-13. *Vented switchgear. Louvered venting showing the damage done by arc-plasma ball during a failure at the primary bus connections (left). Front-vented low-voltage switchgear exposes a worker to an arc-flash hazard even when the doors are closed (right).*

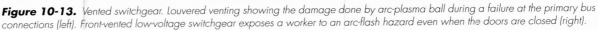

Left: *Courtesy of H. Landis Floyd/DuPont*
Right: *Courtesy of James R. White, Shermco Industries*

one-ninth. Consider using longer tools or test probes, wearing leather gloves (even for 120-volt panelboards), and keeping your arms extended to a comfortable distance.

2. If the equipment fails, it does not matter which task is being performed. Chances are that the worker will be exposed to the full arc flash and its effects. Downsizing arc-rated PPE and equipment usually is not a safe course of action, unless the PPE itself poses a risk or increases the risk.

3. When troubleshooting a problem, the chances of failure are probably higher than normal. That is why you are troubleshooting.

4. The chances of injury increase with the number of workers present. Only the people directly involved with the task should be exposed to the hazards. People who are monitoring or auditing the task should be outside either the limited approach boundary or the AFB, whichever distance is greater.

5. Each layer of nonmelting clothing under arc-rated PPE reduces the arc-flash heat to the body by approximately 50% per layer. The air that is trapped between layers of clothing acts in the same way that feathers act in a down jacket. However, consider the *NFPA 70E* 130.7(C)(9)(a) requirements described in the chapter covering PPE. *NFPA 70E* Section 130.7(C)(9)(a) Layering states, in part:

If nonmelting, flammable fiber garments are used as underlayers, the system arc rating shall be sufficient to prevent breakopen of the innermost arc-rated layer at the expected arc exposure incident energy level to prevent ignition of flammable underlayers. Garments that are not arc rated shall not be permitted to be used to increase the arc rating of a garment or of a clothing system.

Reprinted with permission from NFPA-70E 2012, *Electrical Safety in the Workplace Copyright 2011, National Fire Protection Association, Quincy, MA 02169. This reprinted material is not the complete and official position of the NFPA on the referenced subject, which is represented only by the standard in its entirety.*

The last sentence is important. If the clothing's arc rating is exceeded and the garment breaks open, the underlayers will be exposed and likely could ignite and burn.

6. Incident energy is proportional to the time of exposure. If the exposure time to an arc flash is doubled, the incident energy received by the worker is doubled.

Qualifications for Performing Maintenance on Electrical Equipment

Several factors should be considered when deciding whether to perform electrical power systems maintenance and testing in-house or to use an outside company. Depending on the circumstances of your particular company, the following considerations may or may not be an issue:

1. The cost of test equipment. If the equipment is already purchased, there is no issue. If it has to be purchased, it also has to be maintained, which would include periodic calibration and repair for field damage.

2. Training of the test technician. Tests must be performed in a standardized manner that does not affect the results. Test technicians must be well trained and familiar with the specific test equipment used and the equipment being tested. All technicians must perform testing in the same manner, using the same methods each time. Factors that affect test results must also be part of the knowledge base. Experience goes a long way with this issue.

3. Having the correct test equipment for the tests being performed. Trying to "make do" with certain test equipment does not work well when something more sophisticated or with a larger capacity is needed. Test equipment is like any other tool; it must fit the task.

4. Documentation. Customers require accurate documentation of both the as-found and as-left test results. Handwritten notes or forms are not acceptable for most companies. Several companies offer software for this purpose.

5. Elevated pay levels. Test technicians must be highly trained, motivated, and accurate. Their work will require more training and precision than what is required for the average Electrical Worker.

6. Frequency of use. The old saying, "Use it or lose it," is certainly applicable in testing. To be good at testing, technicians must perform testing work frequently. People tend to lose some expertise when performing tasks infrequently, especially when performing more complex types of testing, such as insulation power factor or calibration of protective relays.

7. Financial exposure. If an incident occurs during testing, the testing company can expect to be sued by the client. The immediate suspicion when a major failure occurs focuses on the testing and maintenance that may have been done in the past. Protective device calibration is always a potential issue.

When outsourcing the organization's electrical maintenance needs to an outside company is contemplated, several items should be considered:

1. Reputation. Does this company have a good reputation in the industry? Contact other customers in similar businesses for which the company has performed maintenance services. Electrical service companies sometimes tend to specialize in certain industries. They can shine in one type of industry but have no practical experience in others.

2. Is the company independent of manufacturers? Companies that are closely associated with manufacturers can be a double-edged sword. On the one hand, they can provide technical expertise that might not be generally available to the industry. On the other hand, they might try to steer work to other parts of their company that might not have the necessary level of expertise. Manufacturers also tend to ask for a higher hourly rate, because they consider themselves "experts." Judge companies by the quality of the technicians who are sent to the site, not their uniforms.

3. When was the company's equipment calibrated? What are the company's requirements for calibration? For example, NETA-member companies are required to calibrate test equipment every 12 months by a National Institute of Standards and Technology (NIST) traceable lab.

4. Which training are the company's technicians required to have? Are they qualified under OSHA's 29 CFR 1910.399 definition of a qualified person, 29 CFR 1910.332, and *NFPA 70E* Article 110.2(D)(1)? OSHA's Multi-Employer Worksite Policy (CPL-2-0.124 Rev. 15.00) ensures that under most circumstances, the premises (host) owner has a shared responsibility with the company that is actually performing the work. If a fatality occurs or OSHA chooses to investigate an accident, using contractors does not absolve the host employer of responsibility or liability. To be qualified, the worker must have the technical skills and expertise to perform the work and the safety training to understand the hazards and avoid them. In addition, *NFPA 70E* Section 205.1, *Qualified Persons*, states:

> Employees who perform maintenance on electrical equipment and installations shall be qualified persons as required in Chapter 1 and shall be trained in, and familiar with, the specific maintenance procedures and tests required.

> *Reprinted with permission from* **NFPA-70E 2012,** Electrical Safety in the Workplace *Copyright 2011, National Fire Protection Association, Quincy, MA 02169. This reprinted material is not the complete and official position of the NFPA on the referenced subject, which is represented only by the standard in its entirety.*

5. Is the electrical service company financially stable? Does it have adequate insurance if something goes wrong? A myriad of horror stories describe small testing companies performing inadequate work, the equipment failing, and the service company closing its doors, only to reopen under a new name.

6. Does the company have the technical expertise to evaluate problems that might arise? NETA-member companies are required to have at least one staff electrical engineer who is also a registered engineer.

Legal Repercussions

Because safety is related to electrical power systems maintenance, it seems reasonable to assume that legal issues could arise if maintenance is not performed. OSHA has not yet taken the stand that not performing maintenance as required by the manufacturer, *NFPA 70B*, or ANSI/NETA MTS-2011 constitutes a willful violation. However, the 2012 edition of *NFPA 70E* requires such maintenance, and OSHA has stated on its website that *NFPA 70E* is "a guide for meeting the requirements of the OSHA electrical regulations." In addition, federal courts have found that *NFPA 70E* is "standard industry practice."

Once a company receives and accepts a willful citation, especially if that citation is received as the result of an accident investigation, its worker's compensation protection no longer shields the company. One definition given by a trial attorney for a willful citation was that it is equal to negligent behavior. OSHA defines a willful citation as a violation that

the employer knowingly commits with plain indifference to the law. The employer either knows that what he or she is doing constitutes a violation, or is aware that a hazardous condition existed and made no reasonable effort to eliminate it.

Safe Versus Unsafe Example

Industrial insurers also require maintenance of electrical power systems and their protective devices. Often the risk an insurer assumes will cover a company not only for loss and repair of equipment, but also for the company's financial losses. An incident can cost the insurer several million dollars. At a Midwest refinery that flooded in 2007, the insurer paid

for loss of profitability at the rate of $4 million per day. The insurer paid this compensation for several weeks while the company made repairs to its electrical systems. Not surprisingly, then, companies may request expedited services to meet an insurer's maintenance requirements.

The importance of maintenance is illustrated by one customer's experience with the service team of Shermco Industries. On a yearly basis, Shermco Industries calibrated a customer's protective relays on a 13,800-volt system. Each year its field service technicians would discuss the maintenance needed on air-magnetic circuit breakers, and each year the customer would decline, saying, "They never operate and should be like brand new." One year, several months after the company's protective relays were serviced, the customer had a failure on one of its 13.8-kilovolt underground feeder cables. The fault cascaded through six levels of circuit breakers before it finally cleared. The arc lasted only two to three seconds, but it destroyed not only the switchgear, but also a 20 cal/cm^2 arc-rated arc-flash suit that was hanging on the wall some 15 feet away. Because this company could not show that it had performed the circuit breaker maintenance required by the insurer, its claim for financial loss was denied. **See Figure 10-14**.

Imagine if an Electrical Worker had initiated an arcing fault in this switchgear. The calculated arc-flash hazard if the immediate upstream circuit breaker would have operated properly is not known. However, it is reasonable to assume that the actual arc-flash incident energy was considerably greater than would have been calculated since the immediate upstream breaker and maybe others upstream did not operate properly.

Lack of Maintenance Example

BACKGROUND

Acceptance testing consists of field tests and inspections that assess the suitability for initial energization of electrical power distribution equipment and systems. This testing is performed to ensure that the equipment and systems are operational and within applicable standards and manufacturer's tolerances as well as within design specifications. The data from acceptance testing is retained to use and compared to future maintenance inspections and testing. In the case of circuit breakers, the circuit breakers are inspected and testing to ensure they perform as specified.

A major customer of Shermco Industries had a new extension built onto its electrical system to supply an additional line. A very well-known manufacturer won

Figure 10-14. *The 13.8-kV breaker and cubicle were damaged by a fault that did not clear (top left). Ceramic portions of arc chutes were damaged by extreme heat (top right). All metal inside the breaker was melted or vaporized (bottom left). An arc-flash suit hanging on the wall was reduced to ashes (bottom right).*

Courtesy of James R. White, Shermco Industries

the project bid, with the proposed work including the installation of the equipment along with all start-up and acceptance testing. The project proceeded smoothly, and the work was completed. Several weeks after the extension was placed in service, a failure occurred in a 13.8-kilovolt underground feeder cable. Medium-voltage circuit breakers failed to open, and approximately $5.2 million was lost between the expense of replacing equipment and lost production. What went wrong?

Shermco Industries' forensic investigation determined that a simple $100, 2-pole, 100-ampere molded-case circuit breaker would not carry the rated current. This breaker tripped in 70 to 90 seconds when 45 amperes was injected into it, which is a characteristic that would be expected from a 15-ampere cir-

cuit breaker. This breaker fed the battery charger that supplied direct current (DC) control power, including tripping power, for the 13.8-kilovolt circuit breakers. **See Figure 10-15**.

The client's maintenance personnel did not check the batteries because they were "brand new." The battery bank slowly lost DC voltage; thus, when the cable failure occurred, no power was available to trip the breaker. The result was damage to a power transformer and a reactor (used to limit the short-circuit current). **See Figure 10-16**. Note that the transformer windings were literally blown away from the core, which is characteristic of large current through-faults.

Was the manufacturer's testing company responsible for this incident? Not really. It turned out that the purchasing agent for the client company did not specify

Figure 10-15. *The low-voltage AC panelboard supplied a battery charger and batteries, which provided the DC power for the medium voltage AC circuit breaker tripping mechanisms.*

Courtesy of James R. White, Shermco Industries

Figure 10-16. *A transformer and a reactor were damaged by through-fault current when MV circuit breakers did not operate due to insufficient DC tripping circuit power.*

Courtesy of James R. White, Shermco Industries

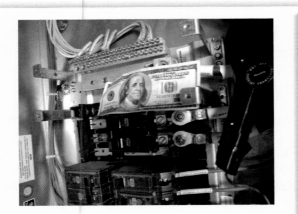

Figure 10-17. *Investigation after the incident showed that the 100-ampere main circuit breaker was not properly calibrated and had nuisance tripped, which then resulted in the battery system losing its charge over time.*

Courtesy of James R. White, Shermco Industries

that these small breakers be tested in the request for quote (RFQ). After all, why spend $150 to test a $100 circuit breaker? When the breaker feeds a critical load, however, that is why a $100 circuit breaker should be tested at a cost of $150. When we assume that devices will function properly or that someone else did his or her job or that the circuit we are about to test is deenergized without testing it, we set ourselves up for a problem. **See Figure 10-17.**

Just because electrical equipment, especially an OCPD, is brand new does not mean it is good. "Brand new" often means that the equipment has not been tested and verified. Acceptance testing is performed to ensure that electrical devices function according to the manufacturer's specifications on all settings and all functions. It comprises a thorough series of tests that detect many problems, such as the one described here. How was this issue resolved? It is probably still in court.

Source: Contributed by Jim White of Shermco Industries

Summary

Worker safety can be enhanced through the implementation of OCPD safe work practices and the performance of regular maintenance on electrical equipment. Various work practices can improve the level of safety when working on electrical equipment; these include not resetting circuit breakers or replacing fuses after a fault until it is safe to reenergize the system, properly evaluating circuit breakers prior to resetting them after a fault interruption, using remote racking devices for circuit breakers, replacing circuit breakers and fuses with devices of the proper type and rating, and testing fuses properly.

Electrical equipment maintenance directly affects worker safety. Poorly maintained equipment can create hazards in electrical equipment. *NFPA 70E* contains requirements for maintaining electrical equipment. Preventive maintenance of all electrical equipment is essential, but it is especially critical for OCPDs, given that the arc-flash hazard can increase if the lack of maintenance delays an OCPD's opening time. If the condition of maintenance is questionable, extra precautions must be taken. In these cases, deenergization should occur prior to working, whenever possible. If that is not possible, additional PPE and more cautious work practices must be considered. The 2012 *NFPA 70E* requires that OCPDs be maintained and that the maintenance, inspections, and tests be documented.

1. After a circuit is de-energized by the automatic operation of a circuit protective device, the circuit shall not be manually reenergized until it has been determined that the equipment and circuit can be _?_ energized.
   a. cautiously
   b. quickly
   c. safely
   d. slowly

2. After interrupting a fault approaching its interrupting rating, a _?_ must be inspected and tested in accordance with the manufacturer's instructions.
   a. circuit breaker
   b. fuse
   c. ground fault protection relay
   d. motor starter

3. The 2012 edition of *NFPA 70E* states that OCPDs shall be _?_ in accordance with the manufacturer's instructions or industry consensus standards.
   a. maintained
   b. replaced
   c. retrofitted
   d. sized

4. For the purposes of *NFPA 70E*, Chapter 2, maintenance shall be defined as preserving or restoring the condition of electrical equipment and installations, or parts of either, for the safety of _?_ who work where they are exposed to electrical hazards.
   a. employees
   b. employers
   c. salespersons
   d. vendors

5. The *NFPA 70E* Technical Committee considers enclosed, energized 600-V or less electrical equipment not likely to pose an electrical hazard if the equipment has been properly _?_.
   a. cleaned
   b. ground fault protected
   c. installed and maintained
   d. maintained

6. When performing the hazard identification and risk assessment for a specific task and the maintenance condition of the OCPD is questionable, assuming that the OCPD _?_ function properly is safer.
   a. shall
   b. should
   c. will
   d. will not

7. When replacing a molded case circuit breaker in a panelboard, besides making sure the circuit breaker is from the same manufacturer, an Electrical Worker should be sure that the replacement circuit breaker has the _?_.
   a. longest warranty
   b. lowest ampere rating for the load
   c. lowest price
   d. proper voltage rating, ampere rating, and interrupting rating

8. *NFPA 70E* Section 130. _?_ directs a shock hazard analysis and an arc flash hazard analysis be done prior to work beginning when the electrical equipment is not in an electrically safe work condition.
   a. 3(A)(1)
   b. 3(A)(2)
   c. 3(B)(1)
   d. 3(B)(2)

9. It is important to remember that garments that are not arc rated shall not be permitted to be used to _?_ the arc rating of a garment or of a clothing system.
   a. decrease
   b. increase
   c. segment
   d. tier

10. Acceptance testing of equipment and systems consists of _?_ for initial energization of power distribution equipment and systems to ensure that electrical devices function according to the manufacturer's specifications and that the systems function as designed.
    a. an owner and vendor meeting
    b. Electrical Workers' qualification testing and skills demonstration
    c. engineer and contractor meeting
    d. field tests and inspections

11 Electrical System Design and Upgrade Considerations

OBJECTIVES

1. Understand the important role that system design can play in eliminating or reducing electrical hazards.
2. Understand that choosing and installing overcurrent protective devices (OCPDs) is an important safety design consideration in regard to arc-flash and arc-blast hazards.
3. Identify other engineering and design practices that can enhance electrical safety for new systems and for upgrading of existing systems.

REFERENCE

1. *NFPA 70E®*, 2012 Edition

An electrician with approximately 10 years of experience was working at a power plant. At about 12:00 p.m., he was electrocuted while replacing a limit switch. The victim had been on vacation up to and including the day before the incident.

Around 11:30 a.m. that day, the victim received a briefing on the need to replace a limit switch. The victim supposedly took a normal lunch break from 11:45 a.m. to 12:15 p.m. At about 12:25 p.m., three workers saw the victim lying face up underneath a conveyor belt. On the ground next to the victim were a pack of cigarettes and two folded one-dollar bills.

The body had no signs associated with a fall and was not in a position that would result from a fall. Examination of the body showed electrical burns on the right hand, aspiration of food into the secondary and tertiary bronchi, a contusion on the left mastoid scalp, and an abrasion of the right mid-tibial area. The medical examiner concluded that the cause of death was electrocution.

The probable sequence of events follows: The victim was standing on the conveyor belt guard installing the new limit switch. Two of the three wires were connected, and the victim was connecting the last wire, the 220-volt energized wire. The wire coming out of the conduit was too short to reach the switch. The victim probably grabbed the wire with his right hand and attempted to pull it farther out of the conduit.

As he did, the bottom of part of his hand contacted the limit switch. When the wire hit the upper part of his palm, a circuit was completed. Due to the relatively low voltage, the victim was not killed instantly; his heart probably went into arrhythmia. He probably felt uncomfortable, decided to get down and have a smoke, and died seconds later.

The major etiologic factor in this fatal incident was the failure of the victim to follow standard procedures for locking out electrical power. The failure apparently resulted from inattention by the victim rather than the difficulty of the procedure or a lack of time to do the job. An explanation for the failure could be a somewhat cavalier attitude of the victim toward relatively small voltages. The research team observed such an attitude in other electricians at the plant.

Source: For details of this case, see FACE Investigation #83PA08. Accessed July 10, 2012.

For additional information, visit qr.njatcdb.org Item #1201.

Introduction

Engineering and system design significantly impacts worker safety. If a system is designed so that an electrical hazard does not exist, then the worker will not have to contend with the hazard. Also, if the use of a specific design technique mitigates an electrical hazard to a much lower level, the workplace, although not hazard free, is still a safer environment. Therefore, using engineering to eliminate hazards or at least create a lower-level hazard is a higher priority than safe work practices. This does not mean designing for electrical safety is more important than safe work practices. Rather, by designing for safety, there is the opportunity to eliminate or minimize work exposure to electrical hazards before a worker ever has a chance to become exposed to them.

This chapter briefly mentions a few techniques for new systems as well as upgrades for existing systems. However, there are many more design techniques to consider and the reader is encouraged to continually strive to search for, investigate, and implement design techniques that result in safer workplaces.

Overcurrent Protective Device Considerations

Choosing and installing overcurrent protective devices (OCPDs) is an important safety design consideration in regard to arc flash and arc blast. Overcurrent protection device selection decisions for new systems and for existing system upgrades may affect arc-flash hazard either positively or negatively. The magnitude of the arcing fault current and the length of time for which the current flows are directly related to the energy released by an arcing fault. OCPDs are the "safety valves" that may limit the magnitude of the fault current that flows and reduce the time duration of the current, thereby mitigating the arc-flash hazard to some extent. Many considerations related to circuit breakers, fuses, and relays can affect electrical safety.

Non-Current-Limiting Overcurrent Protective Devices

There are two broad categories of OCPDs: current limiting and non-current limiting. When the fault current is in the current-limiting range of a current-limiting OCPD, the fault current's magnitude and time duration are shortened, which reduces the energy released in an arcing fault. **See Figure 11-1.** The selection of OCPDs can have a significant impact on the level of arc-flash hazard.

Renewable and Class H fuses are outdated types of fuses that are not current limiting. The use of these types of fuses is not recommended.

The typical molded case circuit breakers, insulated case circuit breakers, and low-voltage power circuit

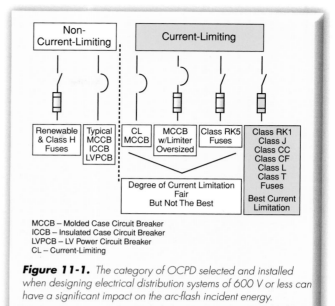

MCCB – Molded Case Circuit Breaker
ICCB – Insulated Case Circuit Breaker
LVPCB – LV Power Circuit Breaker
CL – Current-Limiting

Figure 11-1. *The category of OCPD selected and installed when designing electrical distribution systems of 600 V or less can have a significant impact on the arc-flash incident energy.*

breakers are not listed and marked as current limiting. *NFPA 70E* defines the term *listed* as follows:

Listed. Equipment, materials, or services included in a list published by an organization that is acceptable to the authority having jurisdiction and concerned with evaluation of products or services, that maintains periodic inspection of production of listed equipment or materials or periodic evaluation of services, and whose listing states that either the equipment, material, or service meets appropriate designated standards or has been tested and found suitable for a specified purpose. Informational Note: The means for identifying listed equipment may vary for each organization concerned with product evaluation; some organizations do not recognize equipment as listed unless it is also labeled. The authority having jurisdiction should utilize the system employed by the listing organization to identify a listed product.

Reprinted with permission from NFPA-70E 2012, Electrical Safety in the Workplace *Copyright 2011, National Fire Protection Association, Quincy, MA 02169. This reprinted material is not the complete and official position of the NFPA on the referenced subject, which is represented only by the standard in its entirety.*

Those devices that are not listed and marked as current limiting may not significantly reduce the level of short-circuit currents, and they may take longer to open. They may permit large amounts of energy to be released during an arcing fault. Even if the short-circuit current is in the instantaneous setting range of a circuit breaker, the higher the short-circuit current, the more energy released.

Current-Limiting Overcurrent Protective Devices

Current-limiting OCPDs provide the benefit of reducing the arcing fault energy in higher short-circuit currents by reducing both the current magnitude and its time duration (when the arcing short-circuit current is within their current-limiting range).

Different degrees of current limitation exist. Different OCPDs may become current limiting at different levels of short-circuit current; moreover, once in the current-limiting range, different devices are more current limiting than others. If the arcing short-circuit current is in the current-limiting range of current-limiting fuses, the incident energy released during an arcing fault typically does not increase as the fault current increases. This is an important consideration in mitigating arc flash hazards.

Current-limiting molded case circuit breakers are a better choice than standard molded case circuit breakers. If a circuit breaker is current limiting, it will be marked as "current-limiting." The degree of current limitation is generally moderate but can vary significantly. Underwriters Laboratories (UL) 489, the Molded Case Circuit Breaker Standard, does not establish fixed short-circuit let-through values for peak current (I_p) and ampere^2 seconds (I^2t) for various ampere-rated circuit breakers. (UL tests products and certifies those that meet specific safety standards.) I^2t is proportional to the amount of thermal damage done to the circuit under short-circuit conditions; this measurable value is used to evaluate the fault protection performance of OCPDs. The lower the I^2t that an OCPD lets through, the smaller the amount of thermal energy released into the circuit or arc flash. I_p is the instantaneous peak value of current let through by the OCPD during a short-circuit. Damage to the circuit under short-circuit conditions due to mechanical force is proportional to the square of the maximum instantaneous peak current, or $(I_p)^2$. The lower the value of $(I_p)^2$, the less the mechanical damage done to the circuit.

Standard circuit breakers that incorporate fuses as limiters are another current-limiting alternative. The limiter is solely intended to provide current-limiting short-circuit protection. The fuse limiters normally are oversized to permit the circuit breaker to operate for lower-level short-circuit currents. These fuse limiters give less protection than current-limiting fuses sized to the load, such as when the circuit is a fusible switch system.

The result with the circuit breaker/limiter alternative is typically higher arcing fault energy releases. For instance, the circuit breaker limiter might be sized at 2 to 10 times the equivalent current-limiting fuses that would be used instead of a circuit breaker. As an example, a 600-ampere circuit breaker with fuse limiters may have limiters that are equivalent to fuses of 1,600 amperes or greater. A 600-ampere Class RK1 or Class J fuse would typically provide much lower arc flash incident energy than a limiter that is equivalent to a 1,600-ampere fuse. Properly sized Class RK1 and Class J fuses enter the current-limiting range sooner and let through less fault energy than a 1,600-ampere limiter.

Class RK5 fuses offer a good level of current-limiting protection. A better choice for applications using Class R fuse clips, however, are Class RK1 fuses; these fuses are more current limiting and enter their current-limiting range at lower fault levels. As a consequence, circuits protected by Class RK1 fuses will generally produce a lower arc flash hazard incident energy than circuits protected by Class RK5 fuses.

Class J, Class CF, Class RK1, Class CC, Class L, and Class T fuses offer the best practical current-limiting protection. They generally have a significantly better degree of current limitation than the other alternatives discussed. They also typically enter their current-limiting range at lower currents than the other fuses or limiter alternatives. These types of fuses provide the greatest current limitation for general protection and motor circuit protection.

In evaluating arc flash protection, the overcurrent protective device's I^2t let-through is a direct indicator of the arc flash energy that would be released. **See Figure 11-2**. The UL 248 Fuse Standards maximum

Figure 11-2. *UL standard limits for current-limiting fuses provide insight into the level of arc flash protection. For an arcing fault current at this level, the lower the I²t, the lower the arc flash incident energy.*

UL 248 Fuse Standards Maximum I²t (Ampere² Seconds) Let-Through Limits for 50,000-Ampere RMS Symmetrical Short-Circuit Test

Device Ampere Rating	Class J and Class CF Fuses 600 Volts	Class RK1 Fuses 600 Volts	Class RK5 Fuses 600 Volts
600 A	2,500,000	3,000,000	10,000,000
400 A	1,000,000	1,200,000	5,200,000
200 A	200,000	400,000	1,600,000
100 A	60,000	100,000	500,000
60 A	30,000	40,000	200,000
30 A	7000	10,000	50,000

permitted I^2t let-through limits are compared. The values shown are the maximum limits. Commercially available products may have values lower than those shown in the table, but the products cannot exceed these levels. Among all the products shown, generally the Class J and Class CF fuses have the lowest level of I^2t let-through.

The UL 248 Fuse Standards set short-circuit I^2t let-through limits for current-limiting fuse types such as Classes J, RK1, and RK5. Different limits are set for each fuse's major ampere rating case size, such as 30, 60, 100, 200, 400, and 600. Fuses that are tested and listed as current-limiting are marked "current-limiting."

UL 489 molded case circuit breakers do not have I^2t let-through limits for circuit breakers that are not tested, listed, and marked as "current-limiting." In contrast, circuit breakers that are marked as "current-limiting" have I^2t let-through limits, which is the lower of either the level claimed by the manufacturer or the symmetrical short-circuit calibration wave for a half-cycle without the circuit breaker in the circuit (for the 50,000-ampere rms symmetrical short-circuit test, the I^2t limit permitted by UL 489 for a current-limiting circuit breaker of any ampere rating is 20,750,000 ampere² seconds). UL 489 does not require current-limiting circuit breaker I^2t let-through limits to apply when the circuit breakers are tested under "busbar" test conditions. UL 489 does not set different I^2t let-through limits for different circuit breaker ampere ratings or frame sizes.

OCPDs that have better current-limiting ability may offer better protection. **See Figures 11-3 and 11-4.** Note that the dotted line represents the asymmetrical short-circuit current that could flow with 50,000 symmetrical amperes available: the I_p could reach 115,000 amperes. The UL I_p limit for a 400-ampere Class RK5 fuse is 50,000 amperes; for a 400-ampere RK1 fuse, the I_p limit is 33,000 amperes; and for a 400-ampere Class J fuse, it is 25,000 amperes.

See the chapter, "Incident Energy Varies by Fault Current Magnitude and Duration," for information on the incident energy versus arc fault current for different types of OCPDs.

Overcurrent Protective Devices Life Cycle Consistency

Design criteria for choosing the type of OCPD for new installations or renovations to existing electrical systems should take into account whether the OCPDs will retain their specified fault-clearing operating characteristics over the life cycle of the system. This aspect depends on which type of OCPD is used and whether the appropriate level of maintenance resources, which depends

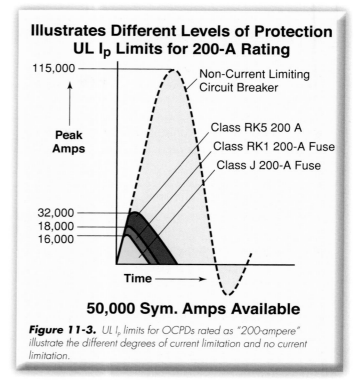

Figure 11-3. UL I_p limits for OCPDs rated as "200-ampere" illustrate the different degrees of current limitation and no current limitation.

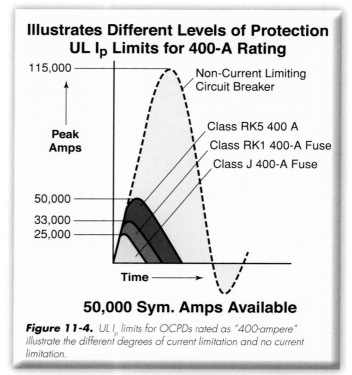

Figure 11-4. UL I_p limits for OCPDs rated as "400-ampere" illustrate the different degrees of current limitation and no current limitation.

on the type of OCPD, will be allocated throughout the electrical system's life cycle.

The reliability of OCPDs in retaining consistent fault-clearing performance over the system life cycle directly impacts arc flash hazards. *NFPA 70E* recognizes this relationship in the third paragraph of 130.5—"The arc flash hazard analysis shall take into consideration the design of the overcurrent protective device and its opening time, including its condition of maintenance"—and in Informational Notes 1 and 2. IN No. 1 advises that inadequate maintenance can result in increased incident energy, while IN No. 2 points to Chapter 2 as the source of safety-related maintenance requirements. 205.4 requires maintaining OCPDs per the manufacturer's instructions or industry consensus standards and states that the inspections, maintenance, and tests must be documented. Other notable sections related to OCPDs include 210.5, 225.1, 225.2, 225.3, and 130.6(L).

The opening time of OCPDs is a critical factor for the resultant arc flash energy released when an arcing fault occurs. The longer an OCPD takes to clear a given arcing fault current, the greater the arc flash hazard. When an arcing fault or any short-circuit current occurs, the OCPD must be able to operate as intended. Therefore, the reliability of OCPDs is critical: they need to open as originally specified; otherwise, the flash hazard can escalate to higher levels than expected.

Two different types of overcurrent protection technology provide different choices in reliability and maintenance requirements and might affect the flash hazard (see the chapter, "OCPD Work Practices and Maintenance Considerations"):

- OCPDs that are reliable and do not require maintenance or minimal maintenance for consistent and reliable fault interruption
- OCPDs that require periodic maintenance according to the manufacturer's instructions and industry standards

Current-Limiting Fuses

Modern fuses are reliable and retain their ability to open as originally designed under fault conditions. **See Figure 11-5**.

When a fuse is replaced, a new factory-calibrated fuse is put into service, and the circuit has reliable protection with performance equal to the original specification. Modern current-limiting fuses do not require maintenance other than visual examination and ensuring that there is no damage to fuses from external thermal conditions (such as conductor terminations), liquids, or physical abuse.

Circuit Breakers

Circuit breakers are mechanical OCPDs that require periodic exercise, maintenance, testing, and possible replacement. A circuit breaker's reliability and operating speed depend on its original specification and its condition. A specific circuit breaker's condition is influenced by the following variables, some of which are not typically recorded and saved:

Figure 11-5. *A range of modern current-limiting fuses are shown: 30 A, 600 V Class CC; 30 A, 600 V Class J; 30 A, 600 V Class RK1; 100 A, 250 V Class RK1; 100 A, 600 V Class RK1; and 1200 A, 600 V Class L.*

Courtesy of Cooper Bussmann, Inc.

- Length of service
- Number of manual operations under load
- Number of operations due to overloads
- Number of fault interruptions
- Humidity
- Condensation
- Corrosive substances in the air
- Vibrations
- Invasion by foreign materials or liquids
- Thermal damage caused by loose connections
- Erosion of contacts
- Erosion of arc chutes

To help keep a circuit breaker within the original specifications, a circuit breaker manufacturer's instructions for maintenance must be followed. Failure to perform periodic maintenance or maintenance after interrupting a fault (*NFPA 70E* 225.3) may result in the circuit breaker requiring longer clearing time or the inability to interrupt overcurrents. Either of these results can drastically affect the potential amount of arc flash energy released.

"Protective Device Maintenance as It Applies to the Arc Flash Hazard," a technical paper by Dennis Neitzel of the AVO Training Institute, is a good resource on this topic. The following excerpts from this paper highlight the importance of circuit breaker maintenance:

> Fuses, although they are protective devices, do not have operating mechanisms that would require periodic maintenance: Therefore, this article will not address them....

Where proper maintenance and testing (on circuit breakers) are not performed, extended clearing times could occur creating an unintentional time delay that will effect the results of flash hazard analysis....

Circuit breakers installed in a system are often forgotten. Even though the breakers have been sitting in place supplying power to a circuit for years, there are several things that can go wrong. The circuit breaker can fail to open due to a burned out trip coil or because the mechanism is frozen due to dirt, dried lubricant, or corrosion. The overcurrent device can fail due to inactivity or a burned out electronic component. Many problems can occur when proper maintenance is not performed and the breaker fails to open under fault conditions. This combination of events can result in fires, damage to equipment or injuries to personnel.

This material is provided courtesy of Dennis K. Neitzel, CPE, Director Emeritus of AVO Training Institute, Inc., Dallas, Texas.

Consideration for Fusible Systems

In addition to current limitation and consistency considerations, which have already been discussed, there are other important considerations for fusible systems.

Rejection-Style Class J, CF, T, R, G, and L Fuses

It is important to ensure that arc flash protection levels are maintained as a facility ages. Class J, CF, T, R, G, and L fuses provide an advantage in that these fuse classes are physically size rejecting. **See Figure 11-6.** Fuses from other classes cannot be inserted into mountings designed for these classes. As a consequence, fuses with lower voltage ratings, lower interrupting ratings, or lower current limitation cannot be accidentally put into service. The designed-in level of arc flash protection will not change in the future as replacement devices are installed. In contrast, circuit breakers with lower

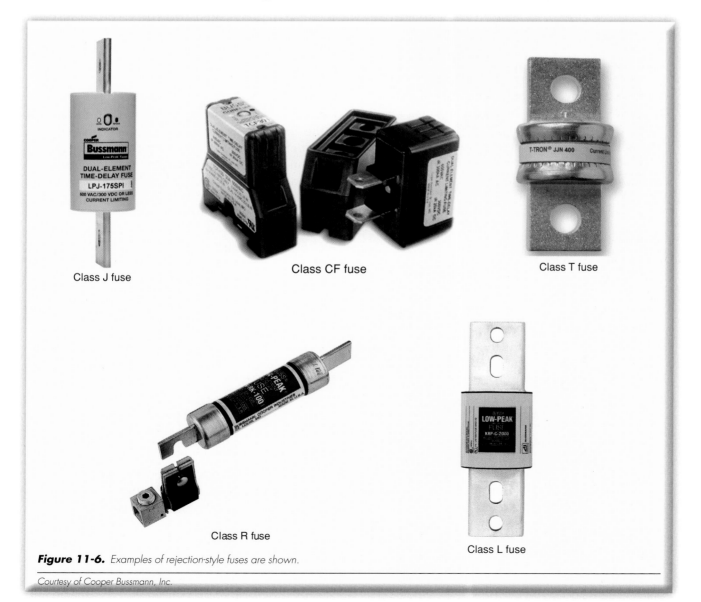

Class J fuse

Class CF fuse

Class T fuse

Class R fuse

Class L fuse

Figure 11-6. *Examples of rejection-style fuses are shown.*

Courtesy of Cooper Bussmann, Inc.

Figure 11-7. *In this fusible panelboard, the branch circuit fuses can be serviced without opening the panel trim.*

Courtesy of Cooper Bussmann, Inc.

Figure 11-8. *Switches with the capability to automatically open when a fuse opens can help reduce the arc flash incident energy for larger ampere-rated circuits.*

Courtesy of Boltswitch

voltage ratings, lower interrupting ratings, and lower current limitation capabilities can mistakenly be used as replacements for superior circuit breakers.

Enhanced Safety of New-Style Fusible Disconnects

In the past decade, the new fusible disconnects have become even safer. **See Figure 11-7.** For instance, the fusible disconnects in the panelboard permit servicing of the fuses without opening the panel trim. The fuses are interlocked with the disconnect handle so the fuses can be inserted or extracted only when the disconnect handle is in the off position.

Switches for Circuits 800 Amperes and Higher

Switches with a shunt-trip that opens all three poles of the switch when the first fuse opens reduce arc flash energy levels. **See Figure 11-8.** This option can be included on new switches or can be retrofitted on some existing switches. Tests have shown that it can reduce the arc flash hazard levels on circuits with large ampere ratings. If a switch is used in this manner, it

should undergo periodic maintenance to ensure consistent arc flash performance.

Reduce Arc Flash Hazard for Existing Fusible Systems

If the electrical system is an existing fusible system, consider replacing or upgrading the existing fuses with fuses that are more current limiting. This step can reduce the arc flash hazard and is a common practice.

Owners of existing fusible systems should consider upgrading Class H, K5, K9, and RK5 fuses to Class RK1 fuses, and should verify that Class J and Class L fuses are the most current-limiting models available. An assessment of many facilities will reveal that the installed fuse types are not the most current limiting, or that fuses were installed decades ago and new fuses with better current limitation are now available. **See Figure 11-9.**

Circuit Breakers

In addition to the types of circuit breakers and consistency considerations that have already been discussed for circuit breaker systems, other important points must be considered in relation to electrical safety.

Adjusting Circuit Breaker Instantaneous Trip Settings

Some circuit breakers have adjustable instantaneous trip settings. **See Figure 11-10.** Such a setting allows adjustment of the fault level at which the circuit breaker will start to operate in its instantaneous mode. This adjustment traditionally is used to help avoid nuisance opening on normal load surges or to improve coordination. If the arcing fault current calculated falls in the current range between the instantaneous trip low and high pickup settings, adjusting the instantaneous trip setting can greatly impact the arc flash incident energy. If the instantaneous trip level is set too high, the circuit breaker might not operate in its instantaneous mode during an arcing fault. The result is a longer opening time, which in turn means that a much higher level of energy may be released during an arcing fault. Conversely, if the instantaneous trip is set low enough to operate for a low arcing fault condition, the arc flash incident energy may be mitigated to a lower level. An engineer

should assess whether the instantaneous trip settings should be adjusted to be as low as possible without incurring nuisance tripping.

One consideration when deciding whether to lower a circuit breaker's instantaneous trip setting is the effect this change will have on selective coordination of the OCPDs. Selective coordination is required for some circuits supplying life-safety loads and is desirable for some circuits supplying mission-critical loads and general-purpose loads.

Sometimes maintenance or service personnel will adjust the instantaneous trip settings without approval. For this reason, it is recommended to perform arc flash hazard analysis based on the highest possible instantaneous trip setting for a given circuit breaker.

Eliminate Short-Time Delay

Some circuit breakers are equipped with a short-time delay setting, which is intended to delay operation of the circuit breaker under fault conditions. **See Figure 11-11.** Short-time delay breakers are used on feeders and mains so that downstream molded case breakers can clear a fault without tripping the larger upstream circuit breaker. Under fault conditions, a short-time delay sensor intentionally delays signaling the circuit breaker to open for the time duration setting of the short-time delay. In many cases, a low-voltage power circuit breaker with a short-time delay setting does not have an instantaneous setting. Therefore, a fault is permitted to flow for an extended time. A low-voltage power circuit breaker

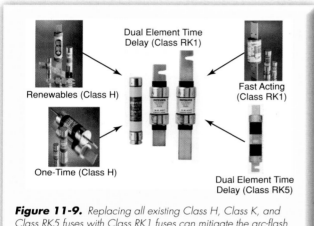

Figure 11-9. *Replacing all existing Class H, Class K, and Class RK5 fuses with Class RK1 fuses can mitigate the arc-flash hazard to lower incident energies in many situations.*

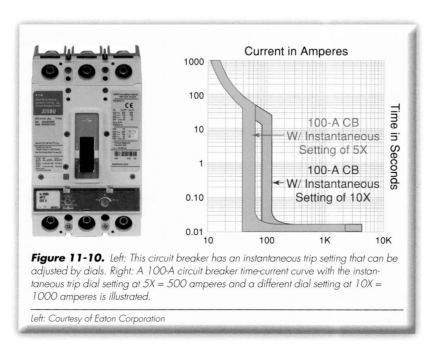

Figure 11-10. *Left: This circuit breaker has an instantaneous trip setting that can be adjusted by dials. Right: A 100-A circuit breaker time-current curve with the instantaneous trip dial setting at 5X = 500 amperes and a different dial setting at 10X = 1000 amperes is illustrated.*

Left: Courtesy of Eaton Corporation

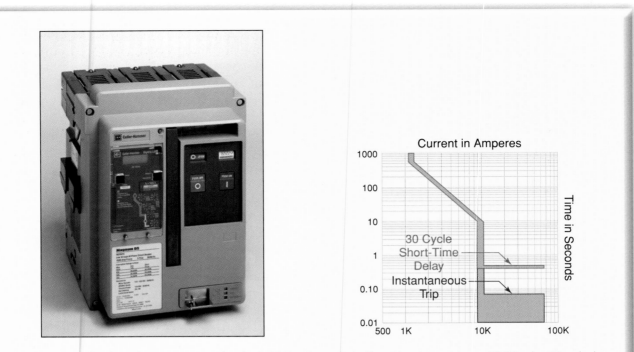

Figure 11-11. *Left: A low-voltage power circuit breaker (LVPCB) is shown. Right: The pink curve represents a LVPCB with a 30-cycle short-time delay setting and no instantaneous trip. The purple curve represents an instantaneous trip for a circuit breaker. A circuit breaker with an instantaneous trip will clear an arcing fault in less time than a circuit breaker with a short-time delay.*

Left: Courtesy of Cooper Bussmann, Inc.

with a short-time delay, and without instantaneous trip, permits a fault to flow for the length of time of the short-time delay setting, which might be 6, 12, 18, 24, or 30 cycles.

An adverse result is associated with using circuit breakers with short-time delay settings. Specifically, if an arcing fault occurs on the circuit protected by a device with a short-time delay setting, a tremendous amount of fault energy might be released while the system waits for the circuit breaker short-time delay to time out. The longer an OCPD takes to open, the greater the flash hazard due to arcing faults. Experience shows that the arc flash hazard increases with the amount of time for which the current is permitted to flow.

System designers and users should understand that using circuit breakers with short-time delay settings could greatly increase the arc flash energy. If an incident occurs when a worker is at or near the arc flash source, that worker might be subjected to considerably more arc flash energy than if an instantaneous trip circuit breaker, a current-limiting circuit breaker, or current-limiting fuses were protecting the circuit.

Arc flash-reducing maintenance switches and zone-selective interlocking, which are covered next, are technologies that permit the use of short-time delays while still being able to mitigate the arc flash incident energy.

Arc Flash-Reducing Maintenance Switching

Arc flash-reducing maintenance switching for circuit breakers, where short-time delay is used (and no instantaneous trip setting), reduces incident energy exposure levels by turning off the short-time delay function and invoking an instantaneous trip while a person is working within the arc flash protection boundary. **See Figures 11-12, 11-13, and 11-14.** When this technology is switched "on," if an arc flash is initiated, the circuit breaker will trip without intentional delay, providing the fastest opening time for the circuit breaker. When the work is completed, the switch is set back to the delay mode to provide the required system selective coordination.

Some systems include a manual switch that the worker uses to turn the bypass on and off. Other systems use automatic means to sense whether the worker is within close proximity of the equipment and

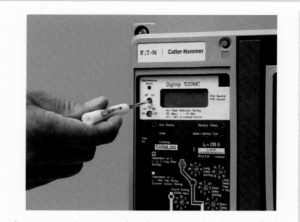

Figure 11-12. *A circuit breaker equipped with an arc flash-reducing maintenance switch is controlled by a screw driver.*

Courtesy of Eaton Corporation

then automatically turn the arc reduction maintenance switch on and off. For instance, as a worker opens an enclosure, a motion sensor would detect the person and invoke an instantaneous trip setting rather than the normal short-time delay.

If a circuit breaker without an instantaneous trip (which uses short-time delay) is placed into service, the 2011 *NEC,* Section 240.87 requires the circuit breaker to be equipped with an arc flash-reducing maintenance switch or some other technology that reduces the incident energy while personnel work on equipment protected by such a circuit breaker.

Zone-Selective Interlocking

While utilizing short-time delay settings may be necessary to achieve selective coordination for a circuit breaker design, short-time delay settings can permit

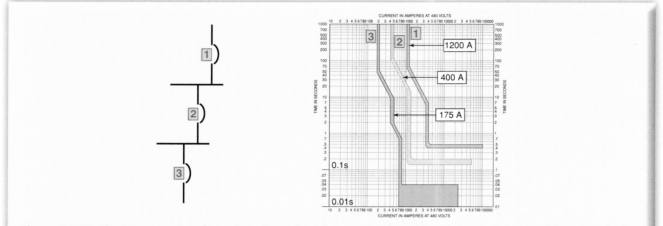

Figure 11-13. *The time current curves for a selectively coordinated system are achieved through the use of short-time delay settings for the 400-ampere and 1200-ampere circuit breakers. Figure 11-14 illustrates the change to the 1200-ampere circuit breaker time-current curve when its arc-flash reduction maintenance switch is turned "on."*

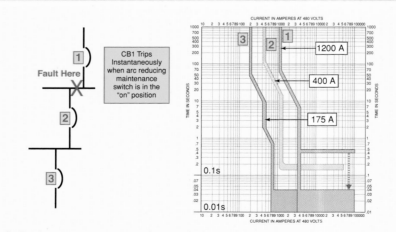

Figure 11-14. *An arc flash-reducing maintenance switch is switched "on," thereby providing an instantaneous trip for the 1200-ampere circuit breaker (number 1) when a worker is exposed to a possible arc-flash hazard. In this situation, the 1200-ampere circuit breaker opens as quickly as possible, reducing the arc-flash hazard. After the worker is no longer exposed to a potential arc-flash hazard, the arc flash-reducing maintenance switch can be turned "off," returning the system to a selectively coordinated state.*

high incident energy levels under arcing fault conditions to occur. A possible solution to this dilemma is to use circuit breakers equipped with zone-selective interlocking. Zone-selective interlocking is an option for some circuit breaker systems where circuit breakers communicate with each other and the time-current functions of the circuit breakers are dynamically modified based on the location of the fault condition. **See Figure 11-15**. Circuit breakers with this option are equipped with communication wiring between the circuit breakers. The benefit is that when a downstream fault is not in a circuit breaker's protection zone, the circuit breaker can be set to operate with a short-time delay to provide selective coordination with downstream circuit breakers. When a fault is within a circuit breaker's protection zone, then that circuit breaker overrides its short-time delay and operates as fast as it can, thereby providing better equipment protection and less incident energy under arcing fault conditions.

Relaying Schemes

Relay schemes can be used on both new installations and existing systems to mitigate arc flash hazard. When an arc flash event occurs that causes the relay to call for the circuit to be interrupted, the relay output signals a properly rated disconnecting means to interrupt the circuit. The disconnecting means can be a vacuum interrupter, circuit breaker, or disconnect switch with a shunt trip.

Overcurrent Relays

Overcurrent relays are used for some low-voltage applications, but are more commonly used on medium- and high-voltage systems. The current and time settings to operate are important factors influencing the arc flash hazard with such systems. Coordination for medium- and high-voltage OCPDs is generally more critical from a business continuity basis, because an outage at this level represents a large amount of power. Thus the objective is to set the relays to minimize the arc flash hazard as well as to provide acceptable coordination levels.

Arc-Flash Relays

Arc-flash relays can be utilized in conjunction with an interrupting disconnect device to mitigate incident energy levels by limiting the time of the fault. Typically these relays monitor for uncharacteristic current, and several light sensors placed at key locations inside an electrical enclosure. Other parameters that could be monitored include sound and pressure. If an arc-flash event occurs within the enclosure, the light, sound, and

pressure all rapidly escalate to high levels. When an uncharacteristic current flow is combined with one or more of these other parameters, the relay reaches its trip point and then signals a disconnecting means to interrupt. This method is an effective arc flash mitigation means for equipment protected by OCPDs that are too large in ampacity to quickly respond to the arcing current.

On a 480-volt service or unit substation with a large ampacity main OCPD, the arc-flash relay technology can be used to signal a medium-voltage disconnect and allow for fast interruption of an arc-flash event. **See Figure 11-16**.

Separate Enclosure for Main Disconnect

Service entrance equipment typically has the most hazardous arc flash hazards due to the line-side connection being protected upstream by a utility OCPD. To reduce the arc-flash hazard exposure, the design may call for the main disconnect to be placed in a separate enclosure from the feeder disconnects or overcurrent protective devices. **See Figure 11-17**. With this configuration, it is possible to open the main disconnect and put the feeder enclosure in an electrically safe work condition. This requires the appropriate personal protective equipment (PPE) for this procedure; however, the arc flash rating of the PPE will be based on the main OCPD, not the utility OCPD. After an electrically safe work condition is achieved, workers do not need PPE and can work in the feeder enclosure without any risk of shock or arc flash. If the main is located in the same line-up, then even with the main disconnect open, there is typically an arc-flash hazard from the energized main disconnect, line-side connections, and conductive parts that are within the enclosure. If work is necessary within the main disconnect enclosure, it will be necessary to deenergize and lock out/tag out any disconnect upstream, which involves having the utility deenergize the service.

A design featuring separately enclosed disconnect supply panels can also be used for industrial control panels (a separately enclosed disconnect is attached on the outside of the industrial control panel enclosure), motor control centers, branch panels, and distribution panels. This arrangement provides an electrical system designed for safer work practices.

If the arc-flash hazard in the feeder enclosure is higher than desired due to the ampacity and characteristics of the main OCPD, an arc flash relay (like that described earlier) can be used to reduce the arc-flash hazard. The arc-flash relay can have optic sensors in the feeder enclosure and use the main disconnect (which needs a shunt-trip) as the device that interrupts the current upon the relay sensing an arc flash.

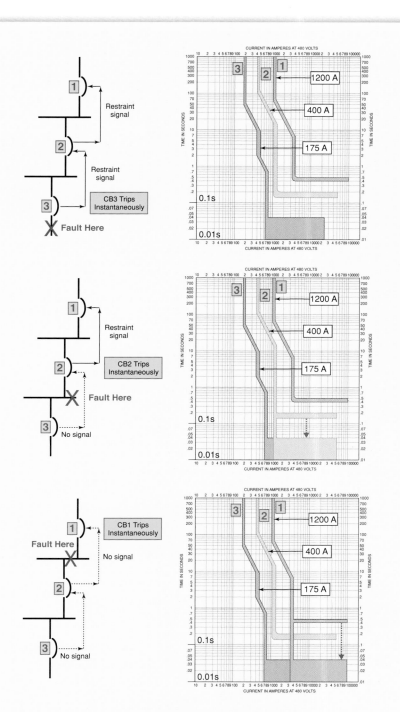

Figure 11-15. *The three fault scenarios for a system demonstrate three circuit breakers equipped with zone-selective interlocking and indicate how they can be selectively coordinated and still help mitigate the arc-flash incident energy to lower levels. Each scenario shows the fault at a particular point in the system and the resulting effective time-current curve for that scenario. Top: The fault occurs on the loadside of circuit breaker 3 (175 A). This fault is in the protection zone for circuit breaker 3, and it operates instantaneously as the time-current curve to the right illustrates. When circuit breaker 3 senses a fault, it sends a restraint signal to circuit breakers 2 (400 A) and 1 (1,200 A); circuit breakers 2 and 1 then function with the short-time delay settings shown by the adjacent time-current curve. Center: For faults between circuit breakers 2 and 3 (circuit breaker 2 protection zone), circuit breaker 3 does not sense the fault current, so it does not send a restraint signal to circuit breaker 2. Circuit breaker 2 does sense the fault current and sends a restraint signal to circuit breaker 1, so that circuit breaker 1 functions with the short-time delay setting shown by the adjacent time-current curve. Circuit breaker 2 will function without any intentional delay; it opens as quickly as possible because it did not receive a restraint signal from circuit breaker 3. Bottom: For faults between circuit breakers 1 and 2 (circuit breaker 1 protection zone), neither circuit breaker 3 nor circuit breaker 2 senses the fault, and no restraint signal is sent to circuit breaker 1. Because circuit breaker 1 senses a fault and does not receive a restraint signal, it will function without any intentional delay. The adjacent time-current curves shows circuit breaker 1 opening as quickly as possible, reducing the arc-flash hazard.*

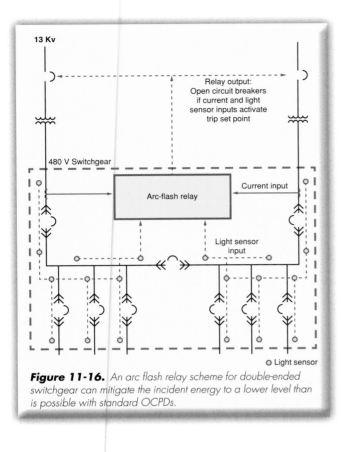

Figure 11-16. *An arc flash relay scheme for double-ended switchgear can mitigate the incident energy to a lower level than is possible with standard OCPDs.*

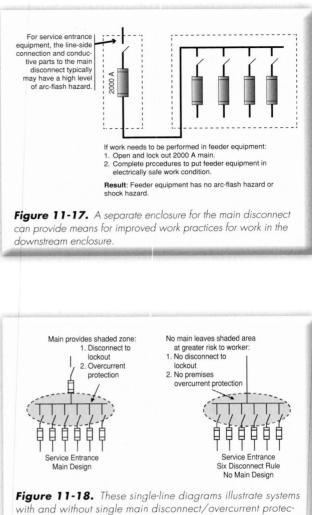

Figure 11-17. *A separate enclosure for the main disconnect can provide means for improved work practices for work in the downstream enclosure.*

Figure 11-18. *These single-line diagrams illustrate systems with and without single main disconnect/overcurrent protection and indicate how each design affects the arc-flash hazard level.*

Specify a Main on Each Service

Generally, a single main service disconnect provides for safer work practices than the six-disconnect rule for service entrances permitted in *NEC* 230.71. The six-disconnect rule is intended to reduce the cost of the service equipment, but this choice typically increases worker exposure to hazards. Without a main OCPD, the main bus and line terminals of the feeders are unprotected. In addition, equipment installed under the six-disconnect rule does not allow the bus to be deenergized without the utility being called to deenergize its supply.

If a worker must work in the enclosure of one of the feeders, the compartment should be placed in an electrically safe work condition. To achieve that end, typically a main disconnect should be locked out. Afterward, no energized conductors will remain in the feeder device compartment. (There may still be an arc-flash hazard due to the main line-side connection unless the main is in a separate enclosure.) **See Figure 11-18.** Note that the system represented by the single-line diagram on the left has a lower arc-flash hazard level than the system represented by the one-line diagram to the right, which has no main overcurrent protection.

If a worker is performing a task within a feeder compartment with an exposed energized conductor, a main OCPD helps protect against arcing faults on the feeder device's line terminals and the equipment's main bus. The arcing fault hazard associated with the main device must be assessed because large ampere-rated OCPDs might permit high arc-flash incident energies. In most cases, the main OCPD can provide better protection than the utility OCPD, which is located on the transformer primary.

High-Speed Fuse Assembly

An engineered assembly installed at the secondary of a large transformer or at other locations that may potentially have high incident energy can greatly mitigate the arc-flash hazards. **See Figure 11-19.** These

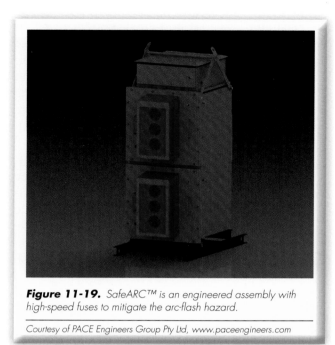

Figure 11-19. *SafeARC™ is an engineered assembly with high-speed fuses to mitigate the arc-flash hazard.*

Courtesy of PACE Engineers Group Pty Ltd, www.paceengineers.com

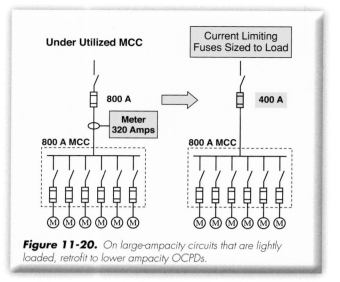

Figure 11-20. *On large-ampacity circuits that are lightly loaded, retrofit to lower ampacity OCPDs.*

assemblies have been used in mining applications, where the transformer secondary has high load current requirements and high arc-flash hazards. They incorporate special-purpose high-speed fuses that are very current limiting. The resulting systems are engineered to carry the normal load currents, but respond very quickly to fault currents. They provide opportunities for electrical design teams on both greenfield and brownfield projects to minimize arc-flash hazards and can also lower a site's carbon footprint through the use of low-impedance transformers. Low-impedance transformers provide greater available fault current, a factor that assists in the arc-flash mitigation operation provided by this assembly.

Sizing Underutilized Circuits with Lower-Ampere-Rated Fuses or Circuit Breakers

When the rated ampacity of a circuit is significantly larger than necessary, the actual load current under the maximum load conditions should be measured, and the most current-limiting fuses should be sized for the load. If an 800-ampere feeder to a motor control center draws only 320 amperes, for example, then 400-ampere current-limiting fuses could be considered. For circuit breakers that incorporate ampere-rated plugs, some benefit might be achieved by lowering the plug rating. A lower circuit breaker plug rating can, in turn, decrease the arc-flash hazard for lower-level arcing fault currents. Downsizing fuses or circuit breakers is a strategy that can be utilized on some new systems or on retrofitted existing systems. **See Figure 11-20**.

Evaluating OCPDs in Existing Facilities for Interrupting Rating

Fuses, circuit breakers, and other OCPDs must be selected so that their interrupting rating is equal or greater than the available short-circuit current for the initial installation as well as over the life of the system. Changes in an electrical system, including changes made to the supplying utility system, can result in higher available short-circuit currents, which may exceed the interrupting rating of existing OCPDs. When a service transformer is increased in kilovolt-ampere size or the transformer is replaced with a lower-impedance transformer, the available short-circuit current might increase.

In addition to *NEC* Section 110.9, which requires fuses and circuit breakers to have adequate interrupting ratings, *NFPA 70E* Section 210.5 and OSHA require that OCPDs have adequate interrupting ratings irrespective of the installation age of the system. §OSHA 29 CFR 1910.302 (b) requires some regulations to be followed irrespective of the original system installation date. One of those regulations is provided here.

> **§OSHA 29 CFR 1910.303(b)(4):**
> *Interrupting rating.* Equipment intended to interrupt current at fault levels shall have an interrupting rating sufficient for the nominal circuit voltage and the current that is available at the line terminals of the equipment. Equipment intended to interrupt current at other than fault levels shall have an interrupting rating at nominal circuit voltage sufficient for the current that must be interrupted.

Whenever system changes occur in a premises or changes are made by the utility that might increase the available short-circuit currents, the existing OCPDs must be reevaluated to determine whether they have sufficient interrupting rating. If a short-circuit analysis or an arc-flash analysis is performed for an existing facility, the employer must evaluate whether all the fuses and circuit breakers have sufficient interrupting ratings for the available short-circuit current at their line terminals. Fuses or circuit breakers that do not have an adequate interrupting rating should be replaced with fuses or circuit breakers that have an adequate interrupting rating.

NFPA 70E defines *interrupting rating* as follows:

> **Interrupting rating.** The highest current at rated voltage that a device is identified to interrupt under standard test conditions.
>
> *Reprinted with permission from* NFPA-70E *2012,* Electrical Safety in the Workplace *Copyright 2011, National Fire Protection Association, Quincy, MA 02169. This reprinted material is not the complete and official position of the NFPA on the referenced subject, which is represented only by the standard in its entirety.*

The device referenced in the definition represents a fuse or circuit breaker. An OCPD that attempts to interrupt a short-circuit current beyond its interrupting rating can rupture violently. This misapplication condition can present an arc flash and arc-blast hazard, and the violent rupturing can initiate an arcing fault in other parts of the equipment. (Another chapter covers calculation of the maximum available short-circuit current.) Modern current-limiting fuses have interrupting ratings of 200,000 and 300,000 amperes, which virtually eliminate this hazard contributor. However, renewable and Class H fuses have only a 10,000-ampere interrupting rating, some Class K fuses have a 50,000-ampere interrupting rating, and Class G fuses have a 100,000-ampere interrupting rating.

Circuit breakers have varying interrupting ratings, so they need to be assessed accordingly. **See Figures 11-21** and **11-22.** Although what these tests depict is a violation of *NEC* Section 110.9, *NFPA 70E* 210.5, and OSHA §1910.303 (b)(4), they emphasize the importance of a proper interrupting rating for arc-flash protection and application of OCPDs. In a fraction of a second, sudden violence can occur.

A short-circuit current can be safety interrupted. **See Figure 11-23.** Note that in this laboratory test the fuses have an interrupting rating greater than the available short-circuit current; therefore, the current is safely interrupted.

"No Damage" Protection for Motor Controllers

Motor starters that are designated as Type 1 protection are susceptible to violent explosive damage from short-circuit currents. They can be the source of an arc flash or can produce ionized gas that initiates an arc fault in an enclosure. If an employee needs to work within an

Figure 11-21. *A laboratory test illustrates what happens when Class H fuses, which have an interrupting rating of only 10,000 amperes, are subjected to a 50,000-ampere fault.*

For additional information, visit qr.njatcdb.org Item #1235.

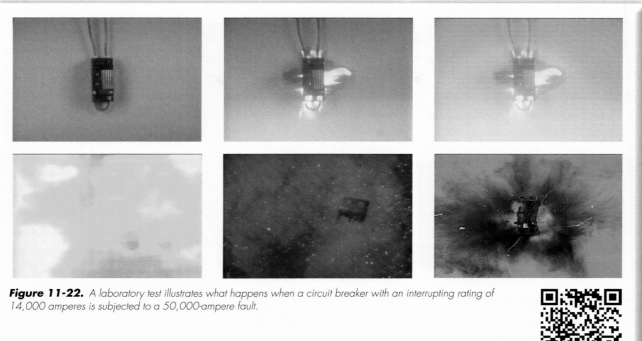

Figure 11-22. *A laboratory test illustrates what happens when a circuit breaker with an interrupting rating of 14,000 amperes is subjected to a 50,000-ampere fault.*

For additional information, visit qr.njatcdb.org Item #1236.

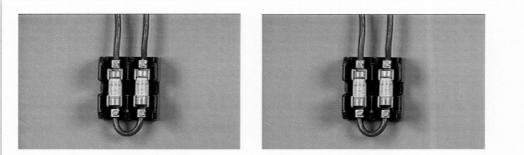

Figure 11-23. *A laboratory test illustrates Class J, low-peak LPJ fuses safely interrupting a 50,000-ampere available short-circuit current. The LPJ fuses have an interrupting rating of 300,000 amperes. Left: Fuses are shown before a laboratory test. Right: Fuses are shown during and after a laboratory test to safely interrupt a short-circuit current.*

For additional information, visit qr.njatcdb.org Item #1237.

enclosure that is not in an electrically safe work condition and that contains a motor starter protected by a Type 1 approach, he or she may be exposed to a serious safety hazard. Specifying Type 2 motor starter protection can reduce the risk, because the level of current limitation typically required to obtain Type 2 protection also provides for excellent arc-flash hazard reduction and minimizes the chance that the starter will be the source of the arc flash or contribute to the initiation of an arcing fault.

Type 1 Protection

IEC 60947-4-1, "Type 1 Protection" (similar to requirements for listing motor starters to UL 508),

requires that, under short-circuit conditions, the contactor or starter cause no danger to persons (with the enclosure door closed) or surroundings, but the contactor or starter might not be suitable for further service without repair and replacement of parts. Note that damage is allowed that may require partial or complete component replacement. It is possible for the overload devices to vaporize and the contacts to weld. Short-circuit protective devices interrupt the short-circuit current but are not required to prevent component damage. **See Figure 11-24.** These still photos were taken from a video of a motor starter tested to Type 1 criteria (with the door open so as to visualize the damage level). The heater elements vaporized and the contacts severely welded, contributing vaporized metal to the atmosphere. If a worker had any unprotected body parts near such an event, he or she might be injured.

Type 2 Protection

Using starters with OCPDs that provide Type 2 protection may mean that the starter is not a potential arc-flash contributor. UL 508E (Outline of Investigation) and IEC 60947-4-1, "Type 2 Protection," require that, under short-circuit conditions, the contactor or starter cause no danger to persons (with the enclosure door closed) or the installation and be suitable for further use. No damage is allowed to either the contactor or the overload relay. Light contact welding is permitted, but contacts must be easily separable. "No damage" protection for the National Electrical Manufacturers Association (NEMA) and IEC motor starters can be provided only by a device that is able to limit the magnitude and the duration of short-circuit current. **See Figure 11-25**.

Fuses that typically meet the requirements for Type 2 "no damage" protection, as demonstrated by the results of the controller manufacturer's testing, include Class J, Class CF, Class CC, and Class RK1 fuses. As mentioned earlier in this chapter, these fuses are very current limiting, which can protect the sensitive controller components.

Selective Coordination

Today, one of the most important parts of any installation is the electrical distribution system. Nothing can stop all activity, paralyze production, create inconvenience, disconcert people, and possibly cause a panic more than a power outage to critical loads.

Selective coordination is considered the act of isolating a faulted circuit from the remainder of the electrical system, thereby eliminating unnecessary power outages. The faulted circuit is isolated by the selective operation of only that OCPD closest to the fault condition. An adequately engineered system enables only the protective device nearest the fault to open, leaving the remainder of the system undisturbed and preserving continuity of service.

Personnel safety is enhanced in a selectively coordinated system because the Electrical Worker is not unnecessarily exposed to arc-flash hazards at upstream panels or switchboards. In a selectively coordinated system, the worker is exposed only at the level of circuit that incurred the problem. In a nonselectively coordinated system, the worker, while troubleshooting

Figure 11-24. *Type 1 protection. The photos were taken before, during, and after testing a motor circuit protector (MCP) intended to provide motor branch-circuit protection for a 10-horsepower (hp) IEC starter with 22,000 amperes of short-circuit current available at 480 volts. The heater elements vaporized, and the contacts were severely welded. This could be a hazard if the door is open and a worker is near.*

Figure 11-25. *These still photos were taken from a video of a motor starter tested to Type 2 criteria (with the door open so as to visualize the lack of a hazard). The photos were taken before, during, and after use of the same test circuit and same type of starter during short-circuit interruption as in Figure 11-24. The difference is that here Cooper Bussmann LPJ-SP Class J current-limiting fuses provide the motor branch-circuit protection. This level of protection reduces the risk for workers.*

the open circuit, is required to work in upstream equipment, where arc-flash energies are often significantly higher. **See Figure 11-26**. In this single-line diagram, the arc-flash energy at a branch-circuit panel is 1.6 cal/cm². If an overcurrent condition on the branch circuit opens only the overcurrent device in the branch-circuit panel, the worker is exposed to 1.6 cal/cm² while troubleshooting the circuit. If the feeder overcurrent device also opens, the worker is forced to work within the feeder panel and is unnecessarily exposed to a higher level of 6.7 cal/cm². If the main also opens unnecessarily, the worker is forced to work within the main panel, thereby exposing the worker to 12.3 cal/cm². Therefore, in this selectively coordinated system example, the worker is exposed to only 1.6 cal/cm², but in the nonselectively coordinated system, the worker is exposed to 12.3 cal/cm².

70E Highlights

NFPA 70E Annex O is helpful for generating designs that minimize exposure of workers to electrical hazards. The title of this annex is "Safety-Related Design Requirements."

Additional Design for Safety Considerations

Beyond overcurrent protective device safety design considerations, there are design techniques for new and existing systems that reduce or eliminate electrical hazards for workers. This section discusses a few of these options; however, a multitude of such design considerations exist.

Remote Monitoring

Specifying remote monitoring of voltage or current, or measurement of other vital electrical parameters, reduces exposure to electrical hazards by transferring the potentially hazardous troubleshooting activity from the actual live equipment to a display visible with the equipment doors closed, thereby enhancing safe work practices. **See Figure 11-27**. Such remote monitoring can be implemented in many ways. For example, the displays may be mounted so that they are readable from the equipment enclosure exterior; the equipment enclosure exterior may support "plug-in" diagnostic displays, instruments, or computers so that troubleshooting can be performed at the equipment with the doors closed; or remote computers may provide network access to the necessary data. These designs reduce the associated electrical hazards and reduce the number of times that required PPE must be worn by Electrical Workers.

Finger-Safe Products and Terminal Covers

One of the best ways to minimize exposure to electrical shock hazard is to use finger-safe products and nonconductive covers or barriers. ("Finger-safe" is a generic term but not a product standard criterion for evaluating electrical products' ability to minimize shock hazard.) Finger-safe products and covers reduce the chance of causing a shock or initiating an arcing fault. If all the electrical components are finger safe or covered, a worker has a much lower risk of making contact with an energized part (shock hazard). The risk of any conductive part falling across bare, energized conductive parts and creating an arcing fault (arc-flash hazard) is greatly reduced.

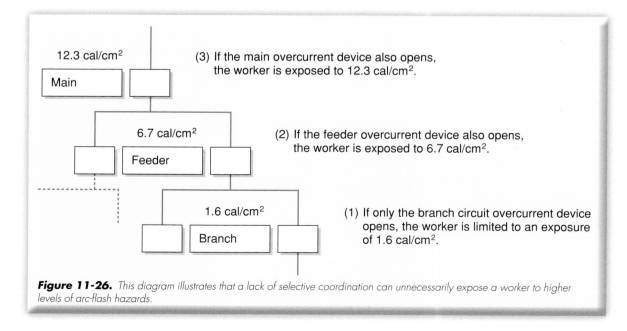

Figure 11-26. *This diagram illustrates that a lack of selective coordination can unnecessarily expose a worker to higher levels of arc-flash hazards.*

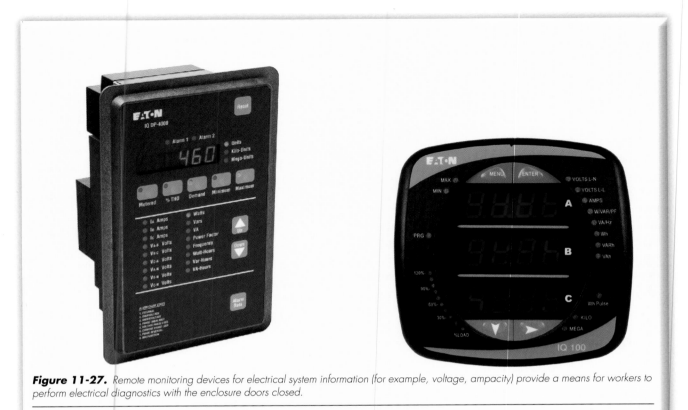

Figure 11-27. *Remote monitoring devices for electrical system information (for example, voltage, ampacity) provide a means for workers to perform electrical diagnostics with the enclosure doors closed.*

Courtesy of Eaton Corporation

Several items have been developed that can help minimize shock hazards and minimize the initiation of an arcing fault. **See Figure 11-28.** All of these devices can reduce the chance that a worker, tool, or other conductive item will come in contact with a live part.

The International Protection (IP) Code rating is an IEC system, detailed in IEC 60529. **See Figure 11-29.** IP2X is often referred to as "finger safe," meaning that a probe the approximate size of a finger must not be able to access or make contact with hazardous energized parts. Principally, IEC 60529 defines the degree of protection provided by an enclosure (barriers/guards) classified under the IP Code and the testing conditions required to meet these classifications. IP20-rated products do not offer any protection against liquids. UL product standards are lagging in the adoption of evaluating enclosures and components for shock hazards; however, some UL product standards are starting to evaluate electrical devices (components) for compliance with IP20. In some cases, the manufacturers of components may make self-certified IP20 claims.

Workers should understand the concept underlying the IP20 rating and its benefits and limitations. Some component products with IP20 claims have a "conditional" IP20 rating. For example, the conductor termination may be considered IP20 rated if the conductor is prepared and installed in the terminal properly. This same terminal may be IP20 rated for the larger AWG conductors for which the terminal is rated, but not IP20 rated for the smaller AWG conductors for which the terminal is rated (a bare finger could contact bare energized metal).

Isolating the Circuit: Installation of an "In-Sight" Disconnect for Each Motor

Electrical systems should be designed to support maintenance, with easy, safe access to the equipment. Their design should provide for isolating equipment for repair purposes, with a disconnecting means for implementing lockout/tagout procedures. A sound design provides disconnecting means at all motor loads, in addition to disconnecting means required within sight of the controller location that can be locked in the open position. Disconnecting means at the motor provide improved isolation and safety for maintenance as well as an emergency disconnect.

Horsepower-rated disconnects should be installed *within sight* (visible and within 50 feet) of every motor

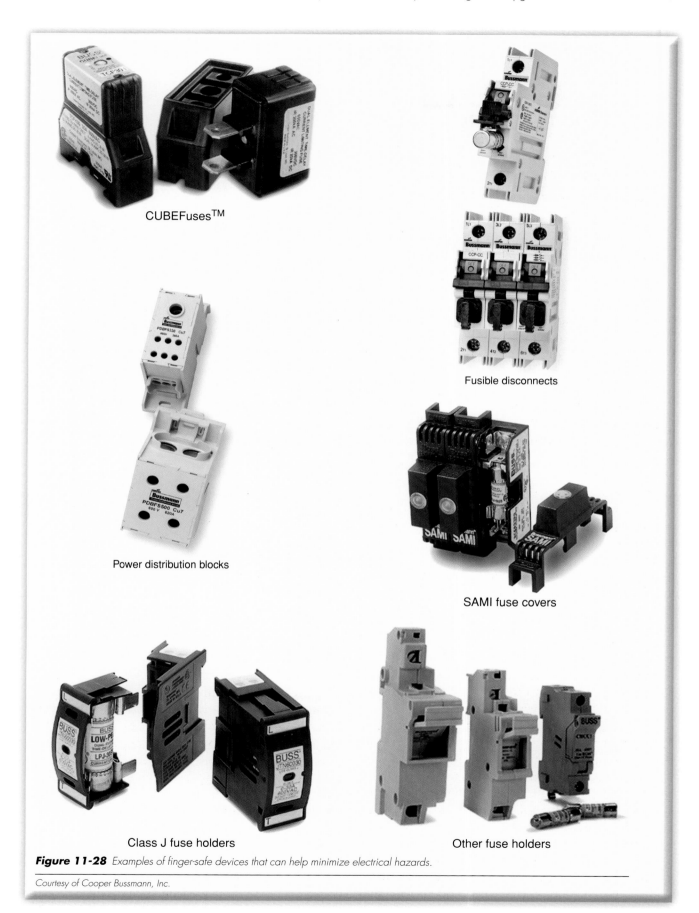

CUBEFuses™

Fusible disconnects

Power distribution blocks

SAMI fuse covers

Class J fuse holders

Other fuse holders

Figure 11-28 *Examples of finger-safe devices that can help minimize electrical hazards.*

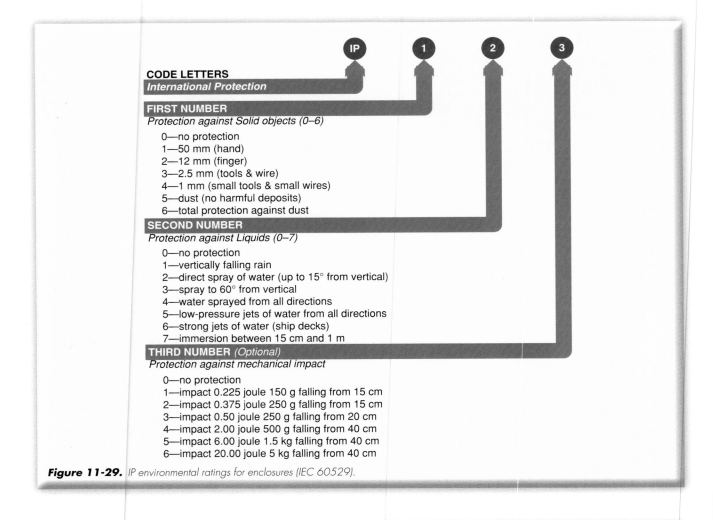

CODE LETTERS
International Protection

FIRST NUMBER
Protection against Solid objects (0–6)

    0—no protection
    1—50 mm (hand)
    2—12 mm (finger)
    3—2.5 mm (tools & wire)
    4—1 mm (small tools & small wires)
    5—dust (no harmful deposits)
    6—total protection against dust

SECOND NUMBER
Protection against Liquids (0–7)

    0—no protection
    1—vertically falling rain
    2—direct spray of water (up to 15° from vertical)
    3—spray to 60° from vertical
    4—water sprayed from all directions
    5—low-pressure jets of water from all directions
    6—strong jets of water (ship decks)
    7—immersion between 15 cm and 1 m

THIRD NUMBER *(Optional)*
Protection against mechanical impact

    0—no protection
    1—impact 0.225 joule 150 g falling from 15 cm
    2—impact 0.375 joule 250 g falling from 15 cm
    3—impact 0.50 joule 250 g falling from 20 cm
    4—impact 2.00 joule 500 g falling from 40 cm
    5—impact 6.00 joule 1.5 kg falling from 40 cm
    6—impact 20.00 joule 5 kg falling from 40 cm

Figure 11-29. IP environmental ratings for enclosures (IEC 60529).

or driven machine. **See Figure 11-30.** The provision for locking or adding a lock to the disconnecting means shall be installed on or at the switch or circuit breaker used as the disconnecting means and shall remain in place with or without the lock installed. The *NEC* defines "in-sight from" as follows:

> Where this *Code* specifies that one equipment shall be "in-sight from," "within sight from," or "within sight of," and so forth, another equipment, the specified equipment is to be visible and not more than 15 m (50 ft) distant from the other.
>
> Reprinted with permission from NFPA-70 2011, National Electrical Code® Copyright 2010, National Fire Protection Association, Quincy, MA 02169. This reprinted material is not the complete and official position of the NFPA on the referenced subject, which is represented only by the Code in its entirety.

An in-sight motor disconnect is more likely to be used by a worker for the lockout procedure to put equipment in an electrically safe work condition prior to doing work on the equipment. Such a disconnect generally is required even if the disconnect within sight of the

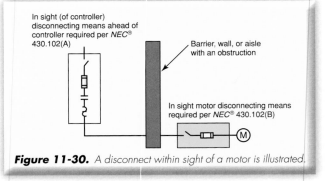

Figure 11-30. A disconnect within sight of a motor is illustrated.

controller can be locked out. Some exceptions exist for specific industrial applications.

Breaking up Large Circuits into Smaller Circuits

In the design phase, if preliminary arc-flash analysis indicates that equipment in high-ampacity circuits is associated with high arc-flash hazards, it may be feasible to divide the loads and supply the various pieces of equipment by multiple smaller-ampacity feeders. In many cases, very large ampere-rated fuses and circuit breakers let through too much energy for a practical PPE arc rating.

A 1,600-ampere circuit, for example, might potentially be broken into two 800-ampere circuits. An analysis of the arc-flash incident energy available in each circuit would then generally indicate that two 800-ampere circuits would be better than one 1,600-ampere circuit.

For specific situations, an arc-flash analysis should be completed, as variables can affect the outcome. This is especially beneficial when using current-limiting protective devices, because the lower ampere-rated devices are typically more current limiting and, therefore, can better reduce the arc-flash hazard. **See Figure 11-31**.

Remote Opening and Closing

Opening and closing large switches and circuit breakers have caused serious arc-flash incidents when these devices failed while being operated. By opening and closing large switches and circuit breakers remotely, a worker might control the operation from a safe distance. If an arc-flash incident should occur in this scenario, the worker will not be exposed to the hazard. **See Figure 11-32**.

Arc-Resistant (Arc-Diverting) Medium-Voltage Switchgear

Arc-resistant switchgear can be installed to withstand internal arcing faults. Arc-resistant equipment typically is designed with stronger door hinges and latches, better door gaskets, and hinged enclosure top venting panels. The underlying concept focuses on diversion of the resultant explosive hot gases and pressure from an internal arcing fault via the hinged enclosure top panels. If the switchgear is installed indoors, then a means of exhausting the hot gases to the outside of the building, such as ducts, is required.

Arc-resistant equipment is rated to withstand specific levels of internal arcing faults with all the doors closed and latched. The rating does not apply with any

door opened or any cover removed. Therefore, arc-resistant equipment does not protect a worker who is performing a task with an open door or panel.

The term "arc-resistant" is a bit misleading. The internal switchgear must withstand an internal arcing fault and, therefore, the sheet metal and other components of the equipment must resist or withstand a specified arcing fault. However, a major feature of this equipment is diversion of the arcing fault by-products (that is, hot ionized gases and blast) via the enclosure top panels. This feature helps to prevent the arcing fault from blowing open the doors or side panels and venting the arcing fault by-products where a worker might be standing. **See Figure 11-33**.

Electrical Motor Operator

Figure 11-32. *A motor operator for remote opening and closing of a circuit breaker can permit a worker to operate a circuit breaker from a long distance.*

Courtesy of Eaton Corporation

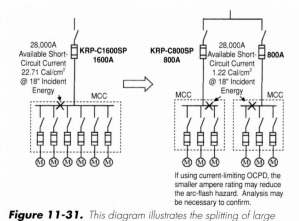

If using current-limiting OCPD, the smaller ampere rating may reduce the arc-flash hazard. Analysis may be necessary to confirm.

Figure 11-31. *This diagram illustrates the splitting of large feeders. Lower ampere-rated devices are typically more current limiting and, therefore, can better reduce the arc-flash hazard.*

Figure 11-33. *This arc-resistant switchgear diverts the arcing fault gases and pressure out the top of the gear if all doors are closed and latched properly.*

Courtesy of Eaton Corporation

Resistance-Grounded Systems

Electrical systems can be designed with resistance-grounded wye systems to increase system reliability and reduce the probability of arc-flash incidents. In this type of installation, a resistor is intentionally inserted between the center point of the wye of the transformer and the ground, so that the ground short-circuit current is limited by the inserted resistor. This type of system can reduce the probability that dangerous or destructive arcing faults will occur. For example, with a resistance-grounded wye system, if a worker's screwdriver slips, simultaneously touching an energized bare-phase termination and the enclosure, a high-energy arc fault would not be initiated. Instead, the fault current would typically be limited to only a few amperes by the resistor intentionally inserted in the ground return path. **See Figure 11-34**. This limited fault current is insufficient to create the vaporized metal that would initiate a 3-phase arcing fault. Such a system does not eliminate the arc-flash hazard, however. If the worker's screwdriver simultaneously touches the energized bare terminations of two phases, a high-level arcing fault might occur because the resistor then is not involved in the current path. If a designer is planning high-resistance/grounded wye systems or retrofitting an existing solidly grounded wye system, he or she must consider the single-pole interrupting capabilities of any circuit breakers and self-protected starters to be installed or already installed. Slash voltage-rated circuit breakers and self-protected starters cannot be used on any system other than solidly grounded wye systems.

Arc Initiator/Absorber Box

An arc initiator/absorber box is a specially designed box made to safely contain an arcing fault when it is used with a relay, light sensors, current sensors, and a shunt-trip main circuit breaker. With this approach, if an arcing fault starts in the switchgear, it is detected, and the special box initiates an arc in the box, which then extinguishes the arcing fault in the switchgear. **See Figure 11-35**.

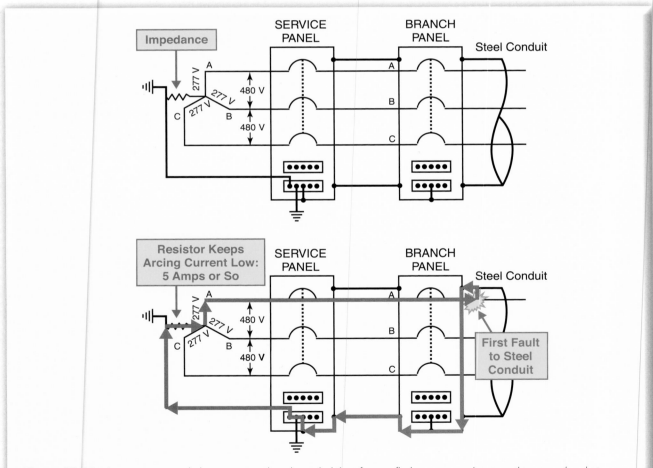

Figure 11-34. A resistance-grounded system can reduce the probability of an arc-flash occurrence; however, there are other design considerations. Top: In a resistance-grounded system, the resistor is positioned between the center tap of the wye transformer and the ground. Center: Because of the resistance, a fault occurring from phase to ground does not cause the OCPD to open. Bottom: The first fault to ground must be removed by Electrical Workers before a second phase goes to ground, or a significant short-circuit current can occur across one pole of the branch-circuit device.

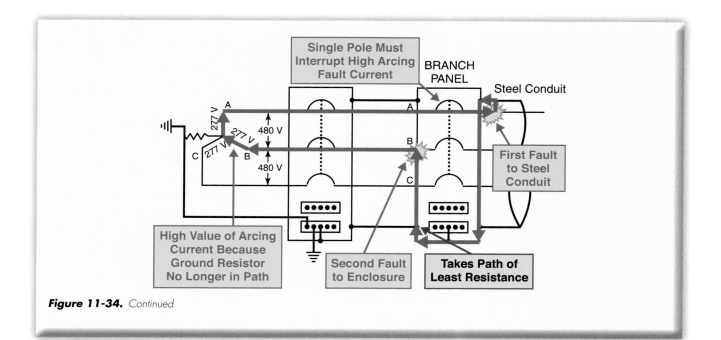

Figure 11-34. *Continued*

Operation

If an arcing fault starts in the switchgear:

1. The current sensor recognizes the arcing-fault signature.
2. Light sensors recognize a flash.
3. The relay processes the inputs and activates the special box to initiate an arcing fault in the specially designed device (this arc is designed to be contained in the box).
4. The relay also signals the main CB to open.
5. The arc initiated in the special box creates low impedance, which results in the fault in switchgear to be extinguished (the voltage at the arcing fault in the switchgear is not sufficient to sustain the arc due to a lower impedance fault in the special box).
6. The main CB opens, which extinguishes the arc in the special box.

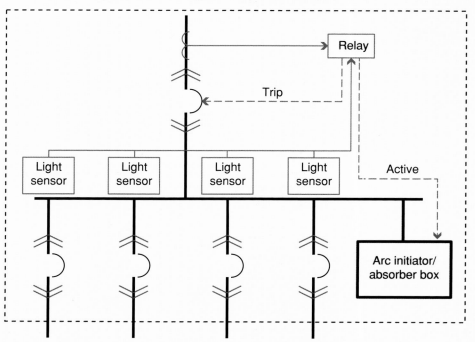

Figure 11-35. *An arc initiator/absorber box can reduce the arc-flash hazard.*

Summary

Electrical system design, whether for a new system or an upgrade, can have a significant impact on personnel safety. The choices made during the design stage play an important role in reducing or eliminating exposure to electrical hazards. Designers must carefully select equipment and circuit designs that provide maximum protection for workers. The selection of the type and characteristics of OCPDs has a significant effect on the arc-flash hazard; in addition, other design techniques can affect the arc-flash hazard and other hazards. Many design techniques other than those presented in this chapter are possible.

REVIEW QUESTIONS

1. By __?__, there is the opportunity to eliminate or minimize work exposure to electrical hazards before a worker ever has a chance to become exposed to them.
   a. designing for safety
   b. performing arc flash hazard analysis
   c. using insulated tools
   d. using PPE

2. __?__ are the "safety valves" that may limit the magnitude of the fault current that flows and reduce the time duration of the current, thereby mitigating the arc-flash hazard to some extent.
   a. Finger-safe products
   b. Overcurrent protective devices
   c. PPE
   d. Transformers

3. The typical molded case circuit breakers, insulated case circuit breakers, and low-voltage power circuit breakers are not listed as __?__ .
   a. current-limiting
   b. electrical equipment
   c. non-current limiting
   d. overcurrent protective devices

4. The longer an OCPD takes to clear a given arcing fault current, the __?__ the arc-flash hazard.
   a. greater
   b. less the incident energy of
   c. lesser
   d. safer

5. __?__ do not require maintenance other than visual examination and protection from damage due to external conditions.
   a. Circuit breakers
   b. Current-limiting fuses
   c. Protective relays
   d. Vacuum interrupters

6. __?__ for circuit breakers, where short-time delay is used (and no instantaneous trip setting), reduces incident energy exposure levels by turning off the short-time delay function and invoking an instantaneous trip while a person is working within the arc-flash protection boundary.
   a. Arc flash-reducing maintenance switching
   b. Electronic sensing
   c. Lubrication
   d. Remote racking

7. Fuses and circuit breakers must be selected so that their interrupting rating is equal to or __?__ the available short-circuit current for the initial installation as well as over the life of the system.
   a. greater than
   b. less than
   c. the same as
   d. weaker than

8. __?__ is a design method that is an effective arc-flash mitigation means for equipment protected by OCPDs that are too large in ampacity to quickly respond to the arcing current.
   a. A ground fault relay
   b. An arc-flash relay
   c. A potential transformer
   d. A Type 2 motor starter

9. __?__ provide a means for workers to perform electrical diagnostics with the enclosure doors closed.
   a. Arc-reduction maintenance switches
   b. Finger-safe products
   c. Insulated tools
   d. Remote monitoring devices

10. Disconnecting means at the motor provide improved __?__ and __?__ for maintenance as well as an emergency disconnect.
   a. fault protection/visibility
   b. isolation/safety
   c. voltage regulation/safety
   d. voltage regulation/visibility

AGREEMENT ESTABLISHING A PARTNERSHIP

BETWEEN

THE OCCUPATIONAL SAFETY AND HEALTH ADMINISTRATION
U.S. DEPARTMENT OF LABOR – OSHA REGION V
COLUMBUS AREA OFFICE

AND

THE NATIONAL ELECTRICAL CONTRACTORS ASSOCIATION (NECA)
CENTRAL OHIO CHAPTER

AND

THE INTERNATIONAL BROTHERHOOD OF ELECTRICAL WORKERS (IBEW)
LOCAL UNIONS 683 & 1105

I. Introduction and Identification of Partners

In an effort to more fully realize the objectives of the Occupational Safety and Health Act of 1970 to provide a safe and healthful work environment for all workers, including those engaged in the electrical construction and maintenance industry, the Central Ohio Chapter of the National Electrical Contractors Association Inc. (NECA) and the International Brotherhood of Electrical Workers Local Unions 683 and 1105 (IBEW), enter into an industry partnership with the United States Department of Labor, Occupational Safety and Health Administration, Region V, Columbus Area Office (OSHA) which has led to this Agreement.

For over 110 years, the NECA/IBEW Labor-Management Partnership has provided its respective members and the entire construction industry with model partnerships designed to meet industry specific issues. These model partnerships are exemplified by our multiple Taft-Hartley Trust Funds and Industry funds. Of these, two NECA/IBEW partnerships deserve special comment.

The first is the National Joint Apprenticeship and Training Committee (NJATC). Through this industry partnership, we have seen the development of a network in excess of 280 Local NECA/IBEW JATCs, where individuals are given the opportunity to become excellent IBEW Journeyman wiremen and IBEW Journeyman installer technicians. Many, as exemplified by the apprentices in the Columbus and Newark JATCs, are completing the program with an Associate Degree.

The second noteworthy partnership is the Council on Industrial Relations (CIR). The CIR is a panel of electrical industry leaders that is authorized to hear and decide/resolve local issues when formally presented to it. As a result, the CIR provides a means for local Labor and Management representatives to settle disputes without strikes or lockouts. Having this industry partnership to assist local parties in resolving their disputes without the emotional strife is a great service. It is because of this NECA/IBEW Partnership our business model proudly bears the moniker of a "strike-lockout proof industry."

The COHNECA/IBEW 683 & 1105 Partnership is pleased to extend the hand of cooperation and consideration to the Columbus OSHA Area Office to form the COHNECA/IBEW 683 & 1105/Columbus OSHA Area Office Partnership, not only for the construction industry, but for general industry as well.

1. Our organizations enjoyed a high level of cooperation with the development of a national Electrical Worker safety curriculum being just one of many examples. The NECA/IBEW/OSHA Partnership builds on this important foundation, and denotes a new era of Labor-Management-Agency cooperation and insight. In the end, meaningful increases in workplace health and safety will benefit the Central Ohio Electrical Workers and their Employers.

Employers who perform electrical construction and maintenance contracting services within the jurisdictional boundaries of (and are signatory to) the International Brotherhood of Electrical Workers (Local Unions 683 and 1105), and are members in good standing of the Central Ohio Chapter of the National Electrical Contractors Association, Inc. shall be eligible to apply to this Partnership.

II. Purpose and Scope

Working as partners and associates, the above parties are committed to achieve measurable, meaningful improvements in electrical worker safety using the following blueprint:

a. Encourage the continuous improvement of the safety culture within the electrical trade.

b. Actively research, share and implement the top safety and health programs for electrical workers. This includes technology, innovations and best practices that provide measurable improvement in electrical worker safety. These will be shared with the NECA/IBEW Partnership members at the quarterly meetings.

c. Continuously develop, build and share improved, effective safety programs specifically for electrical workers.

d. Develop and build, with help from the National Joint Apprenticeship and Training Committee (NJATC), improved, effective and meaningful safety training programs specific enough for the electrical trade, yet broad enough to be effective in every facet of diverse industry.

e. Constantly recognize and promote electrical worker safety excellence.

f. Ensure, through consequential and honest communication, that enforcement policies and practices are consistent, fair and effective.

Counties

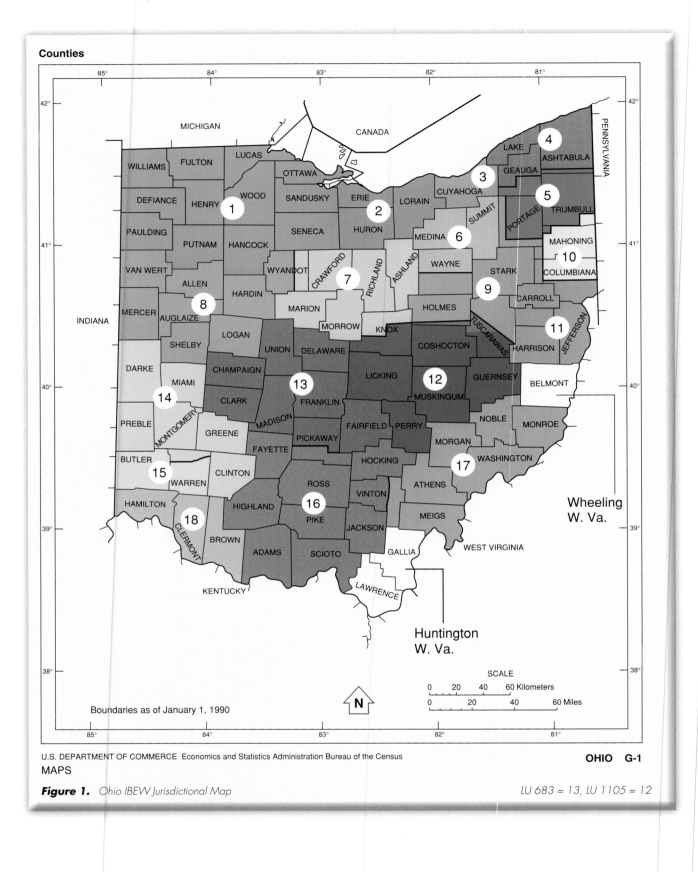

Boundaries as of January 1, 1990

SCALE

U.S. DEPARTMENT OF COMMERCE Economics and Statistics Administration Bureau of the Census
MAPS

OHIO   G-1

Figure 1. *Ohio IBEW Jurisdictional Map*

LU 683 = 13, LU 1105 = 12

III. Goals, Strategies and Measures

Goals

1. Maintain the participating employers' injury illness rates below the Bureau Labor Statistics (BLS) data for NAICS 23821, *Electrical contractors.*
2. Increase the number of and use of safety and health programs and best practices among participants.
3. All parties will work to promote and encourage new participating firms.

Strategies

Strategies to achieve one or all of the above goals:
1. Increase the implementation of effective safety and health program through training, self-inspections, and use of *NFPA 70E Electrical Safe Work practices.*
2. Provide quarterly meeting for Partnership participants and stakeholders.
3. Use Region V fatality data to focus training efforts of participants in order to encourage the use of on-the-job tool box talks.
4. Partnership will be promoted at industry events and functions.

Measures

1. Comparison of DART injury/illness rate of participating employers' to the most recently published BLS data.
2. Document the quarterly meeting and number of participants in attendance.
3. Document the number of employees trained by the electrical training centers in OSHA 10-hour, OSHA 30-hour, and *NFPA 70E Electrical Safe Work practices.*
4. Document the number of meeting to provide safety and health assistance for participating employers.
5. Track the number of new employers who apply and become members to this Partnership.
   These outcomes will be obtained using the following methods:
   A. Adopt and require use of an industry standard checklist designed to exceed OSHA requirements when working energized circuits. This policy will be based on the 2012 National Fire Protection Association® (NFPA®) *70E Standard for Electrical Safety in the Workplace*, which can be found in Appendix A of this partnership. Also refer to Annex E of *NFPA 70E* which provides a typical description of an Electrical Safety Program.
   B. Mandatory OSHA 10-Hour courses for all field employees and mandatory OSHA 30-Hour courses for all field supervisors. All employees will be subject to regular refresher courses through tool box talks or formal classes.
   C. Site specific safety training for new hires.
   D. Regular safety program third-party audits.
   E. Quarterly meetings discussing industry best practices.

All parties will be consulted on a continuing basis for feedback to assess the progression and potential amelioration of these goals.

IV. Performance Measures

1. The measurement system will use OSHA recordable fatalities and accidents identified.
2. Activity measures shall include:
   - The applicable number of employers, supervisors and employees trained.
   - Number of job safety analyses conducted.
   - Number of worksite audits.
   - Increased employee involvement.
   - Enhanced communication between management and employees.

3. Outcome measures will be gathered on an annual basis and will be comprised of data analyzing the number of hours worked, number of injuries, illnesses, and fatalities during the baseline year.

4. All participating firms will keep their OSHA 300 log current and up to date. A copy of the OSHA 300 log will be submitted to the NECA office with the annual recertification and shared with OSHA by June 15th of each year.

   All information will remain confidential as to company name and employee. The only purpose of this data is to assess performance and to target specific training requirements.

V. Annual Evaluation

The annual evaluation shall use the Strategic Partnership Annual Evaluation Format measurement system as specified in Appendix C of CSP 03-02-002, OSHA Strategic Partnership Program for Worker Safety and Health Directive.

All data for the annual evaluation shall be submitted to the NECA/IBEW Partnership no later than August 15th of each year. New applications and re-certifications shall be submitted by June 15th to be considered for the upcoming year. Applications turned in past the June 15th deadline will not be considered for the NECA/IBEW/OSHA Partnership.

It will be the responsibility of the NECA/IBEW Partnership to greater participant data to evaluate and track the overall results and success of the NECA/IBEW/OSHA Partnership. Data collected will be given to OSHA as well.

The OSHA liaison will use the collected data to write and submit the annual evaluation to the Columbus OSHA Area Director. The final draft will be made available to the NECA/IBEW Partnership. The annual evaluation will be sent to the National OSHA Office by December 15 for each calendar year of the NECA/IBEW/OSHA Partnership and Agreement.

VI. NECA/IBEW Partnership Management and Operation

Representative(s) from the NECA/IBEW Partnership will administer this program as outlined herein, and will serve as the primary safety resource, supporting the participating employers and employees. To fulfill this Partnership the NECA/IBEW Partnership will also provide the following services:

1. Act as a liaison for NECA/IBEW members with OSHA. Members will be able to call the NECA/IBEW Partnership with questions and the NECA/IBEW Partnership will contact OSHA for responses, if required. The OSHA primary contact will be appointed by the OSHA Area Director and the NECA/IBEW Partnership will be notified in writing.

2. In concert with The Electrical Trades Center and the Newark JATC Training Center, offer ongoing, quality training on topics of importance of members – specifically the focused areas of fall protection, electrical hazards, trenching, confined spaces etc.

> Although providing mandatory fall protection at the 6' level is not required by the current OSHA construction standards and is not mandated by OSHA as a requirement for participation in any OSHA partnership agreement, the contractors on this partnership are committed to providing a greater level of protection to the employees working at this site and will require protection at the 6' level and above.

3. Provide up-to-date informational materials and brochures to NECA/IBEW Partnership members (from OSHA, OSHA Ohio On-Site Consultation Service, the Bureau of Workers' Compensation (BWC) and other appropriate organizations).

4. Organize and provide to participating employers OSHA's interpretations of major standards, as well as local inspection perspectives.

5. Develop written safety and health policies and programs for participating employers, including emphasis on employer/employee responsibilities.

6. Promote construction safety excellence through an annual NECA Safety Recognition Program.

7. Administer the NECA/IBEW/OSHA Partnership, including, but not limited to, the initial evaluation of potential participating employer applications to determine whether the firm meets the criteria specified within the Partnership. Information considered by the NECA/IBEW Partnership will include pertinent company information as referenced in Section VII (demonstrated safety and health program, training commitments, OSHA citation history, fatalities, injury/illness experience and similar factors). Any and all information garnered by the NECA/IBEW Partnership will be held with the greatest confidentiality.
8. Notify OSHA on a regular and recurring basis with the name(s) of Employers which have met the partnership criteria.
9. Conduct periodic audits to determine the impact and effectiveness of the NECA/IBEW/OSHA Partnership.
10. A representative(s) from the NECA/IBEW Partnership and a representative(s) from OSHA together will conduct random onsite verification of 10 percent or more of participating employers to ensure that participating employers are fulfilling their commitment to the Partnership. During these onsite evaluations, if an OSHA representative observes serious hazards, these hazards will be addressed in a comprehensive and systematic manner, which may include citations.
11. If necessary, terminate employers from the NECA/IBEW/OSHA Partnership if findings indicate unacceptable performance or submission of falsified documentation. (Note: At the discretion of the NECA/IBEW partnership and OSHA, a participating employer may be permitted to correct deficiencies within 30 days of notification and apply to the NECA/IBEW partnership for continued recognition as a partner in good standing before termination would take effect.)

OSHA and the NECA/IBEW Partnership have the discretion to jointly veto Employers from participating for just cause. The OSHA Area Director has the discretion to unilaterally veto Employers from participating in this Partnership for just cause upon providing notification and explanation to the NECA/IBEW Partnership.

VII.   OSHA Benefits

Upon participating Employers' acceptance into this program, OSHA will:
1. Conduct quarterly meetings with the NECA/IBEW partnership and participating Employers to provide information on "what's hot" and to answer general and specific questions.
2. OSHA personnel may be available for a variety of onsite/offsite activities such as review of safety and health management systems, and guidance in conducting workplace audits and evaluations.
3. OSHA seminars, workshops, and other speaking events.
4. Informational material such as safety and health brochures, pamphlets and electronic tools.
5. Through the Columbus OSHA Area Director or designated representative meet with participating employers to address their role in the partnership.
6. Review the participating Employers' records and provide limited onsite training as needed during inspections where minor problems (ex: missing mid-rail) are found.

VIII.   Participating Employer Verification

To become and take full advantage of participating employer status, the Employer must:
1. Submit application and be willing to submit documents to the NECA/IBEW Partnership for review by OSHA (such as the OSHA Log 300 and the company's safety program).
2. Not have had willful or repeat violations in the last three years in Ohio, and nationwide, must not have had serious violations that resulted in a fatality(ies) or catastrophe(s).
3. Have a written safety and health program that complies with ANSI A-10.38 recommendations or OSHA's 1989 Voluntary Guidelines. The program must include active employee involvement.
4. Ensure that all new field Employees receive site-specific training before beginning work, and receive at least a two-hour safety orientation within the first week of hire. Topics for the orientation must include fall protection, electrical, struck-by, trenching/caught-between and personal protective equipment.
5. Ensure that within a year all field employees attend an OSHA 10-Hour training course. Field employees must receive refresher training in the 10-Hour Course every three years.

6. Ensure that within a year all field supervisors attend an OSHA 30-Hour training course. Field supervisors must receive refresher training in the 30-Hour course every three years.
   a. Note: A supervisor is defined as someone who directs/controls work.
7. Have a Days Away Restricted and Transfer (DART) rate for work in Ohio at least 25 percent below the BLS national average for the company's North American Industry Classification Code (NAICS).
   a. The company will submit individual DART rates for the past three years, and will be evaluated on a three-year average.
   b. If a company fulfills all other requirements, but does not have a qualifying DART, it may appeal for inclusion in the Partnership. OSHA and the NECA/IBEW Partnership will review these appeals on a case by case basis, and may allow the company to participate if the improvement can be shown.
8. Show evidence that both employees and supervisors are held accountable for safety.
9. Allow OSHA access to sites for inspection if the Employer has the authority to allow an inspection of the site. OSHA will follow the guidelines for inspections as outlined in the Field Operations Manual (FOM).
10. Recognize that OSHA implements Local Emphasis Programs (LEP) and National Emphasis Programs (NEP) to better manage specified hazards. These specific programs will involve inspections. The NECA/IBEW Partnership will be informed by OSHA of all LEPs and NEPs, and through the Partnership, this information will be shared with all participating Employers.
11. Participate in a site audit by an outside, independent source approved by the NECA/IBEW Partnership. The audit must include an action plan to prevent future hazards, as well as methods to abate current hazards. These audits will be made available to OSHA upon request.
12. Agree that all work on energized circuits will be performed under a standard policy as referenced in Appendix A to be developed and built in accordance with the current National Fire Protection Association publication 70e. This policy will cover hazard/risk evaluation, and procedures including protective barriers and shields, communication, insulated tools and equipment, and personal protective equipment.
13. Be a member in good standing of the National Electrical Contractors Association.

IX.   Employer/Employee Rights

This Partnership Agreement does not preclude employees and/or employers from exercising any right provided under the OSH Act (or federal employees, 29 CFR 1960), nor does it abrogate any responsibility to comply with rules and regulations adopted pursuant to the Act.

Routine Employee involvement in the day to day implementation of worksite safety and health programs is expected to be assured, including Employee participation in Employer self-audits, site inspections, job hazard analysis, safety and health program reviews and near miss investigations.

X.   Termination of this Partnership

Any party may withdraw from this Partnership Agreement by providing written notification to the other parties. Termination will be effective 30 days after receipt of said notification. Furthermore, an individual participating Employer may withdraw from this Agreement by providing written notification to the NECA/IBEW partnership and OSHA. Termination shall be effective 30 days after receipt of notification.

If OSHA chooses to withdraw its participation in the partnership, the entire Agreement is terminated. Either party may also propose modification of amendment of the Agreement.

An individual participating Employer's violation of this Partnership Agreement shall not be grounds for OSHA to terminate this Partnership or the Partnership Agreement.

The NECA/IBEW partnership and OSHA may terminate an individual Employer from the partnership if the Employer fails to meet the qualifications or otherwise violates the terms and conditions of this Partnership.

Any party may propose modification or amendment to this Partnership subject to concurrence by every other party to the Partnership.

Unless modified or superseded, this Partnership Agreement will terminate three years after the signing date. Agreed this 3rd day of December, 2012.

Deborah Zubaty
Area Director
U.S. Department of Labor OSHA
Columbus Area Office

Mario Ciardelli
Business Manager, Financial Secretary
Local Union 683, IBEW
Columbus, Ohio

Brian Damant
Chapter Manager
Central Ohio Chapter, NECA, Inc.
Columbus, Ohio

William Hamilton
Business Manager, Financial Secretary
Local Union 1105, IBEW
Newark, Ohio

Courtesy of the Electrical Trades Center

Appendix A
Energized Checklist

ENERGIZED CHECKLIST

TO BE REVIEWED BY FOREMAN PRIOR TO START OF WORK
VALID FOR NAMED WIREMAN AND DAY ISSUED ONLY

Date: _____   START TIME: _____ A.M.   P.M.

PROJECT _____

BUILDING LOCATION _____

PANEL LOCATION _____

DESCRIPTION OF WORK TO BE PERFORMED _____

_____

SPECIFIC REASON EQUIPMENT CIRCUIT CANNOT BE DE-ENERGIZED: _____

_____

_____

MAXIMUM VOLTAGE PRESENT _____

MAXIMUM VOLTAGE PRESENT _____

PERSONAL PROTECTIVE EQUIPMENT (PPE) WORN/USED:

HARD HAT _____   *INSULATED GLOVES _____

FACE SHIELD _____   *INSULATED SLEEVES _____

FLASH RESISTANT JACKET _____   *INSULATED BLANKET _____

INSULATED MATS _____

ADDITIONAL PRE REQUIRED _____

WIREMAN SIGNATURE _____

WORK AUTHORIZED BY _____ TITLE _____

REQUIRED SECOND WIREMAN ASSISTING SIGNATURE _____

On all energized circuits or equipment carrying four hundred forty (440) volts or over, as a safety measure, two (2) or more journeyman must work together.

OWNER/CUSTOMER REPRESENTATIVE _____

TIME WORK COMPLETED _____ A.M.   P.M.

REVIEW BACK SIDE OF CHECKLIST FOR INFORMATION ON RUBBER GOODS

Rubber goods are classed, color coded, rated and tested at voltages as follows:

Class	Color	Maximum Use Voltage Phase-To-Phase Ac	Maximum Use Voltage Phase-To-Phase DC
00	Beige	500 V	700 V
0	Red	1,000 V	1,500 V
1	White	7,500 V	11,250 V
2	Yellow	17,000 V	25,500 V
3	Green	26,500 V	39,750 V
4	Orange	36,000 V	54,000 V

Rubber goods are required to be labeled as follows:
- Manufacturer's name
- Type I for non-ozone-resistant equipment or Type II for ozone-resistant equipment
- Size (Gloves only).
- Voltage class (0, 1, 2, 3, and 4).
- Color coding according to voltage class.

Each Foreman should ensure that rubber protective equipment under his or her control and care is maintained, inspected, used and tested as per the equipment manufacturer's instructions and as per the appropriate American Society for Testing and Materials (ASTM) standards that are available for review in the Safety Coordinator's office:

11.2 RUBBER GOODS TESTING GUIDELINES

The Site Superintendent should ensure that Foreman and employees inspect all rubber protection products before each use in accordance with the manufacturer's inspection guidelines and the ASTM standard:
- ASTM F 1236 Standard Guide for Visual Inspection of Electrical Protective Rubber Products.

The Safety Coordinator should maintain a master list of all Company rubber protection equipment and set up a testing schedule so that there will be an adequate amount and types of rubber protection equipment and devices on site while tested rubber goods are off site undergoing manufacturer's retesting.

1. Manufacturer's Periodic Electrical Testing Instructions.

Product	Maximum Test Interval, Months	Astm	Notes
Gloves	6	F496	Tested. Unused gloves may be placed into service within 12 months of the previous tests without retesting.
Sleeves	12	F496	Tested. Unused sleeves may be placed into service within 12 months of the previous tests without retesting.
Blanket	12	F479	Tested. Unused blankets may be placed into service within 12 months of the previous tests without retesting.
Mats	-	D178	Only needs to be tested by Manufacturer, when new.
Covers	-	F478	Retested when in-service in section indicates a need.
Line Hoses	-	F478	Retested when in-service in section indicates a need.

This material relates to Chapters 8 and 10, which cover the importance of maintenance for electrical safety. This appendix is reprinted with the permission of the InterNational Electrical Testing Association (NETA) and is extracted from Appendix B of ANSI/NETA MTS-2011, *Standard for Maintenance Testing Specifications for Electrical Power Distribution Equipment and Systems*. Appendix C in this book provides an example of the maintenance and tests recommended for molded-case circuit breakers and insulated-case circuit breakers. For the complete standard, visit www.netaworld.org.

APPENDIX B

Frequency of Maintenance Tests

NETA recognizes that the ideal maintenance program is reliability-based, unique to each plant and to each piece of equipment. In the absence of this information and in response to requests for a maintenance timetable, NETA's Standards Review Council presents the following time-based maintenance schedule and matrix.

One should contact a NETA Accredited Testing Company for a reliability-based evaluation.

The following matrix is to be used in conjunction with Appendix B, Inspections and Tests. Application of the matrix is recognized as a guide only.

Specific condition, criticality, and reliability must be determined to correctly apply the matrix. Application of the matrix, along with the culmination of historical testing data and trending, should provide a quality electrical preventive maintenance program.

MAINTENANCE FREQUENCY MATRIX			
	EQUIPMENT CONDITION		
	POOR	AVERAGE	GOOD
EQUIPMENT RELIABILITY REQUIREMENT LOW	1.0	2.0	2.5
MEDIUM	0.50	1.0	1.5
HIGH	0.25	0.50	0.75

	Inspections and Tests Frequency in Months (Multiply These Values by the Factor in the Maintenance Frequency Matrix)			
Section	Description	Visual	Visual & Mechanical	Visual & Mechanical & Electrical
7.1	Switchgear & Switchboard Assemblies	12	12	24
7.2	Transformers			
7.2.1.1	Small Dry-Type Transformers	2	12	36
7.2.1.2	Large Dry-Type Transformers	1	12	24
7.2.2	Liquid-Filled Transformers	1	12	24
	Sampling	–	–	12
7.3	Cables			
7.3.1	Low-Voltage, Low-Energy	–	–	–
7.3.2	Low-Voltage, 600 Volt Maximum	2	12	36
7.3.3	Medium- and High-Voltage	2	12	36
7.4	Metal-Enclosed Busways	2	12	24
	Infrared Only	–	–	12
7.5	Switches			
7.5.1.1	Air, Low-Voltage	2	12	36
7.5.1.2	Air, Medium-Voltage, Metal-Enclosed	–	12	24
7.5.1.3	Air, Medium- and High-Voltage Open	1	12	24
7.5.2	Oil, Medium-Voltage	1	12	24
7.5.3	Vacuum, Medium-Voltage	1	12	24
7.5.4	Medium-Voltage SF_6 Switches	1	12	24
7.5.5	Cutouts	12	24	24
7.6	Circuit Breakers			
7.6.1.1	Air, Insulated-Case/Molded-Case	1	12	36
7.6.1.2	Air, Low-Voltage Power	1	12	36
7.6.1.3	Air, Medium-Voltage	1	12	36
7.6.2	Oil, Medium-Voltage	1	12	36
	Sampling	–	–	12
7.6.2	Oil, High-Voltage	1	12	12
	Sampling	–	–	12
7.6.3	Vacuum, Medium-Voltage	1	12	24
7.6.4	SF_6	1	12	12
7.7	Circuit Switchers	1	12	12
7.8	Network Protectors	12	12	24
7.9	Protective Relays			
7.9.1	Electromechanical and Solid State	1	12	12
7.9.2	Microprocessor-Based	1	12	12
7.10	Instrument Transformers	12	12	36
7.11	Metering Devices			
7.11.1	Electromechanical and Solid-State	12	12	36
7.11.2	Microprocessor-Based	12	12	36
7.12	Regulating Apparatus			
7.12.1.1	Step-Voltage Regulators	1	12	24
	Sample Liquid	–	–	12

	Inspections and Tests Frequency in Months (Multiply These Values by the Factor in the Maintenance Frequency Matrix)			
Section	Description	Visual	Visual & Mechanical	Visual & Mechanical & Electrical
7.12.1.2	Induction Regulators	12	12	24
7.12.2	Current Regulators	1	12	24
7.12.3	Load-Tap-changers	1	12	24
	Sample Liquid	–	–	12
7.13	Grounding Systems	2	12	24
7.14	Ground-Fault Protection Systems	2	12	12
7.15	Rotating Machinery			
7.15.1	AC Induction Motors and Generators	1	12	24
7.15.2	Synchronous Motors and Generators	1	12	24
7.15.3	DC Motors and Generators	1	12	24
7.16	Motor Control			
7.16.1.1	Motor Starters, Low-Voltage	2	12	24
7.16.1.2	Motor Starters, Medium-Voltage	2	12	24
7.16.2.1	Motor Control Centers, Low-Voltage	2	12	24
7.16.2.2	Motor Control Centers, Medium-Voltage	2	12	24
7.17	Adjustable Speed Drive Systems	1	12	24
7.18	Direct-Current Systems			
7.18.1	Batteries	1	12	12
7.18.2	Battery Chargers	1	12	12
7.18.3	Rectifiers	1	12	24
7.19	Surge Arresters			
7.19.1	Low-Voltage Surge Protection Devices	2	12	24
7.19.2	Medium- and High-Voltage Surge Protection Devices	2	12	24
7.20	Capacitors and Reactors			
7.20.1	Capacitors	1	12	12
7.20.2	Capacitor Control Devices	1	12	12
7.20.3.1	Reactors, (Shunt and Current-Limiting) Dry-Type	2	12	24
7.20.3.2	Reactors, (Shunt and Current-Limiting) Liquid-Filled	1	12	24
	Sampling	–	–	12
7.21	Outdoor Bus Structures	1	12	36
7.22	Emergency Systems			
7.22.1	Engine Generator	1	2	12
	Functional Testing	–	–	2
7.22.2	Uninterruptible Power Systems	1	12	12
	Functional Testing	–	–	2
7.22.3	Automatic Transfer Switches	1	12	12
	Functional Testing	–	–	2
7.23	Telemetry/Pilot Wire SCADA	1	12	12

(continued)

		Inspections and Tests Frequency in Months (Multiply These Values by the Factor in the Maintenance Frequency Matrix)		
Section	Description	Visual	Visual & Mechanical	Visual & Mechanical & Electrical
7.24	Automatic Circuit Reclosers and Line Sectionalizers			
7.24.1	Automatic Circuit Reclosers, Oil/Vacuum	1	12	24
	Sample	–	–	12
7.24.2	Automatic Line Sectionalizers, Oil	1	12	24
	Sample	–	–	12
7.27	EMF Testing	12	12	12

This material relates to Chapters 8 and 10, which cover the importance of maintenance for electrical safety. This appendix is reprinted with the permission of the InterNational Electrical Testing Association (NETA) and is extracted from ANSI/NETA MTS-2011, *Standard for Maintenance Testing Specifications for Electrical Power Distribution Equipment and Systems.* These specifications cover the suggested field tests and inspections that are available to assess the suitability for continued service and reliability of electrical power distribution equipment and systems. The purpose of these specifications is to assure that tested electrical equipment and systems are operational and within applicable standard and manufacturer's tolerances and that the equipment and systems are suitable for continued service. The following is extracted as an example of the maintenance and tests recommended for molded-case circuit breakers and insulated-case circuit breakers. Appendix B in this book provides guidelines on the Frequency of Maintenance. For the complete standard, visit www.netaworld.org.

7. Inspection and Test Procedures

7.6.1.1 Circuit Breakers, Air, Insulated-Case/Molded-Case

1. **Visual and Mechanical Inspection**
   1. Inspect physical and mechanical condition.
   2. Inspect anchorage and alignment.
   3. Prior to cleaning the unit, perform as-found tests, if required.
   4. Clean the unit.
   5. Operate the circuit breaker to insure smooth operation.
   6. Inspect bolted electrical connections for high resistance using one or more of the following methods:
      1. Use of a low-resistance ohmmeter in accordance with Section 7.6.1.1.2.
      2. Verify tightness of accessible bolted electrical connections by calibrated torque-wrench method in accordance with manufacturer's published data or Table 100.12.
      3. Perform a thermographic survey in accordance with Section 9.
   7. Inspect operating mechanism, contacts, and arc chutes in unsealed units.
   8. Perform adjustments for final protective device settings in accordance with coordination study provided by end user.
   9. Perform as-left tests.

2. **Electrical Tests**
   1. Perform resistance measurements through bolted connections with a low-resistance ohmmeter, if applicable, in accordance with Section 7.6.1.1.1.
   2. Perform insulation-resistance tests for one minute on each pole, phase-to-phase and phase-to-ground with the circuit breaker closed, and across each open pole. Apply voltage in accordance with manufacturer's published data. In the absence of manufacturer's published data, use Table 100.1.

3. Perform a contact/pole-resistance test.
*4. Perform insulation-resistance tests on all control wiring with respect to ground. The applied potential shall be 500 volts dc for 300-volt rated cable and 1000 volts dc for 600-volt rated cable. Test duration shall be one minute. For units with solid-state components, follow manufacturer's recommendation.
5. Determine long-time pickup and delay by primary current injection.
6. Determine short-time pickup and delay by primary current injection.
7. Determine ground-fault pickup delay by primary current injection.
8. Determine instantaneous pickup current by primary injection.
*9. Test functions of the trip unit by means of secondary injection.
10. Perform minimum pickup voltage test on shunt trip and close coils in accordance with Table 100.20.
11. Verify correct operation of auxiliary features such as trip and pickup indicators, zone interlocking, electrical close and trip operation, trip-free, antipump function, and trip unit battery condition.
12. Reset all trip logs and indicators.
13. Verify operation of charging mechanism.

3. **Test Values**

   3.1 **Test Values—Visual and Mechanical**
      1. Compare bolted connection resistance values to values of similar connections. Investigate values which deviate from those of similar bolted connections by more than 50 percent of the lowest value. (7.6.1.1.1.6.1)
      2. Bolt–torque levels should be in accordance with manufacturer's published data. In the absence of manufacturer's published data, use Table 100.12. (7.6.1.1.1.6.2)
      3. Results of the thermographic survey shall be in accordance with Section 9. (7.6.1.1.1.6.3)
      4. Settings shall comply with coordination study recommendations. (7.6.1.1.1.8)

   3.2 **Test Values—Electrical**
      1. Compare bolted connection resistance values to values of similar connections. Investigate values which deviate from those of similar bolted connections by more than 50 percent of the lowest value.
      2. Insulation-resistance values should be in accordance with manufacturer's published data. In the absence of manufacturer's published data, use Table 100.1. Values of insulation resistance less than this table or manufacturer's recommendations should be investigated.
      3. Microhm or dc millivolt drop values should not exceed the high levels of the normal range as indicated in the manufacturer's published data. If manufacturer's data is not available, investigate values that deviate from adjacent poles or similar breakers by more than 50 percent of the lowest value.
      4. Insulation-resistance values of control wiring should be comparable to previously obtained results but not less than two megohms.
      5. Long-time pickup values should be as specified, and the trip characteristic should not exceed manufacturer's published time–current characteristic tolerance band, including adjustment factors. If manufacturer's curves are not available, trip times should not exceed the value shown in Table 100.7.
      6. Short-time pickup values should be as specified, and the trip characteristic should not exceed manufacturer's published time-current tolerance band.
      7. Ground fault pickup values should be as specified, and the trip characteristic should not exceed manufacturer's published time-current tolerance band.
      8. Instantaneous pickup values of molded-case circuit breakers should fall within manufacturer's published tolerances and/or Table 100.8.
      9. Pickup values and trip characteristics should be within manufacturer's published tolerances.
      10. Minimum pickup voltage on shunt trip and close coils should be in accordance with manufacturer's published data. In the absence of manufacturer's published data, refer to Table 100.20.
      11. Breaker open, close, trip, trip-free, antipump, and auxiliary features should function as designed.
      12. Trip logs and indicators are reset.
      13. The charging mechanism should operate in accordance with manufacturer's published data.

* Optional.

Tables Referenced by 7.6.1.1

Table 100-1. *Insulation-Resistance Test Values Electrical Apparatus and Systems*

Nominal Rating of Equipment (Volts)	Minimum Test Voltage (DC)	Recommended Minimum Insulation Resistance in Megohms
250	500	25
600	1,000	100
1,000	1,000	100
2,500	1,000	500
5,000	2,500	1,000
8,000	2,500	2,000
15,000	2,500	5,000
25,000	5,000	20,000
34,500 and above	15,000	100,000

In the absence of consensus standards dealing with insulation-resistance tests, the NETA Standards Review Council suggests the above representative values. See Table 100.14 for temperature correction factors.

Test results are dependent on the temperature of the insulating material and the humidity of the surrounding environment at the time of the test.

Insulation-resistance test data may be used to establish a trending pattern. Deviations from the baseline information permit evaluation of the insulation.

Table 100-7. *Molded-Case Circuit Breakers Inverse Time Trip Test*
(At 300% of Rated Continuous Current of Circuit Breaker)

Range of Rated Continuous Current (Amperes)	Maximum Trip Time in Seconds For Each Maximum Frame Rating[a]	
	£ 250 V	251–600V
0–30	50	70
31–50	80	100
51–100	140	160
101–150	200	250
151–225	230	275
226–400	300	350
401–600	- - - - -	450
601–800	- - - - -	500
801–1000	- - - - -	600
1001–1200	- - - - -	700
1201–1600	- - - - -	775
1601–2000	- - - - -	800
2001–2500	- - - - -	850
2501–5000	- - - - -	900
6000	- - - - -	1000

Derived from Table 5-3, NEMA Standard AB 4-2000, *Guidelines for Inspection and Preventative Maintenance of Molded-Case Circuit Breaker Used in Commercial and Industrial Applications.*

a. Trip times may be substantially longer for integrally-fused circuit breakers if tested with the fuses replaced by solid links (shorting bars).

Table 100-8. *Instantaneous Trip Tolerances for Field Testing of Circuit Breakers*

Breaker Type	Tolerance of Settings	Tolerances of Manufacturer's Published Trip Range	
		High Side	Low Side
Electronic Trip Units [1]	+30%	- - - - -	- - - - -
Adjustable [1]	+40% −30%	- - - - -	- - - - -
Nonadjustable [2]	- - - - -	+25%	−25%

NEMA AB4-2009 *Guidelines for Inspection and Preventative Maintenance of Molded-Case Circuit Breaker Used in Commercial and Industrial Applications, Table 4*

1. Tolerances are based on variations from the nominal settings.
2. Tolerances are based on variations from the manufacturer's published trip band (i.e., −25% below the low side of the band; +25% above the high side of the band.

TABLE 100.12.1   Bolt-Torque Values for Electrical Connections
US Standard Fasteners[a]
Heat-Treated Steel–Cadmium or Zinc Plated[b]

See Standard for table.

TABLE 100.12.2   US Standard Fasteners[a]
Silicon Bronze Fasteners[bc]
Torque (Pound-Feet)

See Standard for table.

TABLE 100.12.3   US Standard Fasteners[a]
Aluminum Alloy Fasteners[bc]
Torque (Pound-Feet)

See Standard for table.

TABLE 100.12.4   US Standard Fasteners[a]
Stainless Steel Fasteners[bc]
Torque (Pound-Feet)

See Standard for table.

TABLE 100.20.1   Rated Control Voltages and Their Ranges for Circuit Breakers

See Standard for table.

TABLE 100.20.2   Rated Control Voltages and their Ranges for Circuit Breakers
Solenoid-Operated Devices

See Standard for table.

Index

Note: *Page numbers followed by "f" indicate figures.*